FEEDING BEHAVIOR
Neural and Humoral Controls

Research Topics in Physiology

Charles D. Barnes, *Editor*
Department of Veterinary and Comparative Anatomy, Pharmacology, and Physiology
College of Veterinary Medicine
Washington State University
Pullman, Washington

1. Donald G. Davies and Charles D. Barnes (Editors). Regulation of Ventilation and Gas Exchange, 1978

2. Maysie J. Hughes and Charles D. Barnes (Editors). Neural Control of Circulation, 1980

3. John Orem and Charles D. Barnes (Editors). Physiology in Sleep, 1981

4. M. F. Crass III, and C. D. Barnes (Editors). Vascular Smooth Muscle: Metabolic, Ionic, and Contractile Mechanisms, 1982

5. James J. McGrath and Charles D. Barnes (Editors). Air Pollution—Physiological Effects, 1982

6. Charles D. Barnes (Editor). Brainstem Control of Spinal Cord Function, 1984

7. Herbert F. Janssen and Charles D. Barnes (Editors). Circulatory Shock: Basic and Clinical Implications, 1985

8. Richard D. Nathan (Editor). Cardiac Muscle: The Regulation of Excitation and Contraction, 1986

9. Robert C. Ritter, Sue Ritter, and Charles D. Barnes (Editors). Feeding Behavior: Neural and Humoral Controls, 1986

FEEDING BEHAVIOR

Neural and Humoral Controls

Edited by

Robert C. Ritter **Sue Ritter** **Charles D. Barnes**

Department of Veterinary and Comparative Anatomy,
Pharmacology, and Physiology
College of Veterinary Medicine
Washington State University
Pullman, Washington

1986

ACADEMIC PRESS, INC.
Harcourt Brace Jovanovich, Publishers
Orlando San Diego New York Austin
Boston London Sydney Tokyo Toronto

Academic Press Rapid Manuscript Reproduction

ACADEMIC PRESS, INC.
Orlando, Florida 32887

United Kingdom Edition published by
ACADEMIC PRESS INC. (LONDON) LTD.
24–28 Oval Road, London NW1 7DX

Library of Congress Cataloging in Publication Data

Feeding behavior.

(Research topics in physiology)
Includes index.
1. Ingestion—Regulation. 2. Appetite—Physiological aspects. 3. Neuropsychology. I. Ritter, Robert C. II. Ritter, Sue. III. Barnes, Charles D. IV. Series.
[DNLM: 1. Drinking. 2. Eating. 3. Feeding Behavior—physiology. 4. Neurophysiology. W1 RE235E / WL 102 F295]
QP147.F44 1986 612'.3 86-47858
ISBN 0–12–589060–5 (alk. paper)

PRINTED IN THE UNITED STATES OF AMERICA

86 87 88 89 9 8 7 6 5 4 3 2 1

Contents

Contributors

Numbers in parentheses indicate the pages on which the authors' contributions begin.

Hans-Rudolf Berthoud (67), Laboratory of Regulatory Psychobiology, Department of Psychological Sciences, Purdue University, West Lafayette, Indiana 47907

Gaylen L. Edwards (131), Department of Veterinary and Comparative Anatomy, Pharmacology, and Physiology, College of Veterinary Medicine, Washington State University, Pullman, Washington 99164-6250 and W.O.I. Regional Program in Veterinary Medical Eduction, University of Idaho, Moscow, Idaho 83843

Alan A. Epstein (1), Leidy Laboratory, University of Pennsylvania, Philadelphia, Pennsylvania 19104

J. Gibbs (329), Department of Psychiatry, Cornell University Medical College, and Eating Disorders Institute, New York Hospital-Cornell Medical Center, Westchester Division, White Plains, New York 10605

Harvey J. Grill (103), Graduate Groups of Psychology and Neuroscience, University of Pennsylvania, Philadelphia, Pennsylvania 19104

Annette Kirchgessner (27), Program in Neural and Behavioral Science, Downstate Medical Center of the State University of New York, Brooklyn, New York 11210

Sarah F. Leibowitz (191), The Rockefeller University, New York, New York 10021-6399

Yutaka Oomura (235), Department of Physiology, Faculty of Medicine, Kyushu University, Fukuoka 812, Japan

Daniel Porte, Jr. (315), Seattle Veterans Administration Medical Research Center, Seattle, Washington 98108

Terry L. Powley (67), Laboratory of Regulatory Psychobiology, Department of Psychological Sciences, Purdue University, West Lafayette, Indiana 47907

Robert C. Ritter (131), Department of Veterinary and Comparative Anatomy, Pharmacology, and Physiology, College of Veterinary Medicine, Washington State University, Pullman, Washington 99164-6240 and W.O.I. Regional Program in Veterinary Medical Education, University of Idaho, Moscow, Idaho 83843

Sue Ritter (271) Department of Veterinary and Comparative Anatomy, Pharmacology, and Physiology, College of Veterinary Medicine, Washington State University, Pullman, Washington 99164-6250

Edmund T. Rolls (163), Oxford University, Department of Experimental Psychology, Oxford, England

Anthony Sclafani (27), Department of Psychology, Brooklyn College of the City University of New York, Brooklyn, New York 11210

G. P. Smith (329), Department of Psychiatry, Cornell University Medical College, and Eating Disorders Institute, New York Hospital-Cornell Medical Center, Westchester Division, White Plains, New York 10605

B. Glenn Stanley (191), The Rockefeller University, New York, New York 10021-65399

Anton B. Steffens (315), Department of Animal Physiology, State University of Groningen, 9750 AA Haren, The Netherlands

Jan H. Strubbe (315), Department of Animal Physiology, State University of Groningen, 9750 AA Haren, The Netherlands

Stephen C. Woods (315), Department of Psychology, University of Washington, Seattle, Washington 98195

Preface

This volume, *Feeding Behavior: Neural and Humoral Controls,* is the ninth in the series *Research Topics in Physiology.* Having read the individual chapters many times and discussed their contents with the authors, we are still struck by the depth, breadth, and timeliness of the contributions. Each reading of each chapter seemed to bring a new insight or idea. Nevertheless, it is doubtful that the book could have been completed at all without the diligent and sympathetic efforts of Ms. Connie Bollinger, a scientific editor in our department. Her quick grasp and appreciation of an unfamiliar literature, her superb command of language, and her compulsive attention to detail have made our work relatively painless.

The book is a collection of chapters in which eminent investigators have reviewed their own work and attempted to place it in both historical and current perspective. The collection is clearly not comprehensive; rather it is an attempt to sample some of the changes in our thinking about feeding behavior that have occurred over the past decade. Some of the chapters critically reevaluate the role of well-known neural or humoral substrates in the light of new information. Other chapters describe neural and humoral mechanisms whose participation in control of feeding was either unappreciated or largely unstudied until recently. Finally, a few chapters give us views of the complexity of neurobehavioral integration of feeding through detailed analysis of electrophysiology, neuropharmacology, or ontogeny.

The book begins with an incisive chapter on the ontogeny of ingestive behavior. This chapter critiques findings indicating that ingestion in the newborn is accomplished by a behavior neurologically distinct from that which replaces it in the adult. The second chapter examines the role of the hypothalamus in the control of feeding behavior, with particular focus on how the ventromedial hypothalamus participates in control of feeding. The third chapter elucidates the structure and function of the vagus nerve as it relates to the control of ingestion. The fourth and fifth chapters focus on the recently appreciated participation of the hindbrain in the control of feeding behavior. Chapter 6 reviews the electrophysiological circuitry through which various components of food-motivated behavior and ingestion seem to be executed. In Chapter 7 the authors review the burgeoning array of aminergic and neuropeptidergic influences on feeding. Chapter 8 examines the putative role of electrophysiologically detected central chemoreceptors in control of ingestion. Chapters 9 and 10 focus on controls of feeding provided by changes in energy utilization and storage. The final chapter reviews the role of gastrointestinal hormones in the control of ingestion.

We believe this book will be useful both to specialists in ingestive physiology and to individuals in other areas of biology who are interested in control of food intake. Because the volume evaluates substrates with long histories of investigation as well as areas and phenomena whose association with control of feeding is more recent, it should provide many prescriptions for future work. Certainly we hope that the book will stimulate the thinking of future readers as much as it has our own.

Robert C. Ritter
Sue Ritter
Charles D. Barnes

Chapter 1

THE ONTOGENY OF INGESTIVE BEHAVIORS: CONTROL OF MILK INTAKE BY SUCKLING RATS AND THE EMERGENCE OF FEEDING AND DRINKING AT WEANING

Alan A. Epstein

I. INTRODUCTION

The ingestive behaviors of altricial mammals mature through three transformations. The first and last are obvious: at birth the newborn ends its parasitic intrauterine life and begins suckling; at weaning the juvenile leaves the maternal nipple and becomes completely dependent upon foraging for food and water. The second transformation, midway through the suckling period, might be called the transition from the neonatal to the juvenile period. The changes in ingestive behavior that occur then are almost the equal of those at birth and weaning. They are part of a broad range of developmental changes making this transformation the suckling's debut into behavioral independence. In the rat these changes occur between the 10th and 15th days after birth. The pups are now two to three times larger than at birth. They are no longer in almost continuous REM sleep but begin to exhibit slow-wave sleep (SWS) as well. By day 15 they distribute their sleep time between REM and SWS in an almost mature fashion, and they are awake both behaviorally and electroencephalographically 25% of the time (Jouvet-Mounier et al., 1969). Their eyelids and external meati are open. They can locomote without crawling and can thermoregulate while away from their dam and peers. These changes permit them to spend increasing amounts of time away from the litter, and they no longer depend on being retrieved by the dam. Instead, they actively return to her because her olfactory cues have expanded to include both the saliva on her nipples and the caecotroph she excretes with her feces (Leon and Moltz, 1972). They are drawn back to her by its odor, and, having returned to her ventrum, they locate her nipples by their familiar salivary odor (Teicher and Blass, 1976; Hofer et al., 1976).

This new freedom to leave the dam and to reunite with her is facilitated not only by the pup's newly developed locomotor competence and response to the caecotroph, but also by the changes in ingestive behaviors that are an important part of this mid-suckling transition. Earlier in the suckling period, the watertight seal provided by the pups' saliva (Epstein et al., 1970) permitted them to attach to a maternal nipple whenever the dam was with them, regardless of their state of sleep or wakefulness or the recency of their last feeding. They may have been dislodged from a nipple from time to time by the dam's local movements and by the activity of their peers, but they reattached to it rapidly. Except for these few brief interruptions, they remained attached to a nipple for large portions of the time (80% on day 1, diminishing to 30% on day 15) that the dam is with them (Grota and Ader, 1969). On the 11th and 12th days they begin to lose their tenacity for attachment to a nipple. They detach themselves from the nipple after a milk letdown and shift to another nipple with increasing frequency while suckling (Cramer et al., 1980), and they spend increasing amounts of time off the nipple, reattaching only when unfed. Their attachment behavior has now come under the control of upper gastrointestinal fullness; they are no longer confined to the litter by their avidity for a nipple. The two major phases of suckling behavior, attachment to a nipple and ingestion of milk from it, are now under upper GI control. From birth or shortly thereafter the pups were able to control their milk intake by the vigor of their sucking. Now they have added detachment from the nipple after ingestion of milk and reattachment to it when unfed to their repertory of controls. Attachment behavior is no longer a compulsive necessity dictated by the odor of the nipple but has become an appetitive option opened by the pup's nutritive state.

These facts about nipple attachment behavior were discovered in Blass's laboratory (Hall et al., 1975, 1977). The work that followed (Hall and Rosenblatt, 1977, 1978; Cramer and Blass, 1983) has led to the widespread belief that the milk intake of the newborn suckling rat is determined only by the mother's milk supply (Blass et al., 1979; Martin and Alberts, 1979; Henning, 1981; Stoloff and Blass, 1983). That is, the studies of nipple attachment behavior that suggested its extraordinary independence from the newborn pup's nutritive state led to the idea that the pup itself does not make an important contribution to the control of its own intake. Milk ingestion, it is believed, is controlled by the dam, whose limited milk supply prevents the pup from consuming unlimited amounts. This chapter will review the evidence for the newborn's control and show that the belief in its absence is based on experiments that either did not examine milk ingestion (Hall et al., 1977) or that disturbed the balance struck by evolution between the young pup's real but limited control of its intake and the dam's periodic and limited supply of milk (Hall and Rosenblatt, 1977, 1978). I will then discuss weaning as a transformation during which adult feeding and drinking replace suckling rather than succeed it; the two kinds of ingestive behavior are expressions of separate neurological systems that develop in parallel rather than as stages in the maturation of a single system undergoing continuous linear development.

II. CONTROL OF MILK INTAKE BY THE NEWBORN RAT

A. Deprivation-Induced Intake and Upper GI Fullness

Control of milk intake by newborn rat pups was discovered (Houpt and Epstein, 1973) in experiments showing, first, that as early as the third day after birth, newborn rat pups respond to deprivation of mother's milk by ingesting more of it when reunited with her (Fig. 1). The pups were separated from their dam and cared for by a nonlactating foster mother during the 3 to 5 hours of deprivation. Their increased intakes (cross-hatched bars) were therefore caused specifically by milk deprivation, not by the loss of the other benefits of contact with the dam. Second, we used this phenomenon to show that the pup's intake is controlled by upper GI fullness (Fig. 2). The pups were deprived, then given milk, kaolin (a suspension of clay in isotonic saline), or water by gavage just before being reunited with their dam. Their intakes following prefeeding (cross-hatched bars), when combined with the volume of the preload (dashed outlines), made their total upper GI volume similar to the amount they would have ingested by the end of the refeeding period had they not been prefed (open bars). They seemed, in other words, to be "topping themselves up" to an optimum level of upper GI contents. We obtained this same result with both milk and kaolin preloads but not with water, which empties rapidly from the stomach, and we concluded that the pups were equipped with sensory systems for detection of upper GI distention and that these operated, essentially from birth, to control their milk intake. Houpt and Houpt (1975) and others (Drewett et al., 1974; Freidman, 1975) have con-

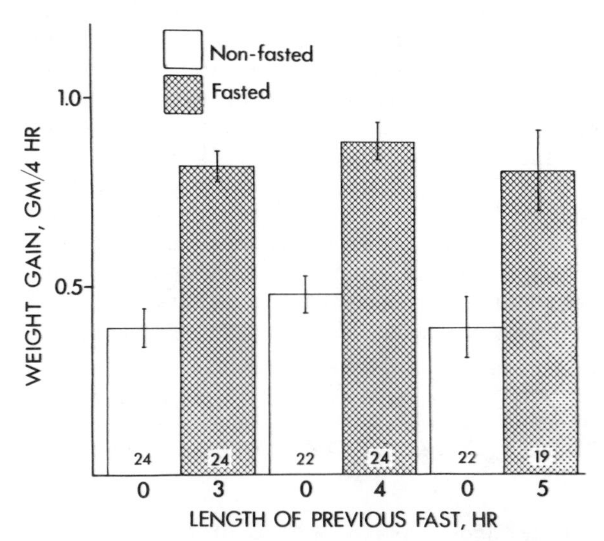

Fig. 1. Increased mother's milk intake (measured as weight gain) of 3- to 7-day-old suckling rats deprived of their dam for 3, 4, and 5 hours. Cross-hatched columns are intakes of deprived pups, open columns are intakes of their non-deprived controls. Numbers within columns are Ns, vertical lines are SEs. From Houpt and Epstein (1973), *Am. J. Physiol*. 225.

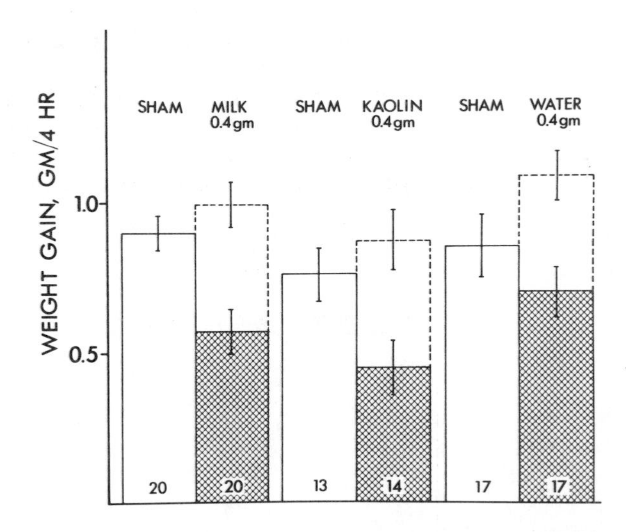

Fig. 2. Suppression of deprivation-induced mother's milk intake by upper GI fill in 3- to 7-day-old sucklings. Open columns (SHAM) are milk intakes of deprived pups that were sham intubated. Cross-hatched columns are intakes of deprived pups given gastric preloads of milk, kaolin, or water. Open columns bounded by dashed lines are volumes of the preloads. Adding preload volume to intake yields the total volume in the upper GI tract of the pup, as shown by the height of the right-hand member of each pair of columns. From Houpt and Epstein (1973).

firmed our finding in rats and extended it to other preload fluids and to older rats. Puppies (Satinoff and Stanley, 1963) and piglets (Stephens, 1975; Houpt et al., 1977) also reduce their mother's milk intake when preloaded by gavage. There is no disagreement about the reality of the phenomenon itself.

B. The Different Controls of Attachment and Ingestion in the Young Pup

Research by Lorenz and others (1982) has added much to our understanding of newborn rats' control of their milk intake. After overnight separation from their dam, pups from 1 to 20 days of age were preloaded with increasing volumes of milk prior to reunion. Their intake from their dam during 2 hours of refeeding, their latencies to reattach to a nipple, and their incidence of attachment during the opportunity to suckle were measured (Fig. 3). The experiment was then repeated at 5,

10, and 20 days of age and in 7- to 9-day-old pups with various denervations of their subdiaphragmatic viscera.

The data in Figure 3 are from non-denervated 1-day-old pups (24 to 48 hours after birth). Orderly suppression of milk intake (decreasing gains in weight during the refeeding period) occurred with increased preloads, and the final volume of the

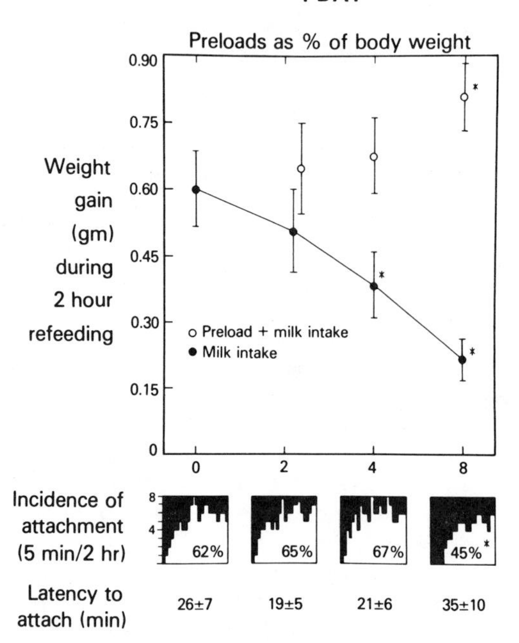

Fig. 3. Mother's milk intake (measured as weight gain), incidence of attachment, and latency to attach during a 2-hour suckling period of 1-day-old pups deprived overnight and given preloads just before refeeding. Results are arranged in vertical columns according to the preload volumes of 0, 2, 4, and 8% of body weight. Upper graph: Milk intake from the nipple (as weight gain in grams ± SE, solid circles connected by solid line). Weight gain from voluntary intake plus preload volume (unconnected open circles). Boxed Histograms: Incidence of attachment. The number of pups (one to eight) attached to a nipple during each successive 5-minute interval of the 2-hour suckling period. White areas are attachment, black areas detachment. Percents are the percentage of pups attached during the entire 2-hour period. Below: Latency to attach in minutes. *Significantly different from 0% preload at $P < 0.05$. From Lorenz, Ellis and Epstein (1981).

prefed pups' upper GI contents approached that of the animals that began the refeeding with empty stomachs and small bowels. They slowed their ingestion of milk to adjust the end-volume of their gut contents as precisely as did older pups and, indeed, adult rats. Lorenz tested his preloaded pups in litters of four, deliberately mixing pups with different preload volumes in the same test litter. Each pup controlled its milk intake according to its upper GI volume at the start of the sucking session, disregarding whether its peers were continuing to ingest (those that received no or small preloads) or had already ceased (those that received the larger preloads). Each pup was individually adjusting its intake (see also Friedman, 1975) so that its total upper GI volume increased to an optimum of approximately 8% of its total body weight.

This work confirms earlier studies of attachment behavior (Hall et al., 1975, 1977). Preloads markedly reducing milk intake had no effect whatsoever on the pups' latencies to attach to a nipple. They remained avid for a nipple even after being filled with the 8% of body weight upper GI load that terminated intake in non-preloaded pups. Their persistent attachment to a nipple throughout the 2-hour suckling period was also very little affected by their upper GI volumes.

Attachment and ingestive behaviors are therefore controlled differently in the newborn rat. By ingestion, the ultimate phase in the sequence of mammalian suckling behaviors, the newborn draws milk from a nipple after having attached to it. It *is* controlled by the animal's nutritive state. By attachment, the penultimate behavior in the sequence, the newborn approaches the dam and identifies and prehends a nipple. It is *not* sensitive, in the rat pup of 10 days of age or less, to the animal's nutritive state (Lorenz et al., 1982).

C. Hyperphagia after Visceral Denervation

Figure 4 shows Lorenz's most remarkable finding. Pups with denervated viscera (vagotomy, VX; or vagotomy plus cordotomy, SCX + VX) are released from the inhibition imposed on their ingestion by gut distention; they become hyperphagic while suckling. The pups were deprived for 5 hours. The surgeries were then performed under hypothermia with rapid rewarming and 5 hours later the animals were fed through anterior oral catheters while away from their dam. The data are therefore the first meals taken after denervation by 10-hour-deprived pups. The vagotomized (VX) animals overate for the first 40 minutes of the 1-hour test. The pups that had lost both vagal and splanchnic innervations (SCX + VX, high thoracic cordotomy severing the ascending projections from the splanchnic bed)

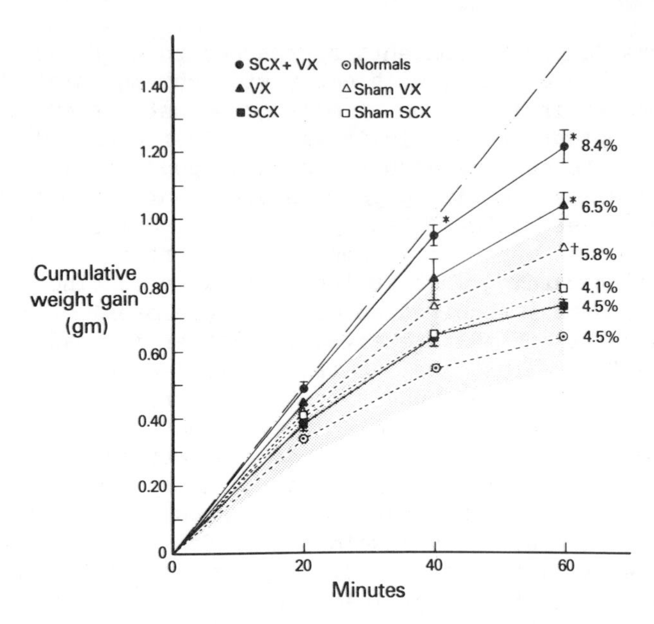

Fig. 4. Cumulative milk intake during a 60-minute feeding period of 10-hour-deprived 7- to 9-day-old pups after complete subdiaphragmatic visceral denervation (SCX + VX = high thoracic spinal cordotomy plus bilateral vagotomy), vagotomy (VX), spinal cordotomy (SCX), or sham vagotomy or sham cordotomy. Unoperated animals are Normals. The interrupted line bisecting the figure is the cumulative rate of milk infusion and would be the rate of intake of a pup that ingested all of the milk offered. The shaded area is the variance of intakes by the normal and sham operated groups. Average weight gain (milk intake) by each group at the end of the 1-hour feed are given at the right as a percentage of each group's average body weight at time 0. *Significantly different from VX and Normals. **Significantly different from SHAM VX. +Significantly different from Normals. All $P < 0.05$. From Lorenz, Ellis, and Epstein, Dev. Psychobiol. 15, copyright © 1982, John Wiley & Sons, Inc. Reprinted by permission of John Wiley & Sons, Inc.

were continuously hyperphagic compared with the normal pups, which ingested 4.5% of their body weight by the end of the 1-hour feeding period. The completely denervated pups (SCX + VX) ate continuously for the first 40 minutes and then showed only a slight and unreliable decline in intake, leading to an 8.4% weight gain. Lorenz has recently confirmed this finding in 10-day-old vagotomized pups that were allowed to suck milk from their dam in a natural litter (1983). He found, moreover, that the hyperphagia could be reduced to normal milk intake by pyloric ligation and to subnormal intake by ligation combined with gastric preloading with water, demonstrating that the satiation of milk intake (presumably mediated through the intact splanchnic afferents) is still possible in the vagotomized pup but requires extremes of gastric distention. Milk is not necessary. Non-nutritive fluids (water in Lorenz's experiments, kaolin in Houpt's earlier work) are effective in demonstrating that mechanical distention is the controlling signal.

The facts, then, are these: Pups of 10 days or less increase their gain of milk from the dam when deprived and reduce their gain when recently fed or filled, and they rely on distention-evoked afferent activity from the upper gut to control their intake. Deprivation will of course reduce the flow of these afferent discharges as the upper GI tract is progressively emptied, and ingestion of milk or preloading with fluid will increase them. The pups appear to use these afferents to adjust their intake of mother's milk so as to maintain upper GI fullness at optimum levels (4 to 6% of body weight).

D. Control of Intake by Changes in Sucking

Neonates can both increase and reduce their own intake. During the 12 to 15 hours a day that the dam is present and tented over them, they are attached almost continuously to one of her nipples, and she lets down milk once every 5 to 20 minutes (Lincoln et al., 1973; Wakerley and Lincoln, 1971). How do they achieve control of their intake of mother's milk? Brake and his colleagues have provided the answer by an informative series of experiments using electromyography of the digastric muscle to monitor sucking behavior in rat pups (Brake et al., 1979; Brake and Hofer, 1980; Brake et al., 1982a,b; Shair et al., 1984). In their typical experiment, a single pup is allowed to attach to a urethane-anesthetized dam (with only a single pup and under anesthesia the dam does not lactate) and its sucking (digastric EMG) is recorded for time spans from 30 minutes to 3 hours. Two kinds of sucking were recorded, *rhythmic runs* and *bursts*. Rhythmic runs (Fig. 5, above) are 5-second or longer episodes of alternating increases and decreases in EMG amplitude occurring at rapid rates. They are sometimes accompanied by rhythmical opening and closing of the mouth and are associated with a series of evenly spaced waves of negative pressure in the oropharynx (Brake et al., 1979). Rhythmic runs in rat pups are the same in all important respects as nutritive sucking as described in human infants (Kron et al., 1967; Wolff, 1978). Bursts (Fig. 5, below) are discrete episodes of arrhythmic high frequency activity unaccompanied by obvious movements of the jaws. Brake and colleagues also recorded, as an artifact of the digastric electrode, movements of the forelimbs, called *treadles*, during which the animals appear to be kneading the area around the nipple (Fig. 5, T). In 3- to 5-day-old pups, 20 to 24 hours of deprivation by separation from the dam produced large increases in rhythmic runs (from complete absence in pups deprived for only 2 to 6 hours to a total of almost 30 minutes of such sucking during a 3-hour session of recording). The rate of treadling was also increased by deprivation. Bursts did not change. Rhythmic runs were also increased by milk ejections from unanesthetized dams (Brake et al., 1982b), and they increased conspicuously in pups having continuous access to milk from an oral catheter. The same selective increase in rhythmic sucking was produced in older pups (10 to 12 days) by 20 to 24 hours of separation from the dam, and this increase was reduced by intragastric preloads of milk (see also Blass et al., 1979). In this experiment the animals were again attached to a nonlactating anesthetized dam, but some of the pups were provided with a continuous supply of milk through an oral catheter. The decrease in rhythmic sucking occurred only in the pups receiving GI loads of milk and was accompanied by decreases in milk intake similar to those reported to us in the 10-day-old (Houpt and Houpt, 1975; Lorenz et al., 1982).

The behavior by which the neonate rat achieves control of its milk intake consists, therefore, of episodes of rhythmic sucking executed while the pup is attached to a nipple. It is

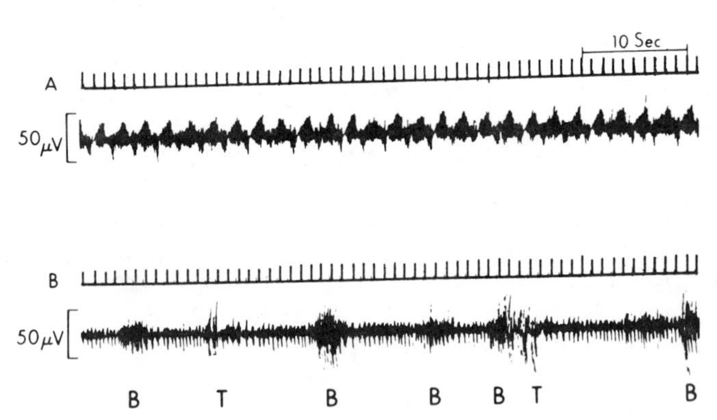

Fig. 5. Electromyographic recordings from the digastric muscle of 20- to 24-hour deprived 10- to 12-day-old pups suckling an anesthetized dam. A rhythmic run is shown in A, bursts (B) and treadles (T) are shown in B. From Brake, Wolfson and Hofer (1979). Copyright 1979 by the American Psychological Association. Reprinted by permission of the author.

the only kind of sucking that increases when the pup is milk-deprived, as it would be if it were dislodged from the nipple and then reattached later or if the dam were to leave the nest for foraging and then return. Rhythmic runs also increase when milk is delivered into the mouth, and this increase is suppressed by upper GI fullness (Brake et al., 1982a). If, as seems very likely, rhythmic runs are elicited by the nipple engorgement and milk leakage that may occur with milk letdowns, then the work of Brake and his colleagues explains the results of Houpt's and Lorenz's studies of intake control in pups 10 days of age or younger. It provides direct evidence for the behavior by which the control is exercised. As the pups age, detachment from the nipple and reattachment under the control of GI afferents (nipple shifting) become additional options for intake control (Cramer et al., 1980).

Of course, the pups must exert their control under the ceiling of the dam's milk supply. They cannot ingest more than she provides. Successful suckling therefore requires a continuous duet between the dam and her litter in which the pups' control of their intake, milk availability from the dam, and her sensitivity to the suckling stimulus are all involved (Friedman, 1975; Lau and Henning, 1984).

E. The Independence of Attachment from Control by Nutritive State

This interplay between the dam and her offspring should not be thought of as occurring without the pups' active participation. This belief arose from the extraordinary experiments of Hall, Cramer, and Blass (1977) showing the independence of nipple attachment behavior from nutritive state (Fig. 6). They positioned three or four pups with their mouths in contact with the ventrum of a nonlactating anesthetized dam in order to make her a passive object of the pups' behaviors and found that 22 hours of separation from the dam (open circles), a severe deprivation in the young pup, produced latencies to attach indistinguishable from those of pups taken from the litter just 15 or 20 minutes before testing (solid circles). Differences in latency based on recency of suckling did not appear until the pups were 11 to 13 days old, after which the latencies diverged so widely that many 20- to 22-day-old pups did not attach at all in the 5-minute test. These findings were confirmed by Lorenz (1982) with his measures of the attachment behavior of the 1-day-old pups (Fig. 3, latency and incidence of attachment). Older pups (10 to 20 days old) were equally avid for attachment over the same range of upper GI volumes, but because they are behaviorally more competent they attached more rapid-

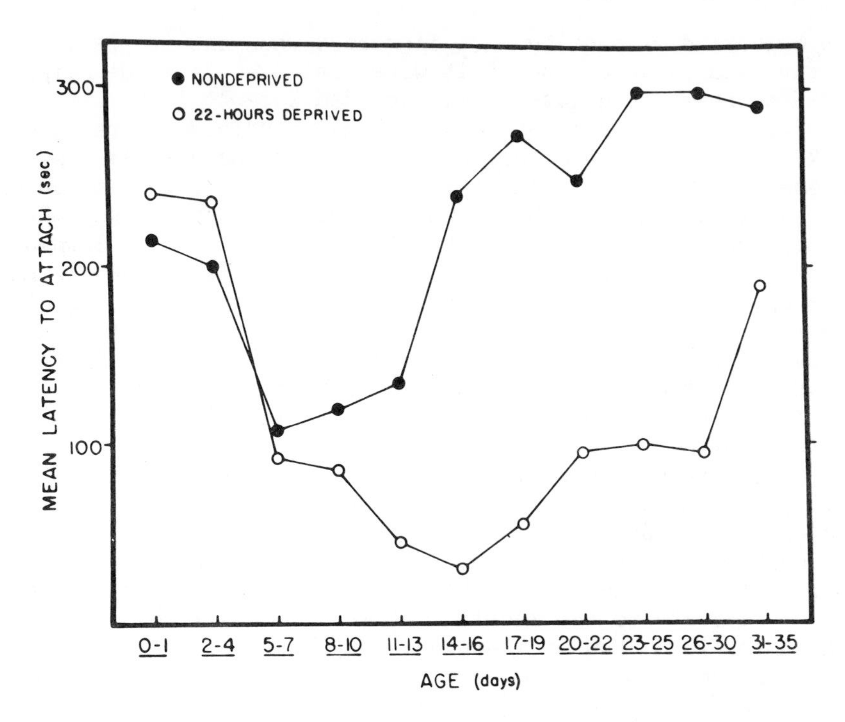

Fig. 6. Mean latency to attach to the nipples of anesthetized rat dams by their deprived and nondeprived pups of increasing age. From Hall, Cramer, and Blass (1979). Copyright 1979 by the American Psychological Association. Reprinted by permission of the author.

ly. Rapid attachment by both fed and unfed pups under 10 days of age was discovered earlier by Hall and his colleagues (1975, 1977) in deprivation studies and seen again by Hall (1979) and by Hall and Rosenblatt (1977, 1978). Again, there is no disagreement about the phenomenon itself.

Teicher and Blass (1976) and Hofer and colleagues (1976) discovered that the young pup's attachment is motivated and guided by exteroceptive odor cues, and Hall, Blass, and Cramer showed that the pup is insensitive to nutritive state. The idea was generalized to all of suckling behavior with no distinction between attachment and ingestive behaviors. Hall, Cramer, and Blass (1977) concluded that suckling by the young pup was "not primarily under interoceptive control." They reinterpreted our prior demonstrations of such control by young pups as the result of "general modulation of pups' behavioral vigor," or, in other words, the result of nonspecific changes in activity.

These experiments of Blass, Hall, and colleagues did not examine ingestion (the pups received no milk) and they were done before Lorenz's and Brakes's studies were published. Lorenz has since shown that upper GI afferents can suppress milk ingestion but are entirely uncoupled from nipple attachment and that latency and incidence of nipple attachment by pups 10 days old or less is as rapid and tenacious immediately after gastric preloading or vagotomy as it is in untreated pups. Brake has further shown that deprivation and upper GI filling modulate rhythmic runs and are therefore behaviorally specific in their effects on suckling behavior; thus, the conclusion of Blass, Hall, and colleagues is unnecessarily vague and imprecise. In a recent review of this issue, Hall (with Williams, 1983) agrees that the Lorenz and Brake evidence demonstrates that young pups attached to a maternal nipple can control their intake by changing the vigor of their sucking.

F. Forced Overingestion through Posterior Oral Catheters

After the studies of attachment behavior by Hall et al., Hall and Rosenblatt (1977, 1978) studied milk ingestion in pups from 5 to 20 days old, but they did so with a method that disturbed the balanced relationship of the dam as milk provider and the pup as milk consumer. They used anesthetized dams, which do not lactate, and they provided milk to the pup through a catheter that pierced the base of the tongue and delivered fluids into the posterior oral cavity at the high rate of 0.1 ml every minute. They found that younger pups (5 and 10 days old) feeding in this way consumed large volumes of milk and were insensitive to deprivation. The pups became engorged with milk and visibly distended, and readily reattached to a nipple until their respiration was embarrassed by reflux of milk out of their nares. These effects were interpreted as dramatic evidence of the young pup's helplessness in limiting its milk intake. They are instead a demonstration of the predicament in which the young pup can be trapped by its avidity for nipple attachment and the irresistibility of its swallowing reflex. The tenacity with which pups of these ages will hold to a nipple was emphasized in the Hall and Rosenblatt work by their observation that pups which had just fallen off a nipple, cyanotic and gasping for breath, with milk running from their mouths and noses, would nevertheless reattach rapidly if held up to the dam's ventrum. The posterior tongue catheter will deliver milk into the back of the mouth and pharynx, from which it must be swallowed at the rate chosen by the experimenter. The particular rate used by Hall and Rosenblatt (0.1 ml/minute) exceeds the rate at which milk is normally provided to young pups by their dam. The pups' control of ingestion by rhythmic sucking is bypassed with this technique, and their only defense against milk engorgement is detachment from the nipple, an infrequent behavior in the pups younger than 10 days old.

Hall and Rosenblatt showed that 10-day-old pups that had been preloaded with 5% of their body weight of either milk or isotonic saline and were then placed on the anesthetized dam and provided with milk through a posterior tongue catheter did detach and then reattach to a nipple without assistance (nipple shifting). They did so at a rate that was almost equal to that of 20-day-olds, and this liberated them from their predicament. By employing this behavioral option they revealed their capacity to control their intake, reducing it by almost 50% as a result of the gastric distention produced by the preloads. But they had only partial control. Whenever a pup lost contact with a nipple and did not reattach for three minutes (nipple *detachment*, not nipple shifting) the experimenter held it at the nipple continuously until it refused to reattach. This refusal to reattach was interpreted as the behavioral sign of the termination of ingestion and the test was ended. But whenever a pup did reattach, it was again trapped in the predicament between avid attachment and forced swallowing, which became increasingly inescapable because nipple-shifting decreased in frequency as milk intake accumulated. The sessions ended with all pups, preloaded or not, grossly distended by stomach volumes of 10% of their body weight. Hall himself in later work (1979) showed that the effects of deprivation are seen with oral catheters only when they are implanted in the anterior oral chamber (just behind the lower incisors under the tongue); for that reason, in my laboratory we implant them only in that position.

Recently, Cramer and Blass (1983) have restored milk letdowns to the anesthetized dam, confirming the essential findings of Houpt and Epstein (1973) and Lorenz et al. (1982). They gave pups access to an anesthetized dam in which the periodic and limited milk supply of the normal dam was imitated by ten oxytocin injections at 4-minute intervals. When suckling from these oxytocin-treated dams, pups of all ages from 5 to 20 days increased their milk intake when deprived and decreased it after gastric preloads of milk. Cramer and Blass agree that changes in sucking behavior while attached to a nipple must be the mechanism of these changes in milk intake. They also find that pups move more frequently to engorged, unsuckled nipples as the length of deprivation increases, except for the 5-day-olds, in which this behavior is rare. Thus, nipple shifting is identified as a behavior available to pups for intake control. But when the anesthetized dams were made supernormal, that is, when the 15-per-hour oxytocin-induced letdowns were continued for 2 hours and a milk-laden, unsuckled dam was introduced for the second hour, young pups could not limit their intake. They remained attached, gorging themselves on milk until they fell off the nipple in respiratory distress, in confirmation of Hall and Rosenblatt. Cramer and Blass interpret this rightly as demonstrating that pups of 10 days or less cannot defend themselves against overingestion, but this is a defense they do not need. Normal dams do not let down at regular 4-minute intervals for as long as 2 hours. The first letdown after attachment is delayed by approximately 10 minutes; others follow at rates that begin in the 4- to 6-minute range but then diminish to 15-20 minutes as the suckling bout continues (Lincoln et al., 1973). Moreover, the dams rarely remain with the litter for 2 hours at a time, even in the daytime. At night they are away for long periods while they forage (Grota and Ader, 1969). The pups themselves are from time to time dislodged from the nipple by their peers or by the mother's grooming of them or herself, and the younger animals take longer to attach. In addition, pups in a normal litter contribute to the timing of their dam's milk letdowns. When they are with an unanesthetized and responsive dam the "chorus" of suckling stimulation they provide makes an important contribution to her oxytocin release (Wakerley and Lincoln, 1973; Wakerley and Drewett, 1975). During a prolonged episode of suckling the pups suck less vigorously as they satiate, making oxytocin release and consequent milk letdown less frequent. This control of letdown frequency by the pups was frustrated in the Cramer and Blass experiment. Their experiment distorted the mutual relations between the dam and her pups that place a ceiling over the amount of milk which can be suckled, and they were misled by the pups' overingestion to the conclusion that young pups control the rate but not the end-volume of their milk intake. Lorenz's experiments show that such control is exercised by pups as young as 1 day old.

The pups are equipped with attachment and sucking behaviors that provide them with a sufficient share of the mother's milk supply to maintain their upper GI volume at between 4% and 6% of their body weight. They simply do not need to be protected from too much milk. The rat dam's milk supply is less than maximum for a litter of normal size. The pups that die in the first 10 days of the suckling period usually do so from starvation, not from milk engorgement.

G. Summary: Attachment and Ingestion Prior to the Mid-Suckling Transition

We have therefore arrived at the following understanding of the young pup's capacities for control of the attachment and ingestive phases of its suckling behavior. Until the 10- to 15-day mid-suckling transition has begun, attachment behavior preliminary to and necessary for suckling is controlled entirely by olfactory and somatosensory cues, the smell of the saliva adhering to the nipple and the feel of the nipple when the pup makes contact with the dam's ventrum. Approach to the dam and nipple attachment (what Hall and Williams refer to as "suckling initiation") are not, in this neonatal period, under the control of the pup's nutritive state, but milk ingestion is. The several behaviors of sucking with which the pup extracts milk from the nipple, rhythmic or nutritive sucking and treadling in particular, are controlled by its nutritive state, specifically by visceral afferents evoked by upper GI fill. This control is bidirectional: When fill is reduced, sucking is increased. When fill is increased, sucking is reduced. It is sensitive both to milk availability and to the pup's GI fill. All attached pups are aroused to vigorous sucking by spontaneous milk letdowns (Wakerley and Drewett, 1975), and pups as young as 1 day of age can appreciate as little as a 2% of body weight difference in the volume of their upper GI contents and adjust their sucking behavior accordingly--it is regulatory. By modulating their sucking behavior in accordance with the volume of their GI fill, they maintain it at an optimum level of between 4% and 6% of their body weight. And modulation of sucking behavior is appropriate in its ontogenetic appearance, being available to the neonate pup, to whom ingesting an adequate share of the mother's milk supply is essential. But, because nipple attachment is insensitive to GI fill, the neonatal pup cannot protect itself from overingestion when it is supplied with excess amounts of milk. Fortunately, this is a laboratory artifact.

The immaturity of the neonatal rat is not expressed in its inability to control its milk intake while attached to a nipple, but in the uncoupling of its need for food from the behaviors that make milk intake possible--its approach to the dam and its readiness to attach to one of her nipples.

During the 10- to 15-day transition, approach to the dam and nipple attachment come under the control of visceral afferents from the upper GI tract, and both phases of suckling (its initiation by approach and attachment, and its culmination by sucking and by swallowing) occur when food is needed. The animal's ingestive behavior is now more adult-like, and weaning has begun.

III. The Emergence of Feeding and Drinking at Weaning

Feeding and drinking are not suckling--the behaviors are different. Adult animals do not suck their food and most of them do not suck water. The materials ingested are different. The adult ingests foods and water as separate commodities, the suckling ingests only milk, and the underlying physiologies are different. Until weaning has begun at the mid-suckling transition, the suckling's approach to the dam, the source of food, is not controlled by hunger as feeding is in the adult. And although milk is accepted as a source of water by thirsty adult rats, young pups do not use their sucking behavior to gain more of it when dehydrated (Bruno et al., 1982; Cramer et al., 1984).

All of these differences should have made it obvious when work on the development of the ingestive behaviors of the rat pup began in the early 1970s that suckling, on the one hand, and feeding and drinking, on the other, had separate neurological mechanisms that underwent parallel ontogenies. But for me, at least, this was not an obvious idea. I began my research on this problem believing that adult ingestive behaviors develop in a serial fashion out of suckling (Houpt and Epstein, 1973) in the same way that walking develops from crawling in the human, and I continued to hold to this analogy after Wirth and I (1976) discovered that rat pups less than one week old would drink water in response to the same challenges that evoke thirst in the adult. The separateness of the suckling and adult ingestion was made obvious to us all by Hall in his important paper of 1979, which revealed the existence of a system for ingestive behavior that was different from suckling and that could be studied from birth.

A. Precocious Feeding by Young Pups Away from Their Dam

Hall had already shown (1975), in experiments that were a technical tour de force, that rat pups which were raised from birth on intragastric feeding and had therefore never suckled were nevertheless remarkably normal in their feeding and drinking behavior at the age of normal weaning. If suckling was a necessary developmental prelude to adult ingestion, then these tube-fed rats should have been unable to eat and drink normally as juveniles. That they were not was an adumbration of the separateness of the two systems for ingestive behavior. In his later work Hall (1980) made the separateness explicit by feeding 1- to 6-day-old pups away from the dam. He infused them with milk through catheters opening into the front of the mouth from which fluids could be either swallowed or rejected, and in later experiments (Hall and Bryan, 1980) he gave them access to puddles of milk at their feet. He found that when the milk was infused they ingested it not by suckling but by licking and chewing movements; when it was puddled onto the floor of the cage, they ingested it by probing into it and swinging their heads back and forth while licking. In addition, the infusions of milk into the mouth did not elicit the extensor stretch aroused by milk letdown and accompanying rhythmic sucking in the pup attached to a nipple. Instead, it aroused a repertory of excited behaviors (mouthing, rolling, curling, crawling, grooming) that do not occur during suckling. Moreover, he found that ingestion away from the dam through anterior oral catheters was responsive to deprivation; this was interpreted as a characteristic of adult feeding. He concluded that presenting milk to the pup while it was separated from its dam elicited intake controlled by an "adult-like ingestive system" present in the infant rat.

This idea made Wirth's and my earlier results understandable. We had subjected rat pups younger than 1 week to cellular dehydration, hypovolemia, and renin release (treatment with isoproterenol) and had found that they drank water when it was offered to them by infusion into the front of their mouths. This was done, of necessity, while the pups were away from their dam. Hall's work made it clear to us that separating rat pups from their dam had freed them to express behavior that normally does not occur until they are weaned.

B. Orexigenic Effect of Norepinephrine OFF but Not ON the Dam

The clearest example from my own work of the separateness of the neurological systems for suckling and adult ingestion is the orexigenic effect of intracranial norepinephrine. In experiments with Ellis and Axt (1984), we injected 2 μg of NE into the third ventricle of the pup (in a benign and rapid procedure that requires no anesthesia and leaves the animal competent for behavior immediately after the injection is made) and then allowed it to ingest milk infused into its anterior oral cavity. The amine elicited an obvious increase in milk intake (Fig. 7, OFF Dam). We found that it occurred abruptly on the tenth day of age and that it was necessary to prefeed the pups to satiation (Fig. 5, milk pre-injection) in order to reveal it. But when Ellis repeated the NE injections and allowed the pups to suckle, they did not ingest more mother's milk (Fig. 7, On Dam). These pups had been prefed to satiety and were the same age as the pups that had expressed the orexigenic effect of NE while feeding away from the dam. This failure to ingest more mother's milk was not the result of the

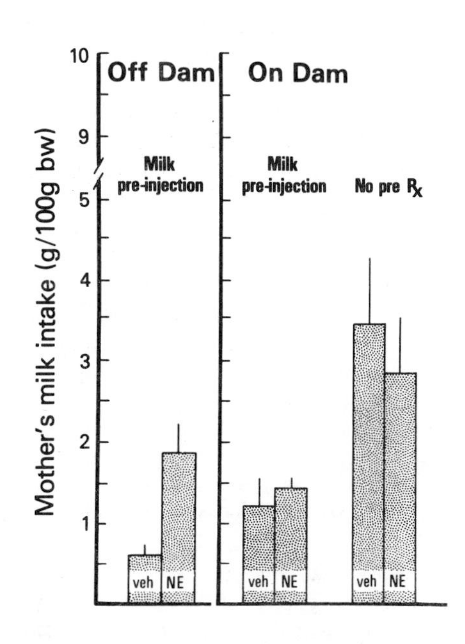

Fig. 7. Lack of effects of intracerebroventricular (i.c.v.) norepinephrine (NE) on suckled milk intake in 11- to 13-day- old pups. On Dam experiments (right) show that pups both satiated on milk (Milk pre-injection) and not satiated (No pre-RX) do not ingest more mother's milk after injection of two μg of NE into their third ventricles. The Off Dam experiment (left) shows that pups of the same age and with the same treatment (pre-satiated) express the orexigenic effect of i.c.v. NE when they are fed while separated from their dams. From Ellis, Axt, and Epstein (1984).

dam's limited milk supply or of the pups' inability to ingest more under the conditions of the sucking or On Dam test. When we repeated the experiment in pups that were not prefed, they ingested twice as much milk (Fig. 7, far right) from their dam. In exact analogy to these results with intracranial norepinephrine, Raskin and Campbell (1981) found that systemic amphetamine is anorexigenic in 5-day-old pups, but only when they are feeding away from the dam. When treated with the drug while they are suckling in a natural litter, they eat more. The drug does not decrease suckling behavior until they are 15 days old.

C. Other Evidence for Separate Systems for Suckling and Adult Ingestion

There are other equally striking examples of the existence of two neural systems for ingestive behavior in the young pup. Ten-day-old rat pups exposed to an arbitrarily flavored milk through a mouth catheter while attached to a nipple do not learn to avoid that flavor after they have been poisoned with lithium chloride. If the pairing of the flavor and illness occurs while they are feeding away from the dam they have no trouble learning. Two weeks later, as weanlings, they avoid diets adulterated with the flavor (Martin and Alberts, 1979). Cholecystokinin does not reduce the intake of young pups ingesting milk while attached and sucking (Blass et al., 1979), but it is satiating when they are feeding independently of the dam (Goldrich et al., 1984).

Together with the results of the norepinephrine and amphetamine work, these results suggest that there are two neurologic systems for ingestive behavior in the brain of the rat pup, and the behavior that the pup is performing (sucking on a nipple held deep in the mouth or licking and chewing food or water brought into the mouth) somehow engages the system appropriate for its control and disengages the other. Sucking on a nipple is, of course, the behavior expressed by the newborn pup, and the system for adult ingestion is "silent" or disengaged during the neonatal period. The adult system can be aroused, however, by electrical stimulation of the lateral hypothalamus in pups as young as 3 days of age (Moran et al., 1983), and it is sufficiently advanced in development to be utilized for consumption of either food or water at very early ages, but only when the pup is required by isolation from its dam to perform acts of adult ingestion (Wirth and Epstein, 1976; Hall, 1979). The adult ingestive system can be damaged selectively by neurotoxins. Treatment of the 3-day-old brain with 6-hydroxydopamine, a selective poison of dopaminergic neurons, has virtually no effect on suckling behavior, but the same treatment administered to adults devastates their ingestive behaviors, disabling them almost as completely as lateral hypothalamic damage (Bruno, 1981; Bruno et al., 1984).

D. Summary: Parallel Ontogenies of Suckling and Adult Ingestion

Rats and other mammals are prepared for their first ingestive behavior by the in utero development of a neurological system for suckling. By using it at birth they complete its development (Dollinger et al., 1978) and it controls their ingestive behavior until weaning begins. However, suckling does not appear to be the developmental prelude to adult ingestion. Instead, a second neurological system for ingestion is nascent in the suckling rat pup (Hall et al., 1977; Blass et al., 1979; Hall and Williams, 1983). It appears to be developing in the neonate, without expression in behavior, in order to replace suckling when the weanling leaves the dam. Each of the systems has its own controls, and we are just learning what they are. Olfactory signals from the dam's nipples are powerful controls of the behaviors by which suckling is initiated, and sensory events in the mouth must be crucial for the control of sucking itself. Olfactory cues also appear to be important for the initiation of feeding in the rat. The odor of the dam's caecotroph attracts the pups and it is the first food they eat, but its odor becomes effective only after they lose their compulsive avidity for the nipple and are able to detach from it. Amines such as norepinephrine and amphetamine control the nascent system for adult ingestion in the neonate just as they do in the adult. So do cholecystokinin and several determinants of thirst, but these are not parts of the mechanism for control of suckling. The systems have several controls in common. Both are aroused by deprivation, and upper GI distention suppresses both, as does the adrenergic activation of the liver. Pups or piglets that are either sucking from a nipple or feeding independently of the dam (or sow) eat more after being deprived of mother's milk, and their intakes are sensitive to their GI volume (Houpt and Epstein, 1973; Houpt et al., 1977). In addition, they will eat less in both settings when treated with intraperitoneal epinephrine (Rodreiguez-Zendejas et al., 1985). The neurological systems are parallel but not completely isolated from each other.

Suckling is not the kindergarten for adult feeding and drinking. Adult feeding behaviors do not develop from suckling like walking develops from crawling. Instead, suckling appears to be the analog of the neonatal walking or stepping done by newborn humans whose head and trunk are supported upright while the plantar surfaces of their feet touch the ground (Zelaso et al., 1972). Such infants make rhythmic stepping movements and extend their legs with sufficient strength to support themselves while they walk forward. However, this behavior disappears from the infant's repertory between 6 weeks to 4 months of age, during which time it flexes its legs when held upright with feet in contact with the ground. Infant neonatal walking is now replaced by crawling, the true prelude to walking. It is not uncommon for vertebrates to have dual neurologic mechanisms for locomotion. Many birds both walk and fly, and some of them do one (usually walk) before the other. Amphibians swim as larvae and walk or hop as adults, and again they do so in developmental sequence (Stehouwer and Favel, 1981).

We are suggesting that the neural mechanisms for ingestion in mammals are similarly organized. There are two of them and they seem to have evolved separately. One, feeding and drinking, is ancient and is common to vertebrates. The other, suckling, is a recent behavior that has been added by evolution to the repertory of newborn mammals. Each mechanism has its own neurology and its own set of afferent and neurochemical controls. Finally, each has its own developmental schedule such that suckling is expressed first and is suppressed at weaning (Williams et al., 1979) as it is replaced by feeding and drinking, which then emerge as the animal's exclusive ingestive behaviors.

ACKNOWLEDGEMENTS

The author's research was supported by NS 03469, HD 17992, and NATO RG/502.

I am grateful to Elliott Blass, Ted Hall, Steve Brake, and Jay Schulkin for their willingness to read this chapter prior to its publication and for the frankness of their criticism.

REFERENCES

Blass, E.M., W. Beardsley, and W.G. Hall. 1979. Age-dependent inhibition of suckling by cholecystokinin. *Am. J. Physiol*. 236: E567-570.

Blass, E.M., and C.P. Cramer. 1982. Analogy and homology in the development of ingestive behavior. In *Changing Concepts of the Nervous System*, ed. A.R. Morrison and P.L. Strick, 503-523. Academic Press, New York.

Blass, E.M., W.G. Hall, and M.H. Teicher. 1979. The ontogeny of suckling and ingestive behaviors. In *Progress in Psychobiology and Physiological Psychology*, ed. J.M. Sprague and A.N. Epstein, 243-299. Academic Press, New York.

Brake, S.C. and M.A. Hofer. 1980. Maternal deprivation and prolonged suckling in the absence of milk alter the frequency and intensity of sucking responses in neonatal rat pups. *Physiol. Behav*. 24: 185-189.

Brake, S.C., D.J. Sager, R. Sullivan, and M.A. Hofer. 1982a. The role of intraoral and gastrointestinal cues in the control of sucking and milk consumption in rat pups. *Dev. Psychobiol*. 15: 543-556.

Brake, S.C., V. Wolfson, and M.A. Hofer. 1979. Electromyographic patterns associated with non-nutritive sucking in 11-13-day-old rat pups. *J. Comp. Physiol. Psychol*. 93: 760-770.

Bruno, J.P. 1981. Development of drinking behavior in preweanling rats. *J. Comp. Physiol. Psychol*. 95: 1016-1027.

Bruno, J.P., L. Craigmyle, and E.M. Blass. 1982. Dehydration inhibits suckling behavior in weanling rats. *J. Comp. Physiol. Psychol*. 96: 405-415.

Bruno, J.P., A.M. Snyder, and E.M. Stricker. 1984. Effect of dopamine-depleting brain lesions on suckling and weaning in rats. *Behav. Neurosci*. 98: 156-161.

Cramer, C.P., and E.M. Blass. 1983. Mechanisms of control of milk intake in suckling rats. *Am. J. Physiol*. 245: R154-R159.

Cramer, C.P., E.M. Blass, and W.G. Hall. 1980. The ontogeny of nipple-shifting behavior in albino rats: mechanisms of control and possible significance. *Dev. Psychobiol.* 13: 165-180.

Cramer, C.P., J.F. Pfister, and E.M. Blass. 1984. Transitions in the dehydration-induced inhibition of milk intake in suckling rats. *Physiol. Behav.* 32: 691-694.

Dollinger, M.J., W.R. Holloway, and V.H. Denenberg. 1978. Nipple attachment in rats during the first 24 hours of life. *J. Comp. Physiol. Psychol.* 92: 619-626.

Drewett, R.F. 1978. Gastric and plasma volume in the control of milk intake in suckling rats. *Q. J. Exp. Psychol.* 30: 755-764.

Drewett, R.F., C. Statham, and J.B. Wakerley. 1974. A quantitative analysis of the feeding behavior of suckling rats. *Anim. Behav.* 22: 907-913.

Ellis, S., K. Axt, and A.N. Epstein. 1984. The arousal of ingestive behaviors by the injection of chemical substances into the brain of the suckling rat. *J. Neurosci.* 4: 945-955.

Epstein, A.N. 1984. The ontogeny of neurochemical systems for control of feeding and drinking. *Proc. Soc. Exp. Biol. Med.* 175: 127-134.

Epstein, A.N., E.M. Blass, M.L. Batshaw, and A.D. Parks. 1970. The vital role of saliva as a mechanical sealant for suckling in the rat. *Physiol. Behav.* 5: 1395-1398.

Friedman, M.I. 1975. Some determinants of milk ingestion in suckling rats. *J. Comp. Physiol. Psychol.* 89: 636-647.

Goldrich, M.S., P.H. Robinson, P.R. McHugh, and T.H. Moran. 1984. Cholecystokinin inhibition of independent milk ingestion in neonatal rats. International Society for Developmental Psychobiology, Baltimore, Maryland.

Grota, L.J., and R. Ader. 1969. Continuous recording of maternal behavior of *Rattus norvegicus*. *Anim. Behav.* 17: 722-729.

Hall, W.G. 1975. Weaning and growth of artificially reared rats. *Science* 190: 1313-1314.

Hall, W.G. 1979. The ontogeny of feeding in rats: ingestive and behavioral responses to oral infusions. *J. Comp. Physiol. Psychol.* 93: 977-1000.

Hall, W.G., and T.E. Bryan. 1980. The ontogeny of feeding in rats. II. Independent ingestive behavior. *J. Comp. Physiol. Psychol.* 94: 746-756.

Hall, W.G., C.P. Cramer, and E.M. Blass. 1975. Developmental changes in suckling of rat pups. *Nature* 258: 318-320.

Hall, W.G., C.P. Cramer, and E.M. Blass. 1977. The ontogeny of suckling in rats: transitions toward adult ingestion. *J. Comp. Physiol. Psychol.* 91, 1141-1155.

Hall, W.G., and J.S. Rosenblatt. 1977. Suckling behavior and intake control in the developing rat pup. *J. Comp. Physiol. Psychol*. 91: 1232-1247.
Hall, W.G., and J.S. Rosenblatt. 1978. Development of nutritional control of food intake in suckling rat pups. *Behav. Biol*. 24: 413-427.
Hall, W.G., and C.L. Williams. 1983. Suckling isn't feeding, or is it? A search for developmental continuities. In *Advances in the Study of Animal Behavior* 13: 219-254, Academic Press, New York.
Henning, S.J. 1981. Postnatal development: coordination of feeding, digestion, and metabolism. *Am. J. Physiol*. 241: G199-214.
Hofer, M.A., H. Shair, and P. Singh. 1976. Evidence that maternal ventral skin substances promote suckling in infant rats. *Physiol. Behav*. 17: 131-136.
Houpt, K.A. and A.N. Epstein. 1973. The ontogeny of food intake in the rat: GI fill and glucoprivation. *Am. J. Physiol*. 225: 58-60.
Houpt, K.A., and T.R. Houpt. 1975. Effects of gastric loads and food deprivation on subsequent food intake in suckling rats. *J. Comp. Physiol. Psychol*. 88: 764-772.
Houpt, K.A., T.R. Houpt, and W.A. Pond. 1977. Food intake controls in the suckling pig: glucoprivation and gastrointestinal factors. *Am. J. Physiol*. 232: E510-E514.
Jouvet-Mounier, D., L. Astic, and D. Lacote. 1969. Ontogenies of the states of sleep in the rat, cat and guinea pig during the first postnatal month. *Dev. Psychobiol*. 2: 216-239.
Kron, R.E., M. Stein, K.E. Goddard, and M.D. Phoenix. 1967. Effects of nutrient upon the sucking behavior of newborn infants. *Psychosom. Med*. 29: 24-32.
Lau, C., and S.J. Henning. 1984. Regulation of milk ingestion in the infant rat. *Physiol. Behav*. 33: 809-815.
Leon, M., and H. Moltz. 1972. The development of the phermonal bond in the albino rat. *Physiol. Behav*. 8: 683-686.
Lorenz, D.N. 1983. Effects of gastric filling and vagotomy on ingestion, nipple attachment, and weight gain by suckling rats. *Dev. Psychobiol*. 16: 496-483.
Lorenz, D.N., S.B. Ellis, and A.N. Epstein. 1982. Differential effects of upper gastrointestinal fill on milk ingestion and nipple attachment in the suckling rat. *Dev. Psychobiol*. 15: 309-330.
Lincoln, D.W., A. Hill, and J.B. Wakerley. 1973. The milk-ejection reflex of the rat: an intermittent function not abolished by surgical levels of anesthesia. *J. Endocrinol*. 57: 459-476.
Martin, L.T. and J.R. Alberts. 1979. Taste aversions to mother's milk: the age-related role of nursing in

acquisition and expression of a learned aversion. *J. Comp. Physiol. Psychol.* 93: 430-445.

Moran, T.H., G.J. Schwartz, and E.M. Blass. 1983. Stimulation-induced ingestion in neonatal rats. *Dev. Brain Res.* 7: 197-204.

Raskin, L.A., and B.A. Campbell. 1981. The ontogeny of amphetamine anorexia: a behavioral analysis. *J. Comp. Phsyiol. Psychol.* 95: 425-435.

Rodreiguez-Zendejas, A.M., G. Chambert, M.C. Lora-Vilchis, and A.N. Epstein. 1985. Ontogeny of adrenaline-induced anorexia in rats. *Am. J. Physiol.* In press.

Satinoff, E., and W.C. Stanley. 1963. Effects of stomach loading on suckling behavior in neonatal puppies. *J. Comp. Physiol. Psychol.* 56: 66-68.

Shair, H., S.C. Brake, and M.A. Hofer. 1984. Suckling in the rat: evidence for patterned behavior during sleep. *Behav. Neurosci.* 98: 366-370.

Stehouwer, D.J., and P.B. Favel. 1981. Sensory interactions with a central motor program in anuran larvae. *Brain Res.* 202: 131-140.

Stephens, D.B. 1975. Effects of gastric loading on the suckling response and voluntary milk intake of neonatal piglets. *J. Comp. Physiol. Psychol.* 88: 796-805.

Stoloff, M.L., and E.M. Blass. 1983. Changes in appetitive behavior in weanling-age rats: transition from suckling to feeding behavior. *Dev. Psychobiol.* 16: 439-453.

Teicher, M.H., and E.M. Blass. 1976. Suckling in newborn rats: eliminated by nipple lavage, reinstated by pup saliva. *Science* 193: 422-425.

Wakerley, J.B., and R.F. Drewett. 1975. Pattern of suckling in the infant rat during spontaneous milk ejection. *Physiol. Behav.* 15: 277-281.

Wakerley, J.B., and D.W. Lincoln. 1971. Intermittent release of oxytocin during suckling in the rat. *Nature (London)* 223: 180-181.

Wakerley, J.B. and D.W. Lincoln. 1973. The milk-ejection reflex of the rat: a 20- to 40-fold acceleration in the firing of paraventricular neurons during oxytocin release. *J. Endocrinol.* 57: 477-493.

Williams, C.L., J.S. Rosenblatt, and W.G. Hall. 1979. Inhibition of suckling in weaning-age rats: a possible serotonergic mechanism. *J. Comp. Physiol. Psychol.* 93: 414-429.

Wirth, J.R., and A.N. Epstein. 1976. The ontogeny of thirst in the infant rat. *Am. J. Physiol.* 230: 188-198.

Wolff, P.H. 1978. Suckling patterns of infant mammals. *Brain Behav. Evol.* 1: 354-367.

Zelaso, P., N. Zelaso, and S. Kolb. 1972. "Walking" in the newborn. *Science* 176: 314-315.

Chapter 2

THE ROLE OF THE MEDIAL HYPOTHALAMUS IN THE CONTROL OF FOOD INTAKE: AN UPDATE

Anthony Sclafani and Annette Kirchgessner

I. INTRODUCTION

For three decades (1940s-1970s) physiological research on feeding behavior centered on the hypothalamus as the major brain region involved in the control of food intake. This emphasis originated from the now-classic observations that damage to the ventromedial hypothalamus (VMH) results in overeating and obesity, while damage to the lateral hypothalamus (LH) results in aphagia and weight loss (Brobeck et al., 1943; Anand and Brobeck, 1951). Subsequent research revealed that complementary effects are produced by electrical stimulation of the hypothalamus; that is, VMH stimulation inhibits feeding in food-deprived animals, while LH stimulation induces feeding in nondeprived animals (Anand and Dua, 1955). Based on these and other findings, the idea developed that the hypothalamus contained two "centers", a VMH satiety center, and an LH feeding center, that determined when and how much animals eat (Stellar, 1954). This "dual-center theory" became widely accepted since it appeared to account for many aspects of feeding behavior and energy regulation.

Within the last 10 years, however, there has been a gradual shift away from the "hypothalamocentric" view of feeding behavior, and more and more attention has focused on extrahypothalamic neural and hormonal feeding mechanisms. This shift in interest has occurred in part because of exciting new findings implicating extrahypothalamic mechanisms in the control of feeding behavior (e.g., brainstem, vagus nerve, gut hormones). More important, though, the validity of the hypothalamocentric theory of feeding has been so seriously challenged by recent research that some investigators no longer believe that the hypothalamus plays a primary role in the control of food intake (e.g., Stricker, 1983). Other researchers, however, still maintain that the medial and lateral hypothalamus are critically involved in food intake regulation (e.g., Hoebel, 1984; Grossman, 1982).

This chapter reexamines the role of the medial hypothalamus in feeding behavior and in particular focuses on the hypothalamic hyperphagia syndrome. Although traditionally produced by electrolytic lesions of the VMH, the hyperphagia syndrome can be produced by a variety of manipulations of the medial hypothalamus (MH); for this reason it will be referred to as the MH rather than the VMH syndrome. Furthermore, the emphasis here will be on the *hyperphagia* syndrome rather than on the *VMH lesion* syndrome. This is an important distinction because lesions of the VMH produce many behavioral, hormonal, and metabolic effects, not all of which are essential components of the hyperphagia syndrome. Several comprehensive reviews of the MH hyperphagia syndrome have been published (e.g., Bray and York, 1979; LeMagnen, 1983; Powley et al., 1980) and the present paper will not provide a complete account of the extensive literature on this topic. Rather, the following discussion reviews recent developments related to the neuroanatomical substrate and functional etiology of the MH hyperphagia syndrome.

II. ANATOMICAL SUBSTRATE OF HYPOTHALAMIC HYPERPHAGIA

Historically, MH hyperphagia was attributed to VMH damage, particularly damage to the ventromedial nucleus (VMN) and its fiber connections. Furthermore, according to the "dual-center theory," electrolytic lesions of the VMH produce hyperphagia by destroying inhibitory connections of the VMN to the lateral hypothalamic "feeding center." Initial studies using the knife cut technique were consistent with this anatomical model. Parasagittal knife cuts between the VMH and LH produced hyperphagia and obesity (Albert and Storlien, 1969; Sclafani and Grossman, 1969; Gold, 1970a), as well as other behavioral changes comparable to those produced by VMH electrolytic lesions (Sclafani, 1971). However, the anatomical evidence for fiber connections between the VMN and LH has been a controversial issue (see Ricardo, 1983), and behavioral findings obtained during the last decade no longer support the VMN-LH anatomical model of MH hyperphagia.

The effectiveness of parasagittal hypothalamic knife cuts in producing hyperphagia and obesity has been found to depend upon their exact position in the medial-lateral plane, as well as in their rostral-caudal extent. Knife cuts through the medial perifornical region, i.e., through the lateral edge of the VMH, produce greater hyperphagia and obesity than do knife cuts through the lateral perifornical region, i.e., through the medial edge of the LH (Sclafani et al., 1973). Also, only parasagittal knife cuts that extend into the anterior hypothalamus just rostral to the level of the VMN produce the hyperphagia syndrome (Gold, 1970b; Sclafani, 1971). In addition to parasagittal knife cuts, coronal knife cuts through the perifornical hypothalamus anterior or posterior to the level of the VMN increase food intake and body weight (Albert et al., 1971; Grossman, 1971; Grossman and Hennessy, 1976). While all of these knife cuts are in a position to sever fiber connections of the VMN, damage to the VMN or its connections no longer appears to be responsible for the hyperphagia syndrome since electrolytic lesions restricted to the VMN fail to increase food intake or body weight (Beven, 1973; Joseph and Knigge, 1968; Gold, 1973). Rather, it appears that traditional VMH lesions produce hyperphagia by destroying fibers of passage just lateral to he ventromedial nucleus. Nevertheless, the VMN may be importantly involved in some of the metabolic changes associated with the VMH lesion syndrome.

Initially it was thought that parasagittal and coronal hypothalamic knife cuts produce hyperphagia by interrupting different neural pathways (Grossman, 1971; Albert et al., 1971), but subsequent research indicated that a common "feeding-inhibitory" pathway is involved. An important tool in this research was the asymmetrical lesion technique developed by Gold (Gold et al., 1972). With this technique a lesion on one side of the brain, such as a unilateral MH lesion, is combined with a different lesion on the contralateral side of the brain, such as a unilateral posterior hypothalamic lesion. The advantage of this asymmetrical technique is that the only neural systems damaged bilaterally are those that pass through the two areas being lesioned, e.g., through the MH and posterior hypothalamus. All other neural systems are only unilaterally damaged, which is important since unilateral MH lesions typically produce little or no hyperphagia. According to Gold et al. (1972, p. 211), the asymmetrical or crossed lesion technique "is especially suited to anatomical loci, such as the hypothalamus, where several neural systems may overlap or lie in close proximity to one another, but where each neural system may project in a different direction. Under these circumstances, crossed lesions should yield more selective bilateral destruction of neural systems, and thereby more selective behavioral deficits than are possible with bilaterally symmetrical lesions."

By use of this technique, it was discovered that a parasagittal MH knife cut on one side of the brain combined with a coronal hypothalamic knife cut on the other side produced a hyperphagia syndrome comparable to that produced by bilateral parasagittal MH knife cuts (Gold et al., 1977; Sclafani and Berner, 1977; Sclafani, 1982). This finding indicated that the two different knife cuts severed the same fiber system, albeit at different loci. It was further discovered that overeating and obesity can be induced by a unilateral parasagittal MH knife cut combined with a contralateral coronal knife cut or lesion in the midbrain (Gold et al., 1977; Kapatos and Gold, 1973; Sclafani and Berner, 1977). This latter finding implicated the brainstem connections of the hypothalamus in the hyperphagia syndrome, as originally proposed by Hetherington and Ranson (1942). Thus, labeling the hyperphagia syndrome as the "medial hypothalamic" syndrome does not fully describe the neurocircuitry involved in this disorder, although for historical reasons this term will be retained.

The anatomical locus of hypothalamic knife cuts effective and ineffective in producing the hyperphagia syndrome indicated that the fibers involved take a longitudinal path in the perifornical region of the hypothalamus and turn medially just rostral to the VMN (Gold et al., 1977; Sclafani and Berner, 1977). The area of the paraventricular hypothalamus (PVH), and specifically the paraventricular nucleus (PVN), was posited to be the hypothalamic "focus" of this pathway (Gold et al., 1977) Consistent with this hypothesis, localized lesions of the PVN produce hyperphagia and obesity ((Heinbecker et al., 1944; Leibowitz et al., 1981; Aravich and Sclafani, 1983). Furthermore, the changes in feeding behavior produced by PVN lesions are comparable in several respects to those produced by parasagittal knife cuts and VMH electrolytic lesions (Aravich and Sclafani, 1983; Sclafani and Aravich, 1983). However, the degree of overeating and obesity obtained with restricted PVN lesions is less than that seen with large parasagittal knife cuts or VMH lesions. Thus the PVN cannot be the only hypothalamic locus involved in the MH hyperphagia syndrome. Rather, as in the case of VMH lesions (Gold, 1973; Hetherington and Ranson, 1942) and MH knife cuts (Sclafani et al., 1973),

PVH lesions become more effective as their size increases and they extend beyond the borders of the PVN (Leibowitz et al., 1981; Aravich and Sclafani, 1983). Whether large lesions are more obesity-promoting than small ones because they damage a single system that is diffusely organized in the medial hypothalamus, or because they disrupt two or more independent systems that separately promote hyperphagia and obesity, is not known for certain, although some evidence suggests that independent systems may be involved (Weingarten and Chang, 1985).

In either case, neurons of the PVN appear to play an important role in the MH hyperphagia syndrome, which is of considerable interest in light of other data linking the PVN to the control of food intake. Microinjections of norepinephrine (NE) into the medial hypothalamus elicit a robust feeding response in satiated animals, and the PVN has been identified as the most sensitive site for this response (Leibowitz, 1978). In some respects the feeding behavior elicited by NE injections is similar to that produced by MH lesions, which led to the hypothesis that NE injections and MH damage may increase food intake by acting on a common feeding system (see Aravich et al., 1982; Herberg and Franklin, 1972). However, hyperphagia-producing knife cuts lateral to the PVN do not block the eating behavior elicited by PVN microinjections of NE (Aravich et al., 1982). This and other findings indicate that although their anatomical substrates overlap, MH hyperphagia and NE-elicited feeding are mediated by different neural systems (see Aravich et al., 1982; Clavier et al., 1983; Leibowitz et al, 1981). Some evidence suggests that the NE feeding system may be involved in the regulation of macronutrient intake, carbohydrate intake in particular (Leibowitz, 1980; Tretter and Leibowitz, 1980); and PVN lesions, irrespective of whether they produce hyperphagia or not, increase carbohydrate intake in rats allowed to self-select their diet (Sclafani and Aravich, 1983).

In addition to norepinephrine, PVN microinjections of other neurochemicals, i.e., enkephalin, neurotensin, cholecystokinin, and neuropeptide Y have been found to modulate food intake (see Hoebel, 1984; Stanley and Leibowitz, 1984). The functional significance of these feeding effects and their relationship to the MH hyperphagia syndrome remain to be established.

While the hypothalamic course of the feeding-inhibitory pathway involved in the MH hyperphagia syndrome has been extensively studied, much less work has focused on its course and destination (or origin) in the brainstem. As noted above, unilateral midbrain knife cuts or lesions can produce hyperphagia when combined with a contralateral knife cut lateral to the PVN. Since the midbrain lesions and cuts were in a position to sever fibers of the ascending ventral noradrenergic bundle (VNAB), and neurochemical destruction of the VNAB was reported to produce overeating and obesity (Ahlskog and Hoebel, 1973), it was proposed that MH hyperphagia results from the destruction to the VNAB (Gold, 1973; Kapatos and Gold, 1973). Subsequent research has discounted this possibility since destroying the noradrenergic input to the PVN produces hypophagia, not hyperphagia (Azar et al., 1984; O'Donohue et al., 1978); the hyperphagia syndrome produced by VNAB lesions is distinct from the MH hyperphagia syndrome (Ahlskog, 1974; Ahlskog et al., 1975; Sahakian et al., 1983; Sclafani and Berner, 1977); and hyperphagia-inducing MH or midbrain knife cuts or lesions produce inconsistent effects on brain norepinephrine content (Grossman et al., 1977; Coscina, 1983).

Another brainstem pathway that has been implicated in the MH syndrome is the ascending serotonergic fiber system that innervates the hypothalamus and forebrain. Neurochemical treatments that deplete brain serotonin levels have been found to increase food intake and body weight (Breisch et al., 1976; Saller and Stricker, 1976; Waldbillig et al., 1981) and therefore it was proposed that MH lesions or knife cuts, which are in a position to sever serotonin-containing fibers, may promote hyperphagia by destroying a serotonergic feeding-inhibitory system (Gold et al., 1977; Sclafani and Berner, 1977). However, recent work indicates the MH lesions and knife cuts, while they result in similar hyperphagia syndromes, produce different effects on brain regional serotonin content (Coscina, 1983). Note also, that serotonin-depleting midbrain lesions have been found to substantially block the development of MH hyperphagia (Coscina and Stancer, 1977), although these lesions presumably involve a serotonin subsystem different from that responsible for the hyperphagic effects produced by serotonin depletion. Finally, as discussed below, recent findings indicate that the brainstem pathway involved in the MH syndrome extends caudally to the origin of the serotonergic fiber systems. Thus, it now appears unlikely that destruction of an ascending serotonergic feeding-inhibitory system is the principal cause of the MH hyperphagia syndrome, although it may contribute to the syndrome.

In recent studies we have explored the caudal extent of the hypothalamic-brainstem pathway responsible for MH hyperphagia. Using the asymmetrical knife cut technique, we combined a unilateral parasagittal MH cut with a contralateral coronal cut in the caudal brainstem, and it was found that hyperphagia and obesity were produced by coronal knife cuts in the ventrolateral pontine or rostral medullary reticular formation (Kirchgessner and Sclafani, 1983; unpublished observations).

The asymmetrical cuts, however, were generally less effective than bilaterally symmetrical MH knife cuts. This may have occurred because (1) the brainstem cuts were not optimally placed and thus spared some of the feeding-inhibitory fibers; (2) more than one neural system may be involved in the MH hyperphagia syndrome, and not all of the relevant fibers extend into the caudal brainstem; (3) the brainstem knife cuts damaged additional neural circuits which interfered with the expression of the hyperphagia syndrome. Further work is required to resolve this issue.

The cells of origin of the axons transected by the hyperphagia-inducing pontine knife cuts were identified by applying horseradish peroxidase (HRP) to the cut fibers (Kirchgessner and Sclafani, 1983). Hypothalamic HRP-labeled neurons were located primarily in the ipsilateral PVN, particularly in the caudal parvocellular part of the nucleus, which is the PVN area implicated in the MH hyperphagia syndrome (Leibowitz et al., 1981; Aravich and Sclafani, 1983). Labeled cells were also found in areas of the lateral hypothalamus and amygdala, which do not appear to be directly related to MH hyperphagia, while the VMN was free of HRP-labeled cells. In light of the finding that the hindbrain cuts severed fibers that originated in the PVN but not other MH areas, the degree of hyperphagia produced by these cuts, while less than that obtained with large bilateral MH cuts or lesions, may be comparable to that produced by restricted PVN lesions.

Posterior to the pontine knife cut, HRP-labeled neurons were found primarily in the caudal medial portion of the nucleus of the solitary tract (NTS). This nucleus is the principal recipient of gustatory and visceral afferent information (Torvik, 1956; Beckstead and Norgren, 1979; Leslie et al., 1982), and projects heavily to the parvocellular division of the PVN by a pathway that is primarily noradrenergic (see Swanson and Sawchenko, 1983). In addition, the NST has well-established connections with other hypothalamic and limbic structures (Ricardo and Koh, 1978). While the ascending axons of these neurons were severed by the hyperphagia-inducing knife cuts, this does not appear to be the cause of the overeating since lesions of the NST and adjacent area postrema (AP) produce hypophagia and weight loss, and severely attenuate the overeating produced by MH knife cuts (Hyde et al., 1982).

Based on these lesion and knife cut results, Figure 1 provides an updated model of the feeding-inhibitory pathway implicated in the MH hyperphagia syndrome. Whether the pathway is an ascending or descending one has not been clearly established, but a descending course appears the more likely. This is suggested by the fact that PVN lesions produce a hyperphagia syndrome comparable to that produced by knife cuts of its lateral and caudal connections. If the pathway were an ascending one, then the only way in which transections of the pathway and PVN lesions could produce similar behavioral effects would

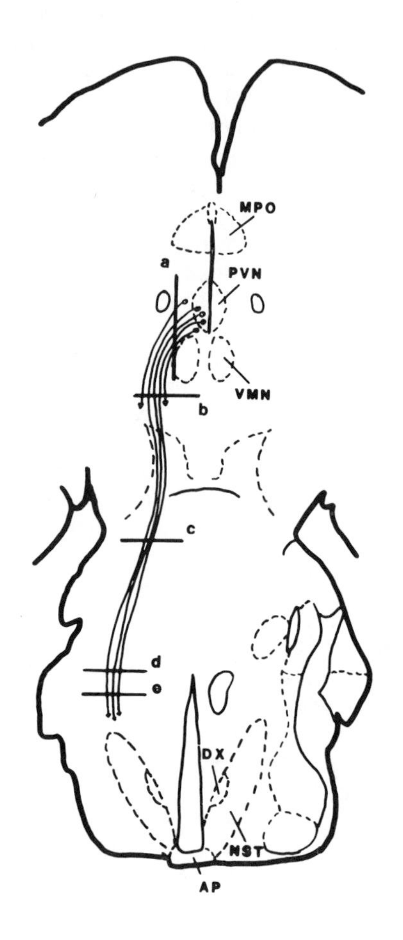

Fig. 1. Schematic representation of the "feeding-inhibitory" pathway implicated in the MH hyperphagia-obesity syndrome. The pathway appears to originate in the PVN and surrounding tissue, and projects laterally and caudally with at least some of the fibers extending to the medulla. The NST and/or DX are likely termination sites for the pathway. The hyperphagia-inducing knife cuts illustrated are as follows: a, parasagittal MH cut; b, posterior hypothalamic coronal cut; c, midbrain coronal cut; d, pontine coronal cut; and e, medullary coronal cut. (For clarity, the pathway and knife cuts are illustrated only one side of the brain.) AP = area postrema; DX = dorsal motor nucleus of the vagus; MPO = medial preoptic area; NST = nucleus of the solitary tract; PVN = paraventricular nucleus; VMN = ventromedial nucleus. The horizontal section was adapted from Paxinos and Watson, *The Rat Brain in Stereotaxic Coordinates*, Academic Press Australia, 1982.

be if the ascending fibers were excitatory to PVN neurons. In this case, though, transection of the relevant PVN efferents would produce effects similar to destroying the PVN cell bodies or their excitatory inputs. Since knife cuts or lesions anterior or dorsal to the PVN do not produce overeating (Gold et al., 1977; Leibowitz et al., 1981), and the PVN-pituitary circuit is not implicated in the hyperphagia syndrome (Cox et al., 1968; Ieni and Gold, 1977; see Gold et al., 1977), the PVN output pathway would appear to take a caudal path.

The exact terminus of the pathway remains to be established, but the NST and dorsal motor nucleus of the vagus (DX) are likely candidates given the caudal PVN's well-established projections to these structures (Swanson and Kuypers, 1980;

Swanson and Sawchenko, 1983). The neurochemical identity of the pathway also remains to be determined, although, as previously discussed (Aravich and Sclafani, 1983), a PVN oxytocinergic pathway is one candidate. Experiments now in progress confirm that hyperphagia-inducing MH cuts and pontine cuts sever descending PVN oxytocinergic fibers, although whether it is the transection of these or other fibers that is responsible for the MH syndrome has not been established (Kirchgessner, Sclafani, and Nilaver, unpublished observations). While it is tempting to conclude that a direct PVN-medullary circuit is involved in the MH syndrome, there is as yet no reason to exclude a multisynaptic system. Also, it must again be emphasized that since restricted PVN lesions or hindbrain cuts produce a less than maximal hyperphagia, additional neural pathways appear to be involved in the MH syndrome.

The functional significance of the PVN-hindbrain feeding-inhibitory pathway remains to be determined, but based on its apparent projection several relationships are suggested. That is, the caudal PVN is in a position to modulate visceral and gustatory afferent information at the level of the NST area, and to influence the activity of parasympathetic and sympathetic functions through its connections with the DX and spinal cord, respectively. That such modulation does, in fact, occur is evidenced by the findings that electrical stimulation of the PVN modulates the activity of NST neurons that relay visceral sensory information (Rogers and Nelson, 1984). These functional connections are of considerable interest in light of the various theories that have been proposed to account for the MH hyperphagia syndrome. As discussed in the next section, MH hyperphagia has been attributed to (1) reduced responsiveness to visceral sensory feedback, (2) an exaggerated responsiveness to orosensory stimuli, and (3) alterations in autonomic motor function and metabolism.

III. FUNCTIONAL ETIOLOGY OF HYPOTHALAMIC HYPERPHAGIA

The nature of the functional disorder responsible for hypothalamic hyperphagia has been extensively examined and discussed, but still remains something of an enigma. As in the case of all lesion syndromes, there is a tendency to assume that the behavioral disorders produced by MH lesions or knife cuts reflect the function of the damaged tissue, rather than the adaptive response of the rest of the brain. Thus, while the pathway transected by the hyperphagia-inducing knife cuts has been referred to as a "feeding-inhibitory pathway" for the sake of discussion, in reality the exact function of this neurocircuit remains to be determined.

Theoretical accounts of the MH hyperphagia-obesity syndrome generally fall into one of two classes. One class attributes the overeating to a disorder in the central regulation of food intake and/or body fat stores, while the other postulates that the overeating is a secondary consequence of autonomically mediated changes in metabolism. In light of the anatomical data reviewed above, the following discussion will focus on three hypotheses of MH hyperphagia that implicate sensory and autonomic functions likely to be disrupted by damage to PVN-brainstem fiber connections.

A. The Satiety Deficit Hypothesis

One early view was that overeating produced by MH damage is due to a deficit in short-term satiety, or more specifically, to a reduced responsiveness to satiety cues generated by ingested food. This interpretation was based on a number of findings, including the failure of MH hyperphagic animals to display increased hunger motivation (Miller et al., 1950; Teitelbaum, 1955), the tendency of hyperphagic rats to overeat by increasing their meal size rather than frequency (Teitelbaum and Campbell, 1958), and the responsiveness of VMH neurons to putative satiety cues such as stomach distention and increased blood glucose levels (Anand et al., 1962, Sharma et al., 1961). While the medial hypothalamus and its fiber connections are still frequently described as constituting a "satiety system," there is substantial evidence against the notion that MH damage impairs short-term satiety.

The increased meal size characteristic of hypothalamic hyperphagia would appear to be *prima facie* evidence for a satiety impairment in MH animals. Enlarged meals, however, are not essential for hyperphagia to occur, since when MH animals are prevented from taking large meals they continue to overeat by increasing their meal number (Sclafani, 1978; Thomas and Mayer, 1978). Analysis of the intrameal feeding pattern of MH hyperphagic rats has also failed to reveal any fundamental disorder in short-term satiety (Sclafani, 1983). It is also relevant to note that enlarged meal size is a feature characteristic of many different animal models of obesity (Sclafani, 1984).

Direct tests of the satiety deficit hypothesis have involved measurements of the MH animal's responsiveness to a variety of feeding challenges. Although some mixed results have been reported, in most cases MH animals have been found to be as or more responsive to the feeding-suppressive effects of intragastric, intraperitoneal, or intravenous injections of nutrients (Fig. 2), to stomach distention, to changes in the caloric density of the diet (when not confounded by palatability changes), and to putative satiety hormones, i.e., cholecystokinin and bombesin (Kulkosky et al., 1976; Liu and Yin, 1974; McHugh and Moran, 1977; Novin et al., 1979; Rowland et al., 1975; Smith et al., 1961; Smutz et al., 1975; Thomas and Mayer, 1968; West et al., 1982). Thus, there is currently little support for the view that MH hyperphagia is due to an insensitivity to the satiating properties of food per se. Furthermore, while the satiety deficit hypothesis originated with the observation that MH hyperphagic animals did not display increased hunger drive (Miller et al., 1950), it is now clear that food motivation is elevated in MH hyperphagic animals (King, 1980; Sclafani, 1978).

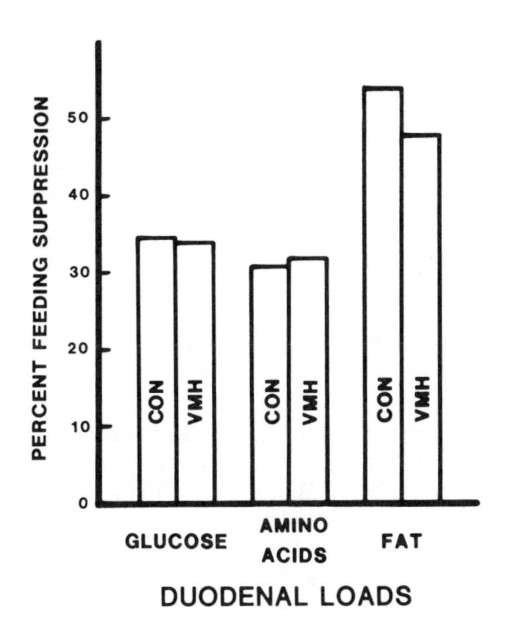

Fig. 2. Feeding-suppressive effects of nutrient loads infused into the duodenum of VMH-lesioned and control rats. The rats were food-deprived, and were duodenally infused with near-isotonic loads of glucose, amino acids, or fat, and their food intake during the subsequent 1 hour was expressed as a percentage of their intake following a saline infusion. The figure illustrates that MH hyperphagic rats do not differ from normal animals in their feeding-suppressive response to duodenal nutrient infusions. Data taken from Novin et al., 1979. Reprinted with permission from *Physiology and Behavior* 22. Copyright 1979, Pergamon Press, Ltd.

B. The Finickiness Hypothesis

One of the most interesting characteristics of MH hyperphagia and obesity, and one that differentiates it from many other models of obesity, is its dependence upon diet palatability (Sclafani, 1984). That is, hypothalamic hyperphagia and obesity can be prevented or reversed by maintaining animals on unpalatable diets, or greatly potentiated by giving highly palatable foods. These and other findings have suggested that the MH hyperphagia syndrome results from an alteration in the animal's responsiveness to the orosensory properties of food.

Although the dietary "finickiness" of MH hyperphagic animals has been extensively studied, it is still not completely understood. Furthermore, it is a matter of dispute whether finickiness is an essential component of the MH hyperphagia syndrome or represents an independent neural disorder. In view of the anatomical evidence linking the MH hyperphagia syndrome with a PVN-hindbrain pathway that possibly involves the NST, the initial gustatory relay station in the brain, the relationship between MH hyperphagia and finickiness will be treated in some detail. It should first be noted that finickiness has two components: an enhanced overeating response to palatable foods (positive finickiness) and an exaggerated undereating response to unpalatable foods (negative finickiness; see Powley, 1977).

1. Are Hyperphagia and Finickiness Common or Separate Disorders?

The study cited most often as demonstrating that finickiness and hyperphagia are distinct disorders is the report of Graff and Stellar (1962). These investigators observed that MH lesions that produced overeating and obesity produced less finickiness than did MH lesions that produced little or no hyperphagia. However, close inspection of their data reveals that (1) contrary to some assertions, as a group the MH obese rats were finicky relative to the unoperated controls; and (2) the finickiness displayed by the nonhyperphagic MH animals was due to negative finickiness to the unpalatable diets (quinine- and NaCl-adulterated chow). Thus, the Graff and Stellar (1962) study demonstrated that negative finickiness can be produced by hypothalamic lesions that do not produce hyperphagia, not that MH hyperphagia can occur without finickiness.

More impressive evidence for a dissociation between hyperphagia and finickiness was obtained in a study by Beven (1973; see Hoebel, 1975). VMH lesions just lateral to the VMN, but not extending into the perifornical component of the medial forebrain bundle (MFB), were observed to produce hyperphagia without finickiness (positive or negative), whereas MFB lesions produced negative finickiness but not hyperphagia. Combined VMH-MFB lesions produced both hyperphagia and finickiness. However, attempts in our laboratory to reproduce these results have not been successful (Sclafani and Aravich, unpublished observations). Furthermore, parasagittal knife cuts through the MH that do not infringe upon the MFB-perifornical region have been consistently found to produce both hyperphagia and finickiness, as do relatively small lesions in the PVH area that also do not invade the MFB-perifornical area (Aravich and Sclafani, 1983; Sclafani, 1971, 1982; Sclafani and Berner, 1977).

Another study frequently cited as evidence that hypothalamic hyperphagia can be dissociated from finickiness is the report of McGinty et al. (1965) that VMH lesions produced hyperphagia in rats that could neither taste nor smell their food. In this experiment animals were prepared with a nasogastric feeding tube and were trained to bar press for intragastrically delivered food. Five out of eight rats with VMH lesions displayed moderate overeating and obesity when feeding themselves in this fashion, although food intake and weight gain were greater when the animals were allowed to eat by mouth. The authors concluded from these findings that while palatable food exaggerates the overeating response, "high palatability of the diet is not necessary for hypothalamic hyperphagia" to occur (McGinty et al., 1965, p. 416).

However, the intragastric feeding technique used by McGinty et al. (1965) has been criticized in that it does not completely eliminate oropharygneal sensations, i.e., thermal sensations produced by a cold diet passing through the nasogastric tube (Holman, 1969). Furthermore, in a replication of the McGinty et al. study, Bauer (1971) found that VMH lesions did not produce hyperphagia in intragastrically feeding rats; in this experiment the diet was kept at room temperature to eliminate thermal cues. Rather, the VMH rats (n = 24) used in this study maintained their food intake and body weight at control levels when feeding themselves intragastrically, although they were hyperphagic and rapidly gained weight when allowed to feed by mouth. The VMH rats were also hyperphagic when half of their food was ingested orally, and half was delivered intragastrically. This later finding is of importance since it indicates that the failure of intragastrically self-feeding VMH rats to overeat is due to the absence of positive orosensory cues rather than to any aversive effect of intragastric food delivery. Thus, contrary to McGinty et al. (1965), Bauer (1971) concluded that positive orosensory cues are necessary for the expression of hypothalamic hyperphagia.

While mixed results have been obtained concerning the relationship between MH hyperphagia and positive finickiness, there is little doubt that negative finickiness can occur in the absence of hyperphagia. The most extreme case of this latter condition would be the LH-lesioned animal (Teitelbaum and Epstein, 1962). Thus, it could be argued that MH lesions produce hyperphagia and negative finickiness by damaging separate, but partially overlapping, neural systems, and that negative finickiness blocks the expression of MH hyperphagia in animals offered unpalatable diets (e.g., Friedman and Stricker, 1976; Stricker, 1983). However, excluding cases of animals with combined MH and LH lesions (see Carlisle and Stellar, 1969; Sclafani, 1971), the negative finickiness of MH animals appears to be primarily a result of their obesity, rather than a direct effect of the neural lesion. That is, several studies report that while obese MH rats are hypophagic and lose weight, relative to controls, when fed unpalatable foods, the same animals display little or no negative finickiness when tested at a nonobese body weight (Ferguson and Keesey, 1975; Franklin and Herberg, 1974; Sclafani et al., 1976). However, since not all forms of animal obesity are associated with negative finickiness (Sclafani, 1984), this characteristic of MH obesity appears to be due to the specific functional disorder responsible for the MH syndrome.

Perhaps the most compelling evidence against the view that negative finickiness represents an independent neural disorder inhibiting the expression of hypothalamic hyperphagia was obtained in a comparative study of the hypothalamic and ovarian obesity syndromes (Gale and Sclafani, 1977). This study demonstrated that, unlike MH hyperphagia, ovarian hyperphagia is not associated with finickiness. That is, whereas a quinine diet blocked MH hyperphagia, and a high-fat diet greatly potentiated it, these diets had essentially no effect, relative to controls, on the overeating and obesity induced by ovariectomy. Most important, though, was the discovery that while the quinine diet blocked MH hyperphagia it did not prevent MH animals from displaying hyperphagia and obesity following ovariectomy (Fig. 3). The fact that in the same animal MH hyperphagia was blocked but ovarian hyperphagia was not indicates that the failure of MH animals to overeat unpalatable foods cannot be attributed to a general overreactivity to such foods.

Thus, although some inconsistent results have been obtained, the weight of the evidence favors the view that finickiness is not a separate disorder, but rather is an intrinsic component of the MH hyperphagia syndrome. MH damage appears to directly or indirectly (see Section III. C) increase the animal's appetite for palatable foods (positive finickiness), and as a consequence of obesity, its rejection of

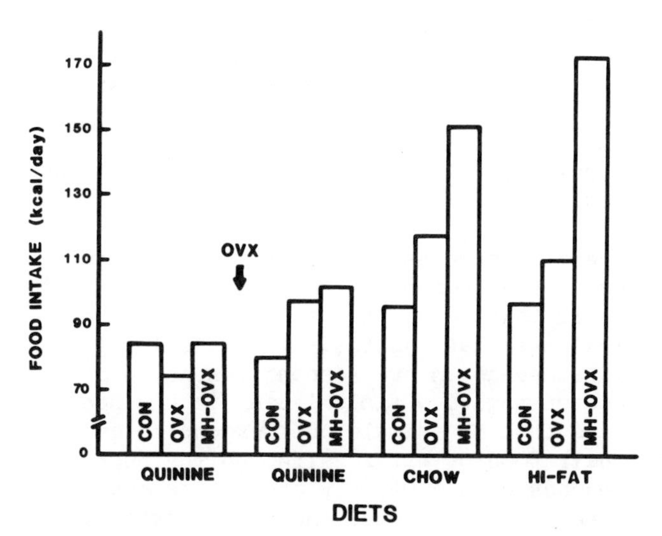

Fig. 3. Effects of diet on the food intake of MH knife cut, ovariectomized rats (MH-OVX), ovariectomized rats (OVX), and control rats (CON). MH knife cuts were placed in the MH-OVX rats 20 days prior to ovariectomy. All rats were initially maintained on a 0.2% quinine chow diet. Following ovariectomy surgery (arrow) the rats were fed the quinine diet, unadulterated chow diet, and high-fat chow diet during successive 20-day periods. The figure shows that the quinine diet blocked hypothalamic hyperphagia, but not ovarian hyperphagia, in MH-OVX rats; and that hypothalamic hyperphagia, but not ovarian hyperphagia, was greatly potentiated by offering the rats chow and high-fat diets. Data taken from Gale and Sclafani, 1977. Copyright 1977 by the American Psychological Association. Reprinted by permission of the author.

unpalatable foods (negative finickiness). Since normal animals tend to overeat and gain weight when offered highly palatable foods, and, when obese, display finickiness to unpalatable food (Maller, 1964; Sclafani and Springer, 1976), MH damage exaggerates rather than fundamentally alters the animal's response to food. Consistent with this interpretation, recent findings indicate that if only one side of the hypothalamus is damaged there is little or no change in food intake when animals are fed a standard chow diet, but their normal hyperphagic response to highly palatable foods, i.e., dietary-induced obesity, is greatly potentiated (Sclafani and Aravich, 1981). Another most interesting finding is that rats with unilateral MH lesions show a preference for palatable foods located in the contralateral sensory field, i.e., the field that projects to the damaged side of the brain (Marshall, 1975). As discussed next, however, the nature of the food cues that promote hyperphagia and finickiness in MH animals is not fully understood.

2. What Diet Qualities Are Responsible for Hypothalamic Finickiness?

It has generally been assumed that MH hyperphagic rats are overresponsive to the orosensory stimuli provided by food, i.e., its taste, smell, texture and temperature. However, a number of findings indicate that hypothalamic finickiness cannot be completely explained as an overresponsiveness to oral stimuli.

In the case of negative finickiness, it has been extensively documented that MH animals overrespond to quinine adulteration of their food or water, although the degree of finickiness depends upon a variety of factors such as body weight, preoperative experience, and the palatability of the unadulterated test substance (see King, 1980; Powley et al., 1980). Yet in two-bottle taste preference tests, MH hyperphagic animals do not differ from controls in their aversion to quinine (Mook and Blass, 1968; Nachman, 1967). This later finding indicates that the quinine finickiness of MH animals cannot be attributed to an alteration in their sensitivity to bitter taste per se. Consistent with this view, recent findings suggest that quinine produces toxic postingestive effects that are largely responsible for the quinine-diet aversions displayed by MH animals as well as normal animals (Aravich and Sclafani, 1980; Heybach and Boyle, 1982; Kratz et al., 1978; Oku et al., 1984; Sclafani et al., 1979). Furthermore, it appears that the ingestion of quinine-adulterated diets leads to a conditioned aversion to bitter tasting foods. Thus, MH rats will overeat a diet made bitter with the nontoxic adulterant sucrose octa-acetate (SOA), but if first fed a quinine diet they will subsequently undereat the SOA diet (Sclafani et al., 1979). This is not to say that rats do not innately dislike bitter-tasting foods, but the profound feeding and weight suppressive effects produced by quinine-adulterated diets appear to represent a combined action of the bitter taste and toxic effects of the drug.

In contrast to this analysis, Weingarten et al. (1983) have recently argued that the orosensory properties of quinine are sufficient to suppress food intake in normal and MH animals. This conclusion was based on the observations that quinine adulteration suppressed sham-feeding in rats with an open gastric fistula, and during the very first exposure to the adulterated diet, suppressed food intake in "real-feeding" rats. It should be noted, though, that these findings were obtained in short-term feeding tests and they do not necessarily explain the long-term suppressive effects of quinine on food intake and body weight. Furthermore, the procedures used by Weingarten et al. do not exclude the possibility that the gastric irritative properties of quinine (Rollo, 1975) may contribute to the suppressive effect of the drug in short-term feeding tests.

In addition to quinine, negative finickiness has been obtained with other adulerants such as sodium chloride and cellulose (Graff and Stellar, 1962; Teitelbaum, 1955). Whether it is the oral effects alone or the combined oral and post-ingestive effects of these adulterants that suppress food intake in MH rats is not known. One well-documented instance of negative finickiness that seems to involve only orosensory cues is the exaggerated feeding response MH rats show to changes in the texture of their chow diet, i.e., powdered vs. pelleted chow (Kramer and Gold, 1980; Sclafani et al., 1979; Teitelbaum, 1955). Interestingly, while MH rats prefer pelleted chow to powdered chow, they show a substantial reduction in their gnawing response to inedible objects (Cox et al., 1967; Sclafani, 1971). Thus, MH damage does not appear to produce a general overreactivity to oral-tactile stimuli, and exactly why MH rats display an exaggerated response to changes in diet texture remains to be determined.

With respect to positive finickiness, most demonstrations of this phenomenon have involved changes not only in the orosensory properties of food, but also in its nutrient composition as well. For example, the high-fat diets widely used to promote MH hyperphagia differ from standard lab chow in taste and texture, but also in caloric density and macronutrient composition. Corbit and Stellar (1964) reported that MH hyperphagic animals overeat a mineral oil/fat diet equivalent in caloric density to lab chow, and they therefore argued that it is the texture of high fat diets, not the caloric density, that is responsible for the finickiness response. This conclusion, however, ignores the fact that the mineral oil/fat diet differed from the chow in its macronutrient composition. Subsequent studies have revealed that (1) MH rats do not overrespond to low-fat diets adulterated with mineral oil (Carlisle and Stellar, 1969); (2) that the overresponsiveness of MH rats to a mineral oil/fat diet is dependent upon the animals having previous experience with a high-fat diet (Coscina et al., 1984); and (3) that the MH rat's preference for high-fat over low-fat diets is not displayed immediately, but develops over time as the animal apparently learns about the postingestive effects of the diets (Carlisle and Stellar, 1969; Sclafani et al., 1983). Thus, the finickiness of MH animals to high-fat diets probably involves a response to the combined effects of the diet's taste, texture, and nutrient composition (see also Kramer and Gold, 1980).

MH hyperphagic rats are also finicky to high-sugar diets, but it seems unlikely that the finickiness is due simply to the sweet taste of the food. In two-bottle taste preference tests MH rats fail to display an enhanced sweet preference for saccharin or sugar solutions (Mook and Blass, 1970; Nachman, 1967). Nor do MH rats typically overconsume saccharin solutions, or highly palatable saccharin-sucrose mixtures in one-bottle tests; in some cases they actually underconsume these solutions relative to controls (Maller, 1964; Sclafani et al., 1973; Sclafani et al., 1976; Sclafani et al., 1981; but see Xenakis and Sclafani, 1982). Furthermore, when tested with pure glucose or sucrose solutions, MH rats usually do not consume more than controls of relatively dilute solutions (less than 20%), although they overconsume concentrated sugar solutions (Beatty, 1973; Sclafani, 1973; Sclafani and Berner, 1977).

The fact that MH rats do not consistently overconsume dilute sugar solutions, saccharin solutions, or sugar-saccharin solutions suggests that it is not sweet taste alone that drives their consumption of calorically rich sugar solutions or high-sugar diets. Instead, it may be the postingestive effects of sugar, either by itself or more likely in conjunction with its sweet taste, that are responsible for the hyperphagic response of MH animals to sugar solutions and diets (see Geiselman and Novin, 1982). In apparent contradiction to this interpretation are findings obtained with sham-feeding rats, i.e., rats fitted with a gastric fistula that allows ingested food to drain from their stomachs. MH rats have been reported to sham-feed more of a concentrated sucrose solution or sweet milk diet than do normal rats (Cox and Smith, 1981; Weingarten, 1982). Since the sham-feeding preparation is thought to eliminate postingestive feedback, these findings were interpreted as strong evidence that MH finickiness is due to an overresponsivity to orosensory stimuli alone. However, recent experiments demonstrate that sham-feeding is not without postingestive effects; rats sham-feeding sugar solutions absorb some of the ingested sugar and show significant increases in blood sugar levels (Grill et al., 1984; Sclafani and Nissenbaum, 1984). Furthermore, while rats with an open gastric fistula consume enormous amounts of concentrated sugar solutions, they do not consume greater than normal amounts of saccharin solutions (Sclafani and Nissenbaum, 1984). In view of these new findings, the sham-feeding data obtained with MH animals cannot be taken as unequivocal evidence for an orosensory interpretation of finickiness.

In summary, while the available evidence indicates that finickiness is an essential component of the MH hyperphagia syndrome, the sensory determinants of this finickiness are not fully understood. Rather than being a response to "taste" stimuli alone, finickiness may involve a synergistic effect of the oral and postingestive feedback provided by food. This concept is discussed further in Section IV.

C. The Autonomic-Metabolic Hypothesis

In contrast to the emphasis given in the past to regulatory deficit interpretations, most recent accounts of the MH syndrome postulate that the hyperphagia and obesity are secondary to autonomically-mediated changes in metabolism. Among the metabolic alterations that accompany MH hyperphagia, the increased insulin secretion produced by VMH lesions has received the most attention, since many of the other metabolic effects appear secondary to hyperinsulinemia (Bray and York, 1979). In fact, several investigators have concluded that hypothalamic hyperphagia is due primarily, if not completely, to hyperinsulinemia, and deny a direct role for the medial hypothalamus in the control of food intake (Bray and York, 1979; LeMagnen, 1983; Stricker, 1983).

1. Evidence for the Autonomic-Metabolic Hypothesis

The evidence supporting the autonomic-metabolic interpretation of MH hyperphagia has been extensively reviewed, and will only be outlined here. (Except where noted, references for the results listed below can be found in Bray and York, 1979, LeMagnen, 1983, or Powley et al., 1980.)

a. VMH Lesions Produce Hyperinsulinemia. That VMH lesions produce hyperinsulinemia is well-documented. Increased insulin release is an early effect of the lesion and occurs before the onset of hyperphagia. Of particular importance, insulin release and fat deposition are elevated even when hyperphagia is prevented or does not spontaneously appear, as in the VMH weanling rat. That MH hyperphagia is a consequence of hyperinsulinemia is suggested by the findings that (1) administration of exogenous insulin to normal rats promotes overeating and obesity, (2) the degree of hyperinsulinemia produced by VMH lesions is positively correlated with subsequent weight gain, and (3) prevention of hyperinsulinemia by experimentally-induced diabetes attenuates or eliminates hypothalamic hyperphagia.

b. Parasympathetic Mediation of VMH Lesion Effects. That the parasympathetic nervous system mediates, at least in part, the effects of MH damage on food intake and insulin release is evidenced by the finding that vagotomy blocks the hyperphagia, obesity, and hyperinsulinemia produced by VMH lesions or MH knife cuts. VMH lesions have also been reported to increase vagal nerve electrical activity (Yoshimatsu et al., 1984) as well as gastric acid release--another sign of increased vagal activity (Weingarten and Powley, 1980). Furthermore, functional denervation of the insulin-secreting ß-cells by pancreatic islet transplantation significantly attentuates MH hyperphagia.

c. Sympathetic Involvement in VMH Lesion Effects. That the sympathetic nervous system is also involved in the MH hyperphagia syndrome is indicated by the findings that VMH lesions decrease (1) pancreatic glucagon levels, (2) salivary gland weight, (3) the free fatty acid (FFA) response to stress, (4) norepinephrine turnover in peripheral tissue, and (5) brown fat metabolism. This lesion-induced reduction in sympathetic activity is thought to promote hyperinsulinemia and lipogenesis, and therefore hyperphagia.

d. Cephalic Phase Hypothesis and VMLH Lesion Effects. These findings indicate that VMH lesions produce a neuroendocrine state conducive to overeating and obesity, but they do not readily account for the diet dependency of the MH hyperphagia syndrome. Some investigators maintain that diet finickiness represents a separate neural disorder and therefore need not be explained by the autonomic-metabolic hypothesis (Stricker, 1983; Friedman and Stricker, 1976). Other workers view finickiness as an integral component of the MH hyperphagia syndrome, and have developed a "cephalic phase" hypothesis to account for both hyperphagia and finickiness within the autonomic-metabolic framework (Powley, 1977; see also LeMagnen, 1983). According to this hypothesis "VMH lesions produce their major effects on feeding behavior by directly heightening the phasic autonomic and endocrine responses triggered by oropharyngeal contact with food stimuli" (Powley, 1977, p. 89). The major evidence in support of the cephalic phase hypothesis is that VMH lesions increase the cephalic insulin and gastric acid response to taste stimuli (Steffens, 1970; Louis-Sylvestre, 1976; Weingarten and Powley, 1980). Also, recent reports indicate that the size of the cephalic insulin response is correlated with diet palatability, and may in turn determine meal size (LeMagnen, 1983). Furthermore, since the vagus nerve mediates the cephalic insulin response, the hyperphagia-blocking effect of vagotomy has been interpreted as support for the cephalic phase hypothesis.

2. Evidence Against the Autonomic-Metabolic Hypothesis

While the evidence outlined above would appear to provide a convincing case for the autonomic-metabolic hypothesis, the degree to which MH hyperphagia is secondary to peripheral neuroendocrine changes rather than a direct effect of the neural lesion is open to question. Note that the emphasis here is on the feeding effects of MH damage rather than the effects of VMH lesions on metabolism and body fat. Clearly, VMH lesions can produce obesity, although to only a mild degree, when overeating fails to appear or is prevented, and this adiposity is a result of neuroendocrine changes. However, as discussed below, whether the same neuroendocrine changes are responsible for the robust hyperphagia and finickiness produced by VMH lesions, PVH lesions, and MH and/or brainstem knife cuts is much less clear.

a. Is Hyperinsulinemia a Necessary or Sufficient Precondition for Hyperphagia? A negative answer is indicated by the following observations:

1) Although experimentally produced diabetes attentuates MH hyperphagia, it does not block it completely, and VMH-diabetic rats maintained at controlled insulin levels consume more food than do similarly treated non-lesioned diabetic rats (Friedman, 1972; Vilberg and Beatty, 1975).

2) Medial hypothalamic manipulations that increase food intake do not invariably produce hyperinsulinemia. In particular, hyperphagia-producing MH knife cuts fail to increase insulin levels when hyperphagia is prevented, although when food is freely available, MH-cut rats show hyperphagia and hyperinsulinemia (Bray et al., 1982; Sclafani, 1981; but see Tannenbaum et al., 1974). Similarly, hyperphagia-inducing PVH lesions do not increase resting or stimulated insulin levels in rats prevented from overeating, and in some cases, in rats allowed to overeat (Sclafani and Aravich, 1983; Steves and Lordon, 1984; Weingarten and Chang, 1985). Also, unlike VMH lesions, PVH lesions in weanling rats do not produce hyperinsulinemia or obesity (Bernardis, 1984), and in adult rats do not produce obesity when hyperphagia is prevented (Weingarten and Chang, 1985). Finally, while it is well established that VMH lesions produce hyperinsulinemia, procaine injections into the VMH, which elicit a short-latency feeding response, actually decrease rather than increase insulin levels (Berthoud and

Jeanrenaud, 1979); and nonirritative VMH lesions, unlike the standard irritative (metal depositing) lesion, are reported not to increase resting insulin levels in food-restricted animals (King et al., 1984).

3) Lesion-induced hyperinsulinemia is not always associated with hyperphagia. This is seen in the VMH lesion weanling rat (Bernardis and Frohman, 1971), although it could be argued that developmental factors limit the expression of hyperphagia in young animals. VMH lesions in adult animals, though, have occasionally been reported to produce hyperinsulinemia in the absence of spontaneous hyperphagia (Hamilton et al., 1976; see also Rabin, 1974). In view of the reports that lesions restricted to the VMN are sufficient to produce hyperinsulinemia (Bernardis and Frohman, 1970, 1971) but not hyperphagia (Beven, 1973; Gold, 1973) it is surprising that hyperinsulinemia in the absence of hyperphagia has not been more frequently reported. Note, however, that most studies of the MH syndrome exclude animals that fail to display hyperphagia, and this obviously reduces the chances of dissociating hyperinsulinemia from hyperphagia. Such a dissociation has been obtained, however, with other brain lesions. In particular, lesions in the area of the suprachiasmatic nucleus (SCN) were found to significantly increase basal insulin levels as well as glucose-stimulated insulin release, but not total daily food intake; the SCN lesions disrupted the day-night distribution of feeding (Yamamoto et al., 1984a,b). Even more surprising is the report that lateral hypothalamic knife cuts producing mild anorexia and weight loss elevated insulin levels relative to weight-matched controls (Grijalva et al., 1980). Also, lesions of the midbrain dorsolateral tegmental area have been found to produce a mild increase in nighttime food intake, but no change in daytime intake; yet basal insulin levels were increased more during the day than during the night (Wellman et al., 1984).

4) While it is well established that administration of exogenous insulin can produce hyperphagia in brain-intact rats, it is also the case that this hyperphagia rarely approaches the magnitude of that produced by MH damage, that insulin-induced feeding is typically associated with severe hypoglycemia whereas MH damage does not have this effect; and that insulin-induced feeding increases meal number more than meal size, whereas MH damage has the opposite effect (LeMagnen, 1983). Furthermore, in contrast to the effects produced by chronic treatment with high doses of insulin, treatment with low doses has been found to suppress food intake and body weight, and it has been argued that insulin normally acts as a satiety signal (Vanderweele et al., 1980). These findings question the valid-

ity of using insulin-induced feeding as evidence for a metabolic explanation of hypothalamic hyperphagia. A related point is that hyperinsulinemia is a common feature of almost all animal models of hyperphagia and obesity, and thus does not readily account for the distinctive features of the MH hyperphagia syndrome (Sclafani, 1984).

b. Is Vagal Hyperactivity a Necessary or Sufficient Precondition for Hypothalamic Hyperphagia? Again the available evidence suggests a negative answer:

1) Following the initial report that vagotomy blocks hypothalamic hyperphagia, several additional studies indicated that this effect is not obtained under all test conditions. In particular, vagotomy performed prior to MH surgery does not in all cases eliminate hypothalamic hyperphagia (see King and Frohman, 1982), although it has been argued that vagal nerve regeneration may mediate this hyperphagia-sparing effect (Fox and Powley, 1984). Even when vagotomy completely reverses hypothalamic hyperphagia and obesity on a chow diet, overeating and obesity can be restored by offering the animals very palatable foods (Sclafani et al., 1981; Fig. 4). Finally, VMH lesions produce a small (20%) but significant increase in food intake in rats in which the pancreatic ß-cells are functionally disconnected from the vagus nerve by means of an islet transplant (Inoue et al., 1978); the rats in this study were fed a chow diet and it is possible that greater hyperphagia would have been obtained if they were fed highly palatable foods.

2) Selective vagotomy of the celiac branch of the vagus nerve attentuates hypothalamic hyperphagia, whereas selective transection of the gastric or hepatic branches does not reduce hyperphagia or obesity (Sawchenko et al., 1981). Since the celiac branch was thought to provide the major innervation to the pancreas, these observations were interpreted as support for the hypothesis that hypothalamic hyperphagia is due to vagally stimulated insulin release. However, recent data indicate that all three vagal branches influence insulin secretion (Berthoud et al., 1983). Furthermore, selective gastric vagotomy was found to produce a greater reduction in vagally stimulated insulin release than did celiac vagotomy. These recent findings question, therefore, the mechanism by which vagotomy reduces MH hyperphagia. Also, while total subdiaphragmatic vagotomy completely reverses MH hyperphagia and obesity on a chow diet, this may be due in part to the aversive effects of the vagotomy that have recently been documented (Bernstein and Goehler, 1983; Sclafani and Kramer, 1985).

3) Cholinergic blocking drugs, which inhibit vagally-stimulated insulin release, only partially inhibit the expression of hypothalamic hyperphagia in long-term tests, and have virtually no effect on the hyperphagic response of MH animals in short-term tests (Carpenter et al., 1979; Sclafani and Xenakis, 1981).

4) While VMH lesions increase gastric acid secretion, a vagally mediated response, hyperphagia-inducing PVN lesions fail to do so (Weingarten and Chang, 1985). On the other hand,

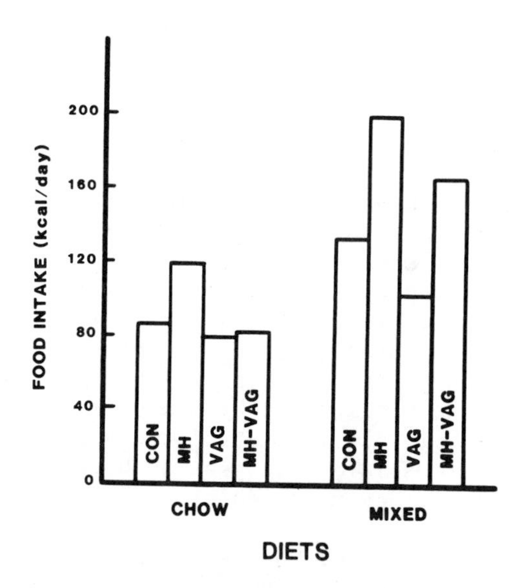

Fig. 4. Effects of diet on the food intake of MH knife cut rats (MH), MH knife cut, vagotomized rats (MH-VAG), vagotomized rats (VAG), and control rats (CON). The vagotomy surgery was performed 10 days after the MH knife cuts, and completely reversed the hyperphagia and obesity on the chow diet induced by the knife cut. The chow intake data depicted here were collected 120 days postvagotomy, at which time the MH rats were obese and only slightly hyperphagic relative to the control rats. When subsequently offered a mixed diet containing chocolate-chip cookies, sweetened milk, high-fat chow, and lab chow, all rats increased their caloric intake, but the two MH groups displayed a greater hyperphagic response than did the brain-intact rats (CON and VAG groups). The figure shows that while vagotomy blocked MH hyperphagia on a chow diet, it did not block MH hyperphagia on the more palatable mixed diet. Data taken from Sclafani et al., 1981. Copyright 1981 by the American Psychological Assn. Reprinted by permission of the author.

PVN lesions, like VMH lesions, have been reported to increase the electrical activity of the pancreatic branch of the vagus nerve (Yoshimatsu et al., 1984); but the same study also reported increased vagal activity following lesions of the dorsomedial hypothalamus (DMH). Yet DMH lesions decrease, rather than increase food intake and body weight (Bernardis, 1973; Dalton et al., 1981). Thus, vagal nerve hyperactivity is not invariably associated with hyperphagia.

5) With respect to the cephalic phase hypothesis, the fact that MH hyperphagia is not completely prevented by vagotomy or treatment with anticholinergic drugs questions the role of the cephalic insulin response in the MH syndrome. It should be noted that it has yet to be demonstrated that an enhanced cephalic insulin response is a direct effect of hyperphagia-producing MH damage, rather than being secondary to basal hyperinsulinemia produced by VMH lesions (Berthoud et al., 1981). In fact, acute deactivation of the VMH with procaine injections increases food intake but actually blocks the cephalic insulin response. (Berthoud et al., 1981). The elevated cephalic gastric acid response may also be secondary to basal hypersecretion, since it is not an immediate effect of the VMH lesions, but, unlike hyperphagia, takes several days to develop (Weingarten and Powley, 1980). Another important point is that cephalic phase responses have not been measured in obesity models other than the MH model, and enhanced cephalic responses may not be unique to the MH syndrome. Recent data indicate that vagal hyperactivity is a contributing factor in the hyperphagia and obesity displayed by the genetically obese fatty (*fa/fa*) rat (Rohner-Jeanrenaud et al., 1983), and it would not be surprising if the fatty rat displayed an increased cephalic insulin response.

c. Is Reduced Sympathetic Activity a Necessary or Sufficient Precondition for Hypothalamic Hyperphagia? Compared to the evidence implicating altered parasympathetic function, the data supporting an important role for reduced sympathetic activity in the hyperphagia-promoting effects of MH damage are less extensive. Furthermore, this aspect of the autonomic hypothesis is seriously questioned by recent findings: (1) chemically-induced sympathectomy does not promote hyperphagia and obesity in normal rats, nor does it reduce the hyperphagia-promoting effects of VMH lesions (Powley et al., 1983), (2) hyperphagia-inducing VMH lesions do not invariably inhibit FFA mobilization (Powley et al., 1983), while lesions or knife cuts in the anterior, lateral or posterior hypothalamus inhibit FFA mobilization but do not produce hyperphagia (Colimbra and Migliorini,

1983; Grijalva et al., 1980), (3) hyperphagia-inducing hypothalamic knife cuts, unlike VMH lesions, do not reduce salivary gland weight (Sclafani, unpublished observations).

In view of these results, it seems unlikely that hyperphagia and finickiness result solely from the autonomic and metabolic alterations produced by MH damage (see also Grijalva and Lindholm, 1982). This is not to say that neuroendocrine changes do not contribute to the syndrome and are not necessary for its maximal expression. According to one estimate, vagally-mediated hyperinsulinemia may account for approximately 40% of the weight gain seen in ad libitum fed VMH-lesioned rats (King and Frohman, 1982). Yet the fact that MH animals continue to overeat, at least to some degree, following vagotomy, pancreatic islet transplantation, experimentally-induced diabetes, and cholinergic blockade, indicates that a significant component of the MH syndrome is a direct result of the central neural lesion rather than a secondary consequence of peripheral neuroendocrine events.

IV. SUMMARY AND FUTURE DIRECTIONS

During the last 40 years the hypothalamus has been lesioned, cut, stimulated, injected, recorded from, and anatomically and biochemically analyzed in an attempt to understand its role in the control of food intake and energy balance. Much has been learned during this time, but our knowledge remains far from complete. This chapter has focused only on the hyperphagia-obesity syndrome produced by MH damage, and there is much more to the involvement of the hypothalamus in feeding behavior, as evidenced by the feeding responses elicited by chemical or electrical stimulation of the MH and LH, the hypophagia syndrome produced by DMH lesions, the altered feeding and endocrine rhythms seen after SCN lesions, and, of course, the classic LH aphagia syndrome. Thus, while a strict "hypothalamocentric" view of feeding behavior is no longer appropriate, the hypothalamus still seems to contain important "crossroads" in the complex neurocircuits controlling ingestive behavior and metabolism.

Concerning the MH hyperphagia syndrome, the present analysis indicates that it results in part from the destruction of a pathway, probably a descending one, that interconnects the PVN with the caudal medulla, with the NST and/or DX as likely destinations. The PVN is not the only hypothalamic area involved, however, since large MH lesions or knife cuts are required to produce maximal hyperphagia and obesity. The relevant "feeding-inhibitory" neurons and fibers may be small in number and functionally identical, but diffusely organized within the anteriomedial hypothalamus. Alternatively, large lesions may be so effective in producing hyperphagia and obesity because they damage different functional systems located within either separate or overlapping areas of the medial hypothalamus. Another interesting puzzle that remains to be solved concerns the relationship between the PVN-medullary neurocircuit implicated in the MH hyperphagia syndrome and PVN noradrenergic and neuropeptide feeding systems, as well as the feeding-related systems of the DMH, SCN, and LH.

While recent research has further defined the anatomy of the MH hyperphagia syndrome, the functional etiology of the syndrome remains something of a mystery. The PVN-medullary pathway described here is in a position to modulate the activity of viscerosensory, orosensory, and autonomic motor systems, and disruptions in all three of these input/output systems have been implicated in the MH hyperphagia syndrome. With respect to viscerosensory input, there is little evidence to support the original short-term satiety deficit hypothesis. Despite their hyperphagia, MH animals seem to be fully responsive to postingestive satiety cues, nor do they appear to suffer from an inability to accurately monitor their long-term caloric intake.

The evidence for an alteration in orosensory function is much stronger, but the nature of the functional changes produced by MH damage remains unclear. On the other hand, the available data suggest that MH animals do not differ from normals in their reactivity to oral stimuli per se.

With respect to the autonomic motor system, the evidence is unequivocal that MH hyperphagia is associated with autonomic and metabolic alterations, but the idea that these are the direct and only cause of hyperphagia is tenuous. In particular, recent evidence suggests that hyperphagia-producing PVN lesions produce minimal disruption to autonomic and metabolic functions. Clearly, additional research is needed to unravel the respective contributions of central regulatory and peripheral neuroendocrine disruptions to the MH hyperphagia syndrome. This research should avoid the use of large MH lesions that most certainly disrupt many different functional systems, but rather should favor the use of techniques (lesion, knife cut, neurochemical) that produce the minimal damage necessary to produce hyperphagia and obesity. Asymmetrical and unilateral lesion/cut procedures, as well as multiple lesion techniques (e.g., VMN plus PVN lesions), should be further exploited in this research. Most important, investigations of the MH hyperphagia syndrome, should concentrate more on non-hyperphagic brain-damaged animals. The fact that some manipulations, e.g., VMN, DMN, or SCN lesions, fail to increase food intake, but nevertheless produce some behavioral and/or neuro-endocrine symptoms characteristic of the MH hyperphagia syndrome, provides important clues as to which changes are primary and which are secondary to the overeating, finickiness, and obesity effects of MH damage.

Finally, one possibility that should be investigated is that MH hyperphagia and finickiness result from a change in the central integration of feedback signals carried by orosensory and viscerosensory systems. While attention has focused primarily on the feeding-inhibitory effects of viscerosensory feedback, which MH damage does not seem to affect, there is some evidence that visceral afferents also provide positive feedback to the brain that stimulates appetite and guides diet selection (Deutsch, 1983; Mather et al., 1978; see also Booth et al., 1972). Thus, palatability may be more a function of the combined action of oral and visceral feedback than of oral feedback alone (see Wyrwicka, 1969). This could explain (1) why MH animals readily overconsume tasty and calorically rich foods (e.g., concentrated sugar solutions), but not tasty and calorically dilute "foods" (e.g., saccharin, saccharin-sugar solutions); (2) why vagotomy, which interrupts visceral sensory and motor function, blocks the MH rat's overconsumption of sugar solutions, but atropine injections, which block only visceral motor function, do not (Sclafani and Xenakis, 1981; Sclafani et al., 1981), and (3) why AP/NST lesions, which primarily affect visceral afferent activity, and vagotomy produce similar effects on the MH hyperphagic response to different diets (Hyde et al., 1982). Since visceral and oral sensory information is integrated at brainstem sites receiving PVN projections (Norgren, 1983; Rogers and Nelson, 1984; Swanson and Sawchenko, 1983), disruption of PVN projections may alter this integration and thereby magnify the animal's ingestive response to palatable foods. This hypothesis is quite speculative given the paucity of information on the positive feedback actions of visceral stimuli, and the complex feeding effects of AP/NST lesions and vagotomy (see Hyde and Miselis, 1983; Kott et al., 1984; Kral, 1981; South and Ritter, 1983), yet represents a new approach to the MH syndrome.

ACKNOWLEDGEMENTS

Preparation of this chapter was supported in part by a grant from the Faculty Research Award Program of the City University of New York. The authors thank Dr. Joseph R. Vasselli for his helpful comments on an earlier draft of this chapter.

REFERENCES

Ahlskog, J.E. 1974. Food intake and amphetamine anorexia after selective forebrain norephinephrine loss. *Brain Res*. 82: 211-240.
Ahlskog, J.E. and B.G. Hoebel. 1973. Overeating and obesity from damage to a noradrenergic system in the brain. *Science* 182: 166-169.
Ahlskog, J.E., P.K. Randall, and B.G. Hoebel. 1975. Hypothalamic hyperphagia: dissociation from hyperphagia following destruction of noradrenergic neurons. *Science* 190: 399-401.
Albert, D.J. and L.H. Storlien. 1969. Hyperphagia in rats with cuts between the ventromedial and lateral hypothalamus. *Science* 165: 599-600.
Albert, D.J., L.H. Storlien, J.G. Albert, and C.J. Mah. 1971. Obesity following disturbance of the ventromedial hypothalamus: a comparison of lesions, lateral cuts, and anterior cuts. *Physiol. Behav*. 7: 135-141.
Anand, B.K. and J.R. Brobeck. 1951. Hypothalamic control of food intake in rats and cats. *Yale J. Biol. Med*. 24: 123-140.
Anand, B.K. and S. Dua. 1955. Feeding responses induced by electrical stimulation of the hypothalamus in cat. *Indian J. Med. Res*. 43: 113-122.
Anand, B.K., G.S. Chhina, and B. Singh. 1962. Effects of glucose on the activity of hypothalamic "feeding centers". *Science* 138: 597-598.
Aravich, P.F. and A.Sclafani. 1980. Dietary preference behavior in rats fed bitter tasting quinine and sucrose octa acetate adulterated diets. *Physiol. Behav*. 25: 157-160.
Aravich, P.F., and Sclafani, A. 1983. Paraventricular hypothalamic lesions and medial hypothalamic knife cuts produce similar hyperphagia syndromes. *Behav. Neurosci*. 97, 970-983.
Aravich, P.F., A. Sclafani, and S.F. Leibowitz. 1982. Effects of hypothalamic knife cuts on feeding induced by paraventricular norepinephrine injections. *Pharmacol. Biochem. Behav*. 16: 101-111.
Azar, A.P., G. Shor-Posner, R. Filart, and S.F. Leibowitz. 1984. Impact of medial hypothalamic 6-hydroxydopamine injections on daily food intake, diet selection, and body weight in freely-feeding and food restricted rats. Paper presented at Eastern Psychological Association Meeting.
Bauer, F.S. 1971. *The Role of Affect in Hypothalamic Hyperphagia*. University Microfilms, No. 72-12, 087, Ann Arbor, Michigan.

Beatty, W.W. 1973. Influence of type of reinforcement on operant responding by rats with ventromedial lesions. *Physiol. Behav.* 10: 841-846.

Beckstead, R.M. and R. Norgren. 1979. An autoradiographic examination of the central distribution of the trigeminal, facial, glossopharyngeal, and vagal nerves in the monkey. *J. Comp. Neurol.* 184: 455-472.

Bernardis, L.L. 1973. Disruption of diurnal feeding and weight gain cycles in weanling rats by ventromedial and dorsomedial hypothalamic lesions. *Physiol. Behav.* 10: 855-861.

Bernardis, L.L. 1984. Paraventricular nucleus lesions in weanling female rats result in normophagia, normal body weight and composition, linear growth and normal levels of several plasma substrates. *Physiol. Behav.* 32: 507-510.

Bernardis, L.L. and L.A. Frohman. 1970. Effects of lesion size in the ventromedial hypothalamus on growth hormone and insulin levels in weanling rats. *Neuroendocrinology* 6: 319-328.

Bernardis, L.L. and L.A. Frohman. 1971. Effects of hypothalamic lesions at different loci on development of hyperinsulinemia and obesity in the weanling rat. *J. Comp. Neurol.* 141: 107-116.

Bernstein, L.L. and L.E. Goehler. 1983. Vagotomy produces learned food aversions in the rat. *Behav. Neurosci.* 97: 585-594.

Berthoud, H.R. and B. Jeanrenaud. 1979. Changes of insulinemia, glycemia and feeding behavior induced by VMH-procainization in the rat. *Brain Res.* 174: 184-187.

Berthoud, H.R., D.A. Bereiter, E.R. Trimble, E.G., Siegel, and B. Jeanrenaud. 1981. Cephalic phase, reflex insulin secretion: neuroanatomical and physiological characterization. *Diabetologia* 20: 393-401.

Berthoud, H.R., A. Niijima, J.-F. Sauter, and B. Jeanrenaud. 1983. Evidence for a role of the gastric, coeliac and hepatic branches in vagally stimulated insulin secretion in the rat. *J. Auton. Nerv. Sys.* 7: 97-110.

Beven, T.E. 1973. *Experimental Dissociation of Hypothalamic Finickiness and Motivational Deficits from Hyperphagia and from Hyperemotionality*. University Microfilms, No. 74-9665, Ann Arbor, Michigan.

Booth, D.A., D. Lovett, and G.M. McSherry. 1972. Postingestive modulation of the sweetness preference gradient in the rat. *J. Comp. Physiol. Psychol.* 78: 485-512.

Bray, G.A., and D.A. York. 1979. Hypothalamic and genetic obesity in experimental animals: an autonomic and endocrine hypothesis. *Physiol. Rev.* 59: 719-809.

Bray, G.A., A. Sclafani, and D. Novin. 1982. Obesity-inducing hypothalamic knife cuts: Effects on lypolysis and blood insulin levels. *Am. J. Physiol.* 243: R445-R449.

Breisch, S.T., F.P. Zemlan, and B.G. Hobel. 1976. Hyperphagia and obesity following serotonin depletion by intraventricular p-chorophenylalanine. *Science* 192: 382-385.

Brobeck, J.R., J. Tepperman, and C.N.H. Long. 1943. Experimental hypothalamic hyperphagia in the albino rat. *Yale J. Biol. Med.* 15: 831-853.

Carlisle, H.J. and E. Stellar. 1969. Caloric regulation and food preference in normal, hyperphagic, and aphagic rats. *J. Comp. Physiol. Psychol.* 69: 107-114.

Carpenter, R.G., B.A. Stamoutsos, L.D. Dalton, L.A. Frohman, and S.P. Grossman. 1979. VMH obesity reduced but not reversed by scopolamine methyl nitrate. *Physiol. Behav.* 23: 955-959.

Clavier, R.M., J.W. Chambers, and D.V. Coscina. 1983. Catecholamine histofluorescence in the paraventricular hypothalamus of rats made hyperphagis by parasagittal knife cuts. *Brain Res. Bull.* 10: 321-325.

Colimbra, C.C. and R.H. Migliorini. 1983. Evidence for a longitudinal pathway in rat hypothalamus that controls FFA mobilization. *Am. J. Physiol.* 245: E332-E337.

Corbit, J.D. and E. Stellar. 1964. Palatability, food intake, and obesity in normal and hyperphagic rats. *J. Comp. Physiol. Psychol.* 58: 63-67.

Coscina, D.V. 1983. Does overeating alter brain neurotransmitters which control feeding? In *Controversies in Obesity*, ed.B.C. Hansen, 41-51. Praeger, New York.

Coscina, D.V. and H.C. Stancer. 1977. Selective blockade of hypothalamic hyperphagia and obesity in rats by serotonin-depleting midbrain lesions. *Science* 195: 416-419.

Coscina, D.V., S. LaCombe, and J.W. Chambers. 1984. A reassessment of greasy dietary texture in promoting hypothalamic obesity. *Soc. Neurosci. Abstr.* 10: 653.

Cox, V.C., J.W. Kakolewski, and E.S. Valenstein. 1967. The relationship between gnawing and food consumption with ventromedial hypothalamic lesions. *Physiol. Behav.* 2: 323-324.

Cox, V.C., J.W. Kakolewski, E.S. and Valenstein. 1968. Effects of ventromedial hypothalamic damage in hypophysectomized rats. *J. Comp. Physiol. Psychol.* 65: 145-148.

Cox, J.E. and G.P. Smith. 1981. Ventromedial hypothalamic lesions produce exaggerated sham feeding. *Soc. Neurosci. Abstr.* 7: 657.

Dalton, L.D., R.G. Carpenter, and S.P. Grossman. 1981. Ingestive behavior in adult rats with dorsomedial hypothalamic lesions. *Physiol. Behav.* 26: 117-123.

Deutsch, J.A. 1983. Dietary control and the stomach. *Prog. Neurobiol.* 20: 313-332.

Ferguson, N.B.L. and R.E. Keesey. 1975. Effect of quinine-adulterated diet upon body weight maintenance in male rats with ventromedial hypothalamic lesions. *J. Comp. Physiol. Psychol.* 89: 478-488.

Fox, E.A., and T.L. Powley. 1984. Regeneration may mediate the sparing of VMH obesity observed with prior vagotomy. *Am. J. Physiol.* 247: R308-R317.

Franklin, K.B.J and L.J. Herberg. 1974. Ventromedial syndrome: the rat's "finickiness" results from the obesity not from the lesions. *J. Comp. Physiol. Psychol.* 87: 410-414.

Friedman, M.I. 1972. Effects of alloxan diabetes on hypothalamic hyperphagia and obesity. *Am. J. Physiol.* 222: 174-178.

Friedman, M.I. and E.M. Stricker. 1976. The physiological psychology of hunger: a physiological perspective. *Psychol. Rev.* 83: 409-431.

Gale, S.K. and A. Sclafani. 1977. Comparison of ovarian and hypothalamic obesity syndromes in the female rat: effects of diet palatability on food intake and body weight. *J. Comp. Physiol. Psychol.* 91: 381-392.

Geiselman, P.J. and D. Novin. 1982. Sugar infusion can enhance feeding. *Science* 218: 490-491.

Gold, R.M. 1970a. Hypothalamic hyperphagia produced by parasagittal knife cuts. *Physiol. Behav.* 5: 23-25.

Gold, R.M. 1970b. Hypothalamic hyperphagia: males get just as fat as females. *J. Comp. Physiol. Psychol.* 71: 347-356.

Gold, R.M. 1973. Hypothalamic obesity: the myth of the ventromedial nucleus. *Science* 182: 488-490.

Gold, R.M., P.M. Quackenbush, G. and Kapatos. 1972. Obesity following combination of rostrolateral to VMH cut and contralateral mammillary area lesions. *J. Comp. Physiol. Psychol.* 79: 210-218.

Gold, R.M., A.P. Jones, P.E. Sawchenko, and G. Kapatos. 1977. Paraventricular nucleus: critical focus of a longitudinal neurocircuitry mediating food intake. *Physiol. Behav.* 18: 1111-1119.

Graff, H. and E. Stellar. 1962. Hyperphagia, obesity, and finickiness. *J. Comp. Physiol. Psychol.* 55: 418-424.

Grijalva, C.V. and E. Lindholm. 1982. The role of the autonomic nervous system in hypothalamic feeding syndromes. *Appetite* 3: 111-124.

Grijalva, C.V., D. Novin, G.A. and Bray. 1980. Alterations in blood glucose, insulin, and free fatty acids following lateral hypothalamic lesions or parasagittal knife cuts. *Brain Res. Bull. 5, Suppl.* 4: 109-117.

Grill, H.J., K.C. Berridge, and D.J. Ganster. 1984. Oral glucose is the prime elicitor of preabsorptive insulin secretion. *Am. J. Physiol.* 246: R88-R95.

Grossman, S.P. 1971. Changes in food and water intake associated with an interruption of the anterior or posterior fiber connections of the hypothalamus. *J. Comp. Physiol. Psychol.* 75: 23-31.

Grossman, S.P. 1982. Mechanisms of appetite regulation. In *Diabetes Mellitus and Obesity*, ed. B.N. Brodoff and S.J. Bleicher, 227-236.

Grossman, S.P. and J.W. Hennessy. 1976. Differential effects of cuts through the posterior hypothalamus on food intake and body weight in male and female rats. *Physiol. Behav.* 17: 89-102.

Grossman, S.P., L. Grossman, and A. Halaris. 1977. Effects on hypothalamic and telencephalic NE and 5-HT of tegmental knife cuts that produce hyperphagia or hyperdipsia in the rat. *Pharmacol. Biochem. Behav.* 6: 101-106.

Hamilton, C.L., P.J. Ciaccia, and D.O. Lewis. 1976. Feeding behavior in monkeys with and without lesions of the hypothalamus. *Am. J. Physiol.* 230: 818-830.

Heinbecker, P., H.L. White, and D. Rolf. 1944. Experimental obesity in the dog. *Am. J. Physiol.* 141: 549-565.

Herberg, L.J. and K.B.J. Franklin. 1972. Adrenergic feeding: Its blockade or reversal by posterior VMH lesions; and a new hypothesis. *Physiol. Behav.* 8: 1029-1034.

Hetherington, A.W. and S.W. Ranson. 1942. The relation of various hypothalamic lesions to adiposity in the rat. *J. Comp. Neurol.* 76: 475-499.

Heybach, J.P and P.C. Boyle. 1982. Dietary quinine reduces body weight and food intake independent of aversive taste. *Physiol. Behav.* 29: 1171-1173.

Hoebel, B.G. 1975. Brain reward and aversion systems in the control of feeding and sexual behavior. In *Nebraska Symposium on Motivation*, ed. J.K. Cole and T.B. Sonderegger, 49-112. University of Nebraska Press, Lincoln, Nebraska.

Hoebel, B.G. 1984. Neurotransmitters in the control of feeding and its rewards: monoamines, opiates, and brain-gut peptides. In *Eating and Its Disorders*, ed. A.J. Stunkard and E. Stellar, 15-38. Raven Press, New York.

Holman, G.L. 1969. Intragastric reinforcement effect. *J. Comp. Physiol. Psychol.* 69: 432-441.

Hyde, T. and R.R. Miselis. 1983. Effects of area postrema/caudal medial nucleus of solitary tract lesions on food intake and body weight. *Am. J. Physiol.* 244: R577-R587.

Hyde, T.M., R. Eng, and R.R. Miselis. 1982. Brainstem mechanisms in hypothalamic and dietary obesity. In *The Neural Basis of Feeding and Reward*, ed. B.G. Hoebel and

D. Novin, 97-114. Haer Institute for Electrophysiological Research, Brunswick, Maine.

Ieni, J.R. and R.M. Gold. 1977. Two satiety systems revealed by hypothalamic knife cuts in hypophysectomized rats. *Brain Res. Bull.* 2: 367-374.

Inoue, S., G.A. Bray, and Y.S. Mullen. 1978. Transplantation of pancreatic b-cells prevents development of hypothalamic obesity in rats. *Am. J. Physiol.* 235: E266-E271.

Joseph, S.A. and J.M. Knigge. 1968. Effects of VMH lesions in adult and newborn guinea pigs. *Neuroendocrinology* 3: 309-331.

Kapatos, G. and R.M. Gold. 1973. Evidence for ascending noradrenergic mediation of hypothalamic hyperphagia. *Pharmacol. Biochem. Behav.* 1: 81-87.

King, B.M. 1980. A re-examination of the ventromedial hypothalamic paradox. *Neurosci. Biobehav. Rev.* 4: 151-160.

King, B.M. and L.A. Frohman. 1982. The role of vagally-mediated hyperinsulinemia in hypothalamic obesity. *Neurosci. Biobehav. Rev.* 6: 205-214.

King, B.M., K.R. Esquerre, and L.A. Frohman. 1984. Obesity in absence of food-restricted hyperinsulinemia in female rats with nonirritative lesions of the ventromedial hypothalamus. *Soc. Neurosci. Abstr.* 10: 1012.

Kirchgessner, A.L. and A. Sclafani. 1983. Hypothalamic-hindbrain feeding inhibitory system: an examination utilizing asymmetrical knife cuts and HRP histochemistry. *Soc. Neurosci. Abstr.* 9: 187.

Kott, J.N., C.L. Ganfield, and N.J. Kenney. 1984. Area postrema/nucleus of the solitary tract ablations: analysis of the effects of hypophagia. *Physiol. Behav.* 32: 429-435.

Kral, J.G. 1981. Vagal mechanisms in appetite regulation. *Intl. J. Obesity* 5: 481-489.

Kramer, T.H. and R.M. Gold. 1980. Facilitation of hypothalamic obesity by greasy diets: palatability vs lipid content. *Physiol. Behav.* 24: 151-156.

Kratz, C.M., D.A. Levitsky, and S. Lustick. 1978. Differential effects of quinine and sucrose octa acetate on food intake in the rat. *Physiol. Behav.* 20: 665-667.

Kulkosky,, P.J., C. Breckenridge, R. Krinsky, and S. Woods. 1976. Satiety elicited by the C-terminal octapeptide of cholecystokinin-pancreaozymin in normal and VMH-lesioned rats. *Behav. Biol.* 18: 227-234.

Leibowitz, S.F. 1978. Paraventricular nucleus: a primary site mediating stimulation of feeding and drinking. *Pharmacol. Biochem. Behav.* 8: 163-175.

Leibowitz, S.F. 1980. Neurochemical systems of the hypothalamus in control of feeding and drinking behavior and water-electrolyte excretion. In *Handbook of the Hypothalamus: Vol. 3. Behavioral Studies of the Hypothalamus,* ed. P.J. Morgane and J. Panksepp, 299-437. Marcel Dekker, New York.

Leibowitz, S.F., N.J. Hammer, and K. Chang. 1981. Hypothalamic paraventricular nucleus lesions produce overeating and obesity in the rat. *Physiol. Behav.* 27: 1031-1040.

LeMagnen, J. 1983. Body energy balance and food intake: a neuro-endocrine regulatory mechanism. *Physiol. Rev.* 63: 314-386.

Leslie, R.A., D.G. Gwyn, and D.A. Hopkins. 1982. The central distribution of the cervical vagus nerve and gastric afferent projections in the rat. *Brain Res. Bull.* 8: 37-43.

Liu, C.M. and T.H. Yin. 1974. Caloric compensation to gastric loads in rats with hypothalamic hyperphagia. *Physiol. Behav.* 13: 231-238.

Louis-Sylvestre, J. 1976. Preabsorptive insulin release and hypoglycemia in rats. *Am. J. Physiol.* 230: 56-60.

Maller, O. 1964. The effect of hypothalamic and dietary obesity on taste preferences in rats. *Life Sci.* 3: 1281-1291.

Marshall, J.F. 1975. Increased orientation to sensory stimuli following medial hypothalamic damage in rats. *Brain Res.* 86: 373-387.

Mather, P., S. Nicolaidis, and D.A. Booth. 1978. Compensatory and conditioned feeding responses to scheduled infusions in the rat. *Nature* 273: 461-463.

McGinty, D., A.N. Epstein, and P. Teitelbaum. 1965. The contribution of oropharyngeal sensations to hypothalamic hyperphagia. *Animal Behav.* 13: 413-418.

McHugh, P.R. and T.H. Moran. 1977. An examination of the concept of satiety in hypothalamic hyperphagia. In *Anorexia Nervosa*, ed. R.A. Vigersky,67-73. Raven Press, New York.

Miller, N.E., C.J. Bailey, and J.A.F. Stevenson. 1950. Decreased "hunger" but increased food intake resulting from hypothalamic lesions. *Science* 112: 256-259.

Mook, D.G. and E.M. Blass. 1968. Quinine-aversion thresholds and "finickiness" in hyperphagic rats. *J. Comp. Physiol. Psychol.* 65: 202-207.

Mook, D.G. and E.M. Blass. 1970. Specific hungers in hyperphagic rats. *Psychon. Sci.* 19: 34-35.

Nachman, M. 1967. Hypothalamic hyperphagia, finickiness, and taste preferences in rats. *Proceedings of the 75th Annual Convention of the American Psychological Association* 2: 127-128.

Norgren, R. 1983. Afferent interactions of cranial nerves involved in ingestion. In *Vagal Nerve Function: Behavioral and Methodological Considerations*, ed. J.G. Kral, T.L. Powley and C.McC. Brooks, 67-77. Elsevier Press, New York.

Novin, D., J. Sanderson, and M. Gonzalez. 1979. Feeding after nutrient infusions: effects of hypothalamic lesions and vagotomy. *Physiol. Behav.* 22: 107-113.

O'Donohue, T.L., W.R. Crowley, and D.M. Jacobowitz. 1978. Changes in ingestive behavior following interruption of a noradrenergic projection to the paraventricular nucleus: histochemical and neurochemical analysis. *Pharmacol. Biochem. Behav.* 9: 99-105.

Oku, J., G.A. Bray, and J.S. Fisler. 1984. Effects of oral and parenteral quinine on rats with ventromedial hypothalamic knife cut obesity. *Metabolism* 33: 538-544.

Paxinos, G. and C. Watson. 1982. *The Rat Brain in Stereotaxic Coordinates.* Academic Press, New York.

Powley, T.L. 1977. The ventromedial hypothalamic syndrome, satiety, and a cephalic phase hypothesis. *Psychol. Rev.* 84: 89-126.

Powley, T.L., C.A. Opsahl, J.E. Cox, and H.P. Weingarten. 1980. The role of the hypothalamus in energy homeostasis. In *Handbook of the Hypothalamus, Vol. 3, Behavioral Studies of the Hypothalamus*, ed. P.J. Morgane and J. Pansepp, 211-298. Marcel Dekker, New York.

Powley T.L., M.C. Walgren, and W.B. Laughton. 1983. Effects of guanethidine sympathectomy on ventromedial hypothalamic obesity. *Am. J. Physiol.* 245: R408-R420.

Rabin, B.M. 1974. Independence of food intake and obesity following ventromedial hypothalamic lesions in the rat. *Physiol. Behav.* 13: 769-772.

Ricardo, J.A. 1983. Hypothalamic pathways involved in metabolic regulatory functions, as identified by track-tracing methods. In *Advances in Metabolic Disorders: Vol. 10. CNS Regulation of Carbohydrate metabolism*, ed. A.J. Szabo, 1-30. Academic Press, New York.

Ricardo, J.A., and Koh, E.T. 1978. Anatomical evidence of direct projections from the nucleus of the solitary tract to the hypothalamus, amygdala, and other forebrain structures in the rat. *Brain Res.* 153: 1-26.

Rogers, R.C. and D.O. Nelson. 1984. Neurons of the vagal division of the solitary nucleus activated by the paraventricular nucleus of the hypothalamus. *J. Auton. Nerv. Sys.* 10: 193-198.

Rohner-Jeanrenaud, F., A.-C. Hochstrasser, and B. Jeanrenaud. 1983. Hyperinsulinemia of preobese and obese fa/fa rats is partly vagus nerve mediated. *Am. J. Physiol.* 244: E317-E1322.

Rollo, I.M. 1975. Drugs used in the chemotherapy of malaria. In *The Pharmacological Basis of Therapeutics*, ed. L.S. Goodman and A. Gilman, 1045-1068. Macmillan, New York.

Rowland, N., M.-J. Meile, and S. Nicolaidis. 1975. Metering

of intravenously infused nutrients in VMH-lesioned rats. *Physiol. Behav.* 15: 443-448.

Sahakian, B.J., T.W. Robbins, R.J. Deeley, B.J. Everitt, L.T. Dunn, M. Wallace, and W.P.T. James. 1983. Changes in body weight and food-related behaviour induced by destruction of the ventral or dorsal noradrenergic bundle in the rat. *Neuroscience* 10: 1405-1420.

Saller, C.F. and E.M. Stricker. 1976. Hyperphagia and increased growth in rats after intraventricular injection of 5,7-dihydroxytryptamine. *Science* 192: 385-387.

Sawchenko, P.E. R.M. Gold, and J. Alexander. 1981. Effects of selective vagotomies on knife cut-induced hypothalamic obesity: differential results on lab chow vs high-fat diets. *Physiol. Behav.* 26: 293-300.

Sclafani, A. 1971. Neural pathways involved in the ventromedial hypothalamic lesion syndrome in the rat. *J. Comp. Physiol. Psychol.* 77: 70-96.

Sclafani, A. 1973. Deficits in glucose appetite and satiety produced by ventromedial hypothalamic lesions in the rat. *Physiol. Behav.* 11: 771-780.

Sclafani, A. 1978. Food motivation in hypothalamic hyperphagia rats reexamined. *Neurosci. Biobehav. Rev.* 2: 339-355.

Sclafani, A. 1981. The role of hyperinsulinemia and the vagus nerve in hypothalamic hyperphagia reexamined. *Diabetologia* 20: 402-410.

Sclafani, A. 1982. Hypothalamic obesity in male rats: comparison of parasagittal, coronal, and combined knife cuts. *Behav. Biol.* 34: 201-208.

Sclafani, A. 1983. Intrameal eating patterns of hypothalamic hyperphagic rats. Paper presented at Eastern Psychological Association Meeting.

Sclafani, A. 1984. Animal models of obesity: classification and characterization. *Intl. J. Obesity.* In press.

Sclafani, A. and P.F. Aravich. 1981. Hyperphagia and obesity produced by unilateral hypothalamic knife cuts in the rat. *Soc. Neurosci. Abstr.* 7: 28.

Sclafani, A. and P.F. Aravich. 1983. Macronutrient self-selection in three forms of hypothalamic obesity. *Am. J. Physiol.* 244: R686-R694.

Sclafani, A. and C.N. Berner. 1977. Hyperphagia and obesity produced by parasagittal and coronal hypothalamic knife cuts: further evidence for a longitudinal feeding inhibitory pathway. *J. Comp. Physiol. Psychol.* 91: 1000-1118.

Sclafani, A. and S.P. Grossman. 1969. Hyperphagia produced by knife cuts between the medial and lateral hypothalamus in the rat. *Physiol. Behav.* 4: 533-537.

Sclafani, A. and T.H. Kramer. 1985. Aversive effects of

vagotomy in the rat. A conditioned taste aversion analysis. *Physiol. Behav.* In press.

Sclafani, A. and J.W. Nissenbaum. 1984. Is gastric sham-feeding really sham feeding? *Soc. Neurosci. Abstr.* 10: 1008.

Sclafani, A. and D. Springer. 1976. Dietary obesity in adult rats: similarities to hypothalamic and human obesity syndromes. *Physiol. Behav.* 17: 461-471.

Sclafani, A and S. Xenakis. 1981. Atropine fails to block the overconsumption of sugar solutions by hypothalamic hyperphagic rats. *J. Comp. Physiol. Psychol.* 95: 708-719.

Sclafani, A., C.N. Berner, and G. Maul. 1973. Feeding and drinking pathways between the medial and lateral hypothalamus in the rat. *J. Comp. Physiol. Psychol.* 85: 29-51.

Sclafani, A., D. Springer, and L. Kluge. 1976. Effects of quinine adulterated diets on the food intake and body weight of obese and non-obese hypothalamic hyperphagic rats. *Physiol. Behav.* 16: 631-640.

Sclafani, A., P.F. Aravich, and J. Schwartz. 1979. Hypothalamic hyperphagic rats overeat bitter sucrose octa acetate diets, but not quinine diets. *Physiol. Behav.* 22: 759-766.

Sclafani, A., P.F. Aravich, and M. Landman. 1981. Vagotomy blocks hypothalamic hyperphagia on a chow diet and sucrose solution, but not on a palatable mixed diet. *J. Comp. Physiol. Psychol.* 95: 720-734.

Sclafani, A., P.F. Aravich, and S. Xenakis. 1983. Macronutrient preferences in hypothalamic hyperphagic rats. *Nutr. Behav.* 1: 233-251.

Sharma, K.N., B.K. Anand, S. Dua, and B. Singh. 1961. Role of the stomach in regulation of activities of hypothalamic feeding centers. *Am. J. Physiol.* 201: 593-598.

Smith, M.H., R. Salisbury, and H. Weinberg. 1961. The reaction of hypothalamic-hyperphagic rats to stomach preloads. *J. Comp. Physiol. Psychol.* 54: 660-664.

Smutz, E.R., E. Hirsch, and H.L. Jacobs. 1975. Caloric compensation in hypothalamic obese rats. *Physiol. Behav.* 14: 305-309.

South, E.H. and R.C. Ritter. 1983. Overconsumption of preferred foods following capsaicin pretreatment of the area postrema and adjacent nucleus of the solitary tract. *Brain Res.* 288: 243-251.

Stanley, B.G., and S.F. Leibowitz. 1984. Neuropeptide Y: Stimulation of feeding and drinking by injection into paraventricular nucleus. *Life Sci.* 35: 2635-2642.

Steffens, A.B. 1970. Plasma insulin content in relation to blood glucose level and meal pattern in the normal and hypothalamic hyperphagia rat. *Physiol. Behav.* 5: 147-151.

Stellar, E. 1954. The physiology of motivation. *Physiol. Rev.* 61: 5-22.
Steves, J.P. and J.F. Lorden. 1984. Vagotomy abolishes obesity in rats with lesions of the paraventricular nucleus. *Soc. Neurosci. Abstr.* 10: 652.
Stricker, E.M. 1983. Brain neurochemistry and the control of food intake. In *Handbook of Behavioral Neurobiology: Vol. 6. Motivation*, ed. E. Satinoff and P. Teitelbaum, 329-366. Plenum Press, New York.
Swanson, L.W. and H.G.J.M. Kuypers. 1980. The paraventricular nucleus of the hypothalamus: cytoarchitectonic subdivisions and organization of projections to the pituitary, dorsal vagal complex, and spinal cord as demonstrated by retrograde fluorescence double-labeling methods. *J. Comp. Neurol.* 194: 555-570.
Swanson, L.W. and P.E. Sawchenko. 1983. Hypothalamic integration: organization of the paraventricular and supraoptic nuclei. *Ann. Rev. Neurosci.* 6: 269-324.
Tannenbaum, G.A., G. Paxinos, and D. Bindra. 1974. Metabolic and endocrine aspects of the ventromedial hypothalamic syndrome in the rat. *J. Comp. Physiol. Psychol.* 86: 404-413.
Teitelbaum, P. 1955. Sensory control of hypothalamic hyperphagia. *J. Comp. Physiol. Psychol.* 48: 156-163.
Teitelbaum, P. and Campbell, B.A. (1958). Ingestion patterns in hyperphagic and normal rats. *J. Comp. Physiol. Psychol.* 51: 135-141.
Teitelbaum, P. and A.N. Epstein. 1962. The lateral hypothalamic syndrome. Recovery of feeding and drinking after lateral hypothalamic lesions. *Psychol. Rev.* 69: 74-90.
Thomas, D.W. and J. Mayer. 1968. Meal taking and regulation of food intake by normal and hypothalamic hyperphagic rats. *J. Comp. Physiol. Psychol.* 66: 642-653.
Thomas, D.W. and J. Mayer. 1978. Meal size as a determinant of food intake in normal and hypothalamic obese rats. *Physiol. Behav.* 21: 113-117.
Torvik, A. 1956. Afferent connections to the sensory trigeminal nuclei, the nucleus of the solitary tract and adjacent structures. *J. Comp. Neurol.* 106: 51-141.
Tretter, J.R. and S.F. Leibowitz. 1980. Specific increases in carbohydrate consumption after norepinephrine (NE) injections into the paraventricular nucleus (PVN). *Soc. Neurosci. Abstr.* 6: 532.
Vanderweele, D.A., F.X. Pi-Sunyer, D. Novin, and M.J. Bush. 1980. Chronic insulin infusion suppresses food ingestion and body weight gains in rats. *Brain Res. Bull.* 5: Suppl. 4, 7-12.
Vilberg, T.R. and W.W. Beatty. 1975. Behavioral changes

following VMH lesions in rats with controlled insulin levels. *Pharmacol. Biochem. Behav.* 3: 377-384.

Waldbillig, R.J., T.J. Bartness, and B.G. Stanley. 1981. Increased food intake, body weight, and adiposity in rats after regional neurochemical depletion of serotonin. *J. Comp. Physiol. Psychol.* 95: 391-405.

Weingarten, H.P. 1982. Diet palatability modulates sham feeding in VMH-lesion and normal rats: implications for finickiness and evaluation of sham-feeding data. *J. Comp. Physiol. Psychol.* 96: 223-233.

Weingarten, H.P., and P.K. Chang. 1985. Comparison of the metabolic and behavioral disturbances following paraventricular- and ventromedial-hypothalamic lesions. *Brain Res. Bull.*. In press.

Weingarten, H.P. and T.L. Powley. 1980. Ventromedial hypothalamic lesions elevate basal and cephalic phase gastric acid output. *Am. J. Physiol.* 239: G221-G229.

Weingarten, H.P., P.K. Chang, and K.R. Jarvie. 1983. Reactivity of normal and VMH-lesion rats to quinine-adulterated foods: negative evidence for finickiness. *Behav. Neurosci.* 97: 221-233.

Wellman, P.J., M. Elissalde, P.A. Watkins, and A. Pinto. 1984. Hyperinsulinemia and obesity in the dorsolateral tegmental rat. *Physiol. Behav.* 32: 1-4.

West, D.B., R.H. Williams, D.J. Braget, and S. Woods 1982. Bombesin reduces food intake of normal and hypothalamically obese rats and lowers body weight when given chronically. *Peptides* 3: 61-67.

Wyrwicka, W. 1969. Sensory regulation of food intake. *Physiol. Behav.* 4: 853-858.

Xenakis, S. and A. Sclafani. 1982. The dopaminergic mediation of a sweet reward in normal and VMH hyperphagic rats. *Pharmacol. Biochem. Behav.* 16: 293-302.

Yamamoto, H., K. Nagai, and H. Nakagawa. 1984a. Bilateral lesions of the suprachiasmatic nucleus enhance glucose tolerance in rats. *Biomedical Res.* 5: 47-54.

Yamamoto, H., K. Nagai, and H. Nakagawa. 1984b. Role of the suprachiasmatic nucleus in glucose homeostasis. *Biomedical Res.* 5: 55-60.

Yoshimatsu, H., A. Niijima, Y. Oomura, K. Yamabe, and T. Katafuchi. 1984. Effects of hypothalamic lesion on pancreatic nerve activity in the rat. *Brain Res.* 303: 147-152.

Chapter 3

PARTICIPATION OF THE VAGUS AND OTHER AUTONOMIC NERVES IN THE CONTROL OF FOOD INTAKE

Terry L. Powley and Hans-Rudolf Berthoud

I. INTRODUCTION: THE AUTONOMIC NERVOUS SYSTEM AND FEEDING

Neural analyses of feeding behavior have proceeded from a number of different perspectives. Historically, much research has emphasized the role of hypothalamic circuitry in ingestive responses (e.g., Stellar, 1954). More recently, neural explanations of feeding have also focused on mechanisms in the caudal brainstem (e.g., see chapters by Grill and R. Ritter in this volume). In the last few years, another area of experimentation has emphasized the dynamic and vital roles of the peripheral branches of the autonomic nervous system (ANS) in the control of feeding (e.g., Grijalva and Lindholm, 1982; Novin and VanderWeele, 1977; Powley, 1977; Powley and Opsahl, 1976). We review some of the conclusions of this latter approach here. The parasympathetic vagus is reviewed first; several general issues concerning the roles of the sympathetic nervous system and feeding then follow. Of course, treating the two branches independently, a strategy convenient for review purposes and often dictated by experimental practicalities, runs the risk of distorting our understanding of the participation of the ANS in feeding. The same caution also clearly applies to considerations of the peripheral autonomic pathways without simultaneous examination of the central autonomic mechanisms.

II. THE PARASYMPATHETICS AND THE CONTROL OF FEEDING

A. The Vagus Nerve

1. The Organization of the Vagus

The vagus, the tenth cranial nerve, is a mixed sensory-motor trunk. Named for its extensive or *wandering* pattern of innervation, the vagus provides the parasympathetic innervation for most of the gut. The main divisions of the nerve, passing caudally, are the pharyngeal, laryngeal, esophageal, and abdominal branches (Fig. 1). (The vagus also gives off auricular, cardiac, and pulmonary branches, but these presumably are less central to the physiology of food intake.) In effect, the vagus provides the innervation of most of the digestive tube, including the regions that control swallowing, propulsion, storage, digestion, and absorption of nutrients. Given this strategic pattern, one is compelled to conclude *a priori* that the vagus must be critically important in food intake. This conclusion is supported by functional analyses discussed below.

The sensory fibers in the vagus innervate the luminal surface and the wall of the gastrointestinal tract from the pharynx to at least the midpoint of the large bowel. These fibers are collected into the trunks of the left and right vagi and distributed distally over the several different branches of the vagi (Fig. 1). The cell bodies of the sensory neurons innervating the gastrointestinal tract are found in the nodose ganglia (estimated in the cat, for example, at approximately 30,000 somata in total: Jones, 1937, 29,600; Foley and Dubois, 1937, 26,000) just external to the jugular foramen. As they enter the brainstem, the central processes of the afferents separate from the motor component of the vagus, form the sensory roots, course to a dorsolateral position on the surface of the medulla, and travel horizontally through the medulla to the solitary tract (Fig. 1), where they then terminate in the commissural and medial subnuclei of the nucleus of the solitary tract (NST).

These vagal afferents relay a number of types of information. Best defined are a series of chemoreceptive inputs from the stomach and small intestine (Davison, 1972; Mei, 1985), as well as mechanoreceptor activities and thermoreceptor inputs

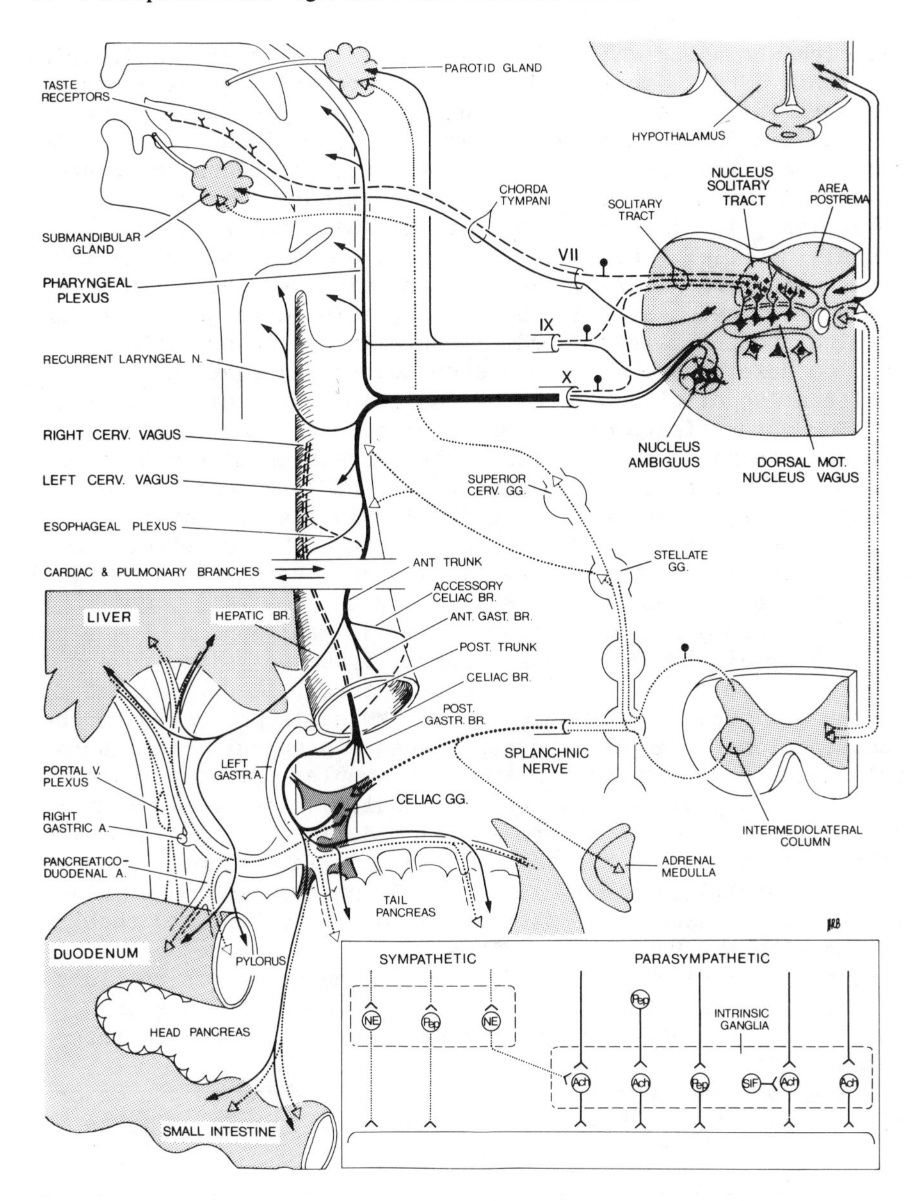

Fig. 1. Schematic representation of the food processing axis and its innervation by the parasympathetic (solid lines) and sympathetic (broken lines) branches of the autonomic nervous system. For clarity, the stomach has been disconnected from the lower esophageal and pyloric sphincters and removed. Note that some vagal fibers (celiac and accessory branches) pass through the celiac ganglia and mix with sympathetic fibers

from the whole length of the gastrointestinal tract (Mei, 1985; Ouazzani, 1984; Paintal, 1973). Electrophysiological evidence suggests that an organotopic organization may exist for the cell bodies in the nodose ganglion (e.g., Davison, 1972a), and some recent tracer evidence suggests that afferents contained in the different vagal branches may terminate somewhat selectively in different zones of the NST commissural and medial subnuclei (Leslie et al., 1982; Norgren and Smith, 1983; Rogers and Hermann, 1983; Shapiro and Miselis, 1985). Little evidence is yet available as to the precise arrangement of the terminals of these first-order inputs in the subnuclei of the NST; to date, there is no evidence suggesting that the different modalities or sensory qualities have unique termini.

The motor fibers of the vagus originate from two major nuclear groups in the medulla. The largest group consists of the dorsal motor nucleus of the vagus (DMN), a long, spindle-shaped nucleus oriented longitudinally in the dorsomedial medulla oblongata. Few estimates of the number of efferents originating in the DMN are available, and, being based on Nissl staining techniques, they are probably too low (e.g., cat: approx. 3,000 per side, Mohiuddin, 1953; man: approx. 9,400 per side, Etemadi, 1961; and rat: approx. 5,000 per side, Lu and Sakai, 1984). In a recent retrograde tracer study, Fox and Powley estimated approximately 6000 DMN cells projected to the abdomen alone. The other group of vagal preganglionic fibers originates from the nucleus ambiguus complex (retrofacial nucleus, n. ambiguus, and n. retroambigualis) distributed longitudinally through the ventrolateral medulla (Fig. 1). Motor fibers of the DMN course through the medulla oblongata in a ventrolateral direction; fibers of the nucleus ambiguus (NA) group form a sharp genu just dorsal to the nucleus and then course with the DMN fibers. These efferents emerge on the ventrolateral side of the medulla, forming the motor rootlets of the vagus, and then combine with the sensory fibers to constitute the vagal trunks that emerge through the lacerated foramen (Fig. 1).

The two motor nuclei of the vagus correspond to two major functional groupings. The motor neurons of the NA in large

---

to innervate the liver, pancreas, and small intestine (see text for details). The insert shows the organization of the different types of peripheral efferent transmission and possible interactions of the two autonomic divisions. Abbreviations: A = artery, Ach = cholinergic, Ant = anterior, BR = branch, CERV = cervical, GG = ganglion, N. = nerve, NE = noradrenergic, Pep = peptides, POST = posterior, V = vein.

part innervate the striated musculature of the throat and esophagus (as well as the cardiac muscle); many of their respective fibers course through the pharyngeal and laryngeal branches of the vagus. The NA is part of the "special visceral efferent" column of the medulla and is technically not autonomic, although its inclusion with the other vagal mechanisms is justified because it receives a heavy projection of second-order (and perhaps first-order) vagal afferents and sends it axons through the vagus. The NA complex has a relatively precise topographic organization corresponding to the major striated muscle groups it innervates (e.g., Davis and Nail, 1984; Gacek, 1975). In contrast, the efferent somata of the dorsal motor nucleus innervate the smooth muscle of the distal gastrointestinal (GI) tract as well as a number of secretomotor structures in the viscera. The functional or topographic organization of the DMN has not been fully determined. In general, although some evidence of organotopic patterns within the DMN has been presented, little of the data can yet support a very punctuate organization. In a recent experiment, we have found that the DMN is organized into a set of longitudinal cell columns corresponding to the separate subdiaphragmatic branches of the vagus (Fox and Powley, 1985).

For more details, two particularly comprehensive and useful reviews of the organization of the vagus are Mitchell and Warwick's 1955 paper and Gabella's 1976 monograph.

2. Roles of the Vagus in Food Intake

a. Principles. There is no *single* role for the vagus in feeding. The nerve is not a feeding circuit. Rather, consistent with the fact that it provides the parasympathetic innervation to a substantial part of the digestive tube, it participates in and affects feeding in a number of ways. We review the better-known roles of the different branches below.

The classification summarized in the left panel of Figure 2 provides a framework for reviewing the vagal roles in feeding. This figure identifies three general ways in which the vagus participates in the control of feeding behavior. First, the vagal branches innervating the pharynx, larynx, and esophagus organize the ingestive sequences associated with both swallowing and ejecting materials. Both afferents and efferents of the tenth nerve have critical roles in these fixed action patterns of feeding. Second, the vagus is a critical link by which the CNS sequences, phases, and paces the transport, digestion, and absorption of energy. The battery of vago-vagal reflexes and their central relays bring these responses under CNS control. These vago-vagal reflexes serve not only to execute various digestive responses, but also to modulate the rates of the processes, thus affecting the partitioning of body energy and, ultimately, modulating the signals that determine feeding behavior. Third, there is also evidence that the vagus provides a major afferent link whereby energy needs and related signals originating in the gastrointestinal tract modulate ingestive mechanisms directly. The right-hand panel of Figure 2 adds some details to the basic organization.

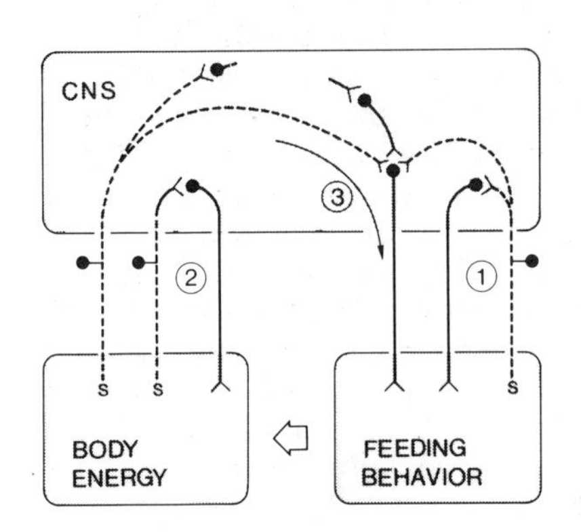

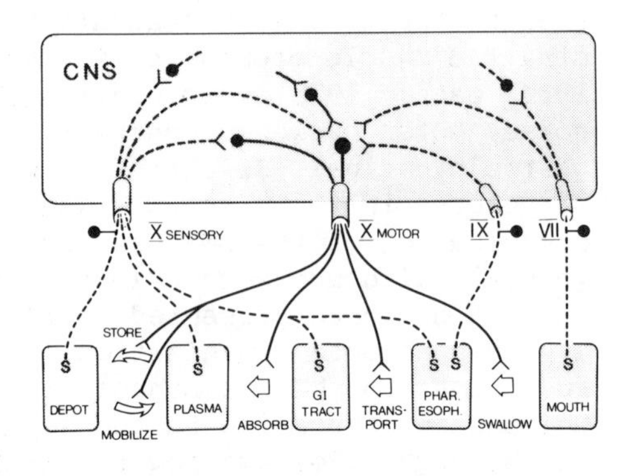

Fig. 2. Left Panel: Highly schematic diagram of the ANS as it informs the CNS through its sensory fibers (broken lines; sensory receptors = S) and influences feeding behavior and the handling of body energy through its motor fibers (solid lines). While the arrangement shown could represent the sympathetic organization as well as the parasympathetic organization, only the latter is illustrated with labels and discussed in detail here. The arabic numbers identify three different classes of sensory-motor reflex loops affecting feeding behavior. (1) represents the reflexes, essential for food intake, associated with ingestion and swallowing in the oropharynx and esophagus. (2) represents vago-vagal reflexes primarily concerned with the coordinated handling of nutrient processing by the abdominal organs of digestion, but that may ultimately also affect feeding through the third loop. (3) represents the classical feedback mechanisms consisting of an afferent vagal limb, e.g., a glucose sensor, and directly modulating the food intake motor command. In general, experimentation has not yet established whether reflex loops 2 and 3 share common afferents or utilize separate afferent channels. The sympathetic branch of the ANS has mechanisms analogous to (2) and (3), but not to (1).

Right Panel: Same schematic as in left panel, but showing some additional details concerning the sensory and motor functions of the vagus as well as the ninth and seventh nerves along the different steps of the food processing chain.

Few of the known vagal effects on feeding are understood in real detail. In many of the cases surveyed below, investigators have identified either feeding effects associated with vagal stimulation or vagal lesions, or changes in feeding that seem correlated with neurophysiological changes of vagal neurons. Interpretations of the experimental observations are necessarily limited by the complexity of the system and the problems inherent in making inferences about normal functions from lesions, stimulation procedures, or neurophysiological correlations.

b. The Pharyngeal and Laryngeal Divisions. Through these two sets of branches, the vagi participate directly and immediately in feeding responses. In particular, they participate in organizing and executing the fixed action patterns of ingestion and rejection recently characterized by Grill and Norgren (1978; see also Grill, this volume). After an animal has masticated its food (or taken a bite small enough to swallow), the bolus is transferred to the back of the mouth. Upon contacting the receptors at the back of the mouth and the pharynx, an appropriate bolus triggers the swallowing motor program, beginning with the pharyngeal phase of deglutition.

An appropriate bolus stimulates mechanoreceptive and chemosensory vagal afferents (working in conjunction with afferents of the glossopharyngeal and trigeminal). These afferent fibers course primarily through a ramus of the superior laryngeal branch (Jean, 1984; Miller, 1982; Norgren, 1983). Elimination of these afferents yields dysphagia. Stimuli of appropriate size, texture, chemical composition, and temperature are swallowed and moved through the esophagus to the stomach by the well-defined motor programs of deglutition. These programs involve integrated sequences of respiratory, cardiovascular, gustatory, and mechanoreceptor patterns; correspondingly, the entire NST probably plays a significant role in the response. Inappropriate stimuli trigger gagging rejection responses. In the extreme case, these rejection responses can include retching or vomiting and reverse peristalsis, which also involve vagal motor pathways. (For extensive reviews, see Dubner et al., 1978; Miller, 1982; Sessle and Hannam, 1976).

Afferents of the pharyngeal and laryngeal branches of the vagus terminate predominantly in caudal and medial subnuclei of the NST. The vagal efferents executing the programs of swallowing and gagging appear to be located in the NA; the effectors of the pharyngeal and palatal responses are situated rostrally and the effectors of the laryngeal reponses extend more caudally within the nucleus. Spontaneously occurring achalasia has been correlated with neuropathologies in the

NA complex, and this disorder of ingestion has also been produced by experimental lesions of the complex (see below).

c. The Esophageal Division of the Vagus. The vagus also innervates the intrinsic ganglia of the esophagus. In the thorax, the left and right vagi course onto the esophagus and run with it to the stomach. As the vagi travel with the esophagus, they rotate with the GI tract from their left-right orientation to a ventral-dorsal orientation, usually assuming the latter position before the level of the diaphragm. In some species, the thoracic vagi form a thoracic plexus where fibers are mingled, while in others, the vagi seem to organize their crossed innervation as a series of discrete communicating branches (see Prechtl and Powley, 1985a, for a fuller description). Apparently along the entire course, and certainly along the entire subdiaphragmatic course (Prechtl and Powley, 1985a), small fascicles separate from the main vagal trunks to innervate the esophagus. Perhaps because these fascicles have not been individually identified and because the esophageal division does not consist of a major branch or bundle, the details of esophageal innervation are not fully worked out. (See Hudson and Cummings, 1985).

The esophageal division of the vagus contains a variety of afferents including mechanoreceptors and thermoreceptors (Miller, 1982; Ouazzani and Mei, 1982; Satchell, 1984). Vagal efferents to the esophagus arise from the n. ambiguus (Holstege et al., 1983; Hudson and Cummings, 1985) and apparently from the dorsal motor nucleus of the vagus (DMN) as well (Fryscak et al., 1984). In general, the NA, particularly the retrofacial subnucleus, innervates the striated musculature of the esophagus, and the DMN appears to innervate the more distal smooth muscle portions of the esophagus (Fryscak et al., 1984; Holstege et al., 1983).

As in the case of the pharynx and larynx, the vagal innervation of the esophagus participates directly in the execution of feeding. From their operating characteristics, it is generally inferred that the stretch and distension receptors provide requisite information about the progress of the food bolus through the upper GI tract. The waves of peristalsis associated with deglutition are phased and in part paced by the vagal efferents. Presumably reflecting the functions of these DMN neurons innervating the esophagus, achalasia has been associated with spontaneous (Cassela et al., 1964; Clifford et al., 1980; Clifford et al., 1973) and experimental (Higgs et al., 1965) lesions of the DMN as well as the NA. Perhaps in part because they eliminate much of this esophageal innervation, radical and "high" subdiaphragmatic vagotomies can also produce a dysphagia (Orlando and Bozymski, 1973). This interpretation is not certain, however, because such surgeries probably compromise the vascularity of the esophagus as well, and dysphagia apparently occurs in only a small percentage of individuals subjected to radical vagotomies.

d. The Abdominal Division of the Vagus. The abdominal division of the vagus consists of several discrete branches, each corresponding to a longitudinal column of motor neurons in the dorsal motor nucleus (Fox and Powley, 1985). In the rat, the animal used in most of the experimental analyses of food intake, these branches are--roughly in order of emergence, moving caudally--the hepatic, accessory celiac, celiac, ventral gastric, and dorsal gastric branch (Fig. 1). (In addition, the subdiaphragmatic vagal trunks are responsible for the more distal of the esophageal fascicles, as discussed above.) In general, these branches innervate the GI tract in regions most actively involved in digestion and absorption, and they innervate the viscera that participate most directly in the metabolism of energy. The pattern suggests, as mentioned above, that the vagus must be pivotal in the monitoring and organization of energy handling. Consistent with such an idea, a variety of data suggests that many of the roles of the abdominal vagus in feeding are indirect: the vagus helps manage the fluxes of energy in the body and thereby modulates the neural, humoral, and hormonal loops that influence feeding (Fig. 2).

Some of the participation of the vagus in feeding behavior can be best appreciated by examining the available information on the separate trunks. Other roles of the vagus in feeding can be best summarized in terms of effects associated with the entire abdominal division of the vagus.

1) The hepatic branch. The main hepatic branch is the first of the abdominal branches to leave the vagus below the diaphragm. It separates from the ventral trunk of the vagus and runs in the omentum towards the hilus of the liver, splitting into several fascicles as it travels (Prechtl and Powley, 1985). As it leaves the ventral trunk, the hepatic branch consists of approximately 3000 fibers, of which fewer than 1% are myelinated (Prechtl and Powley, 1985b). Traveling along the hepatoesophageal artery at this level, it intermingles with other fascicles (possibly of sympathetic origin). In some mammalian species a plexus is formed by these fascicles. Nearly all of the physiological analyses described below have assumed that this branch innervates the liver and that it is the exclusive vagal innervation of the liver. However, two cautions should be mentioned: (1) It has certainly not been established that this branch innervates only the liver; in fact, the evidence indicates that it projects to other tissues

as well. In several species, a branch from the vagal hepatic plexus has been shown in dissection studies to project to the duodenum and pancreas (Jackson, 1948). Such projections are supported by functional studies measuring duodenal motor activity or pancreatic insulin secretion following thoracic or cervical vagal trunk stimulation in animals with severed gastric and celiac branches (Berthoud et al., 1983; Matsuo and Seki, 1978; Stavney et al., 1963). The hepatic branch also appears to innervate the biliary system and the ventral lobe of the pancreas (Griffith, 1969; Stavney et al., 1963). (2) There are almost certainly other vagal inputs to the liver. In man, fascicles from the posterior vagus to the liver have been identified by dissection (McCrea, 1925; Teitelbaum, 1932). On this latter point, Magni and Carobi (1983) have recently identified a "right hepatic branch" in the rat that they trace from the fascicles of the posterior (or right) gastric branch, and Niijima (1983) describes a hepatic branch joining the posterior vagal trunk via the celiac ganglion.

Afferents in the main hepatic branch have been shown to relay many types of information, including chemoreceptive (Niijima, 1969; Niijima et al., 1983), osmoreceptive (Adachi, 1984a; Rogers and Novin, 1980), and thermosensitive (Adachi, 1984b), to the medulla. For a more detailed analysis of the sensory functions of the liver, see Sawchenko and Friedman (1969) and Lautt (1983). Hepatic afferents appear to terminate predominantly in the commissural NST. The components of these afferents that carry glucose-sensitive information have been shown to converge on a population of glucose-sensitive cells in the NST (Adachi et al., 1984). Those in the main branch project almost entirely to the left side of the medulla (Adachi, 1981; Kahrilas and Rogers, 1984). Their distribution as well as their second- and third-order central projections are reviewed by Norgren (1983), Hermann et al. (1983), and others.

Relatively few efferents are found in the main hepatic branch of the vagus (Fox and Powley, 1985; Magni and Carobi, 1983; Rogers and Hermann, 1983). Cell bodies of the hepatic efferents are found primarily on the left side of the DMN, in the longitudinal column of somata associated with the anterior gastric vagus (Fox and Powley, 1985). In their retrograde tracer study, Magni and Carobi (1983) associate a group of cells in the right DMN with their right hepatic branch; these latter cells seem to occupy a column homotypical to that of the main branch on the left.

While the sensory mechanisms associated with the liver could easily be involved in feeding, and offer a theoretical substrate for some of the physiological and behavioral observations discussed below, the relationship remains highly inferential. For most of the cases, it is equally possible that the hepatic afferents participate in metabolic or neuroendocrine responses with only the most indirect consequences for feeding behavior. Examples of a number of vago-vagal physiological reflexes with hepatic afferent limbs, which may conform to this idea, have been described (Granneman and Friedman, 1980, Lee, 1985; Niijima, 1980; Sakaguchi and Yamaguchi, 1979a, 1979b; Sakaguchi and Miyaoka, 1981; Sakaguchi, 1982).

A variety of experiments attempting to more directly implicate the hepatic afferents and efferents in feeding have examined the effects of selective hepatic branch vagotomies. The observation that hepatic vagotomy produced a modest increase in daylight food intake of rats (Friedman and Sawchenko, 1984) suggests that the hepatic branch may participate in normal ad libitum food intake. Females compensated for this modest increase by less night feeding, thus yielding an attenuated day/night differential, while males did not compensate for the increase by altered night feeding, thus yielding a mild hyperphagia and increased growth. The effect occurred within hours of the nerve resection, appeared permanent, and was not produced by selective section of the remaining abdominal branches (Friedman and Sawchenko, 1984; but see also Bellinger and Williams, 1981; Contreras and Kosten, 1981; Louis-Sylvestre et al., 1980; and Tordoff et al., 1982). As concluded by the investigators, this effect of hepatic vagotomy on feeding may operate through a number of indirectly metabolic sequelae that ultimately modulate feeding.

Geary and Smith (1983) have shown that the reduction of feeding or "satiety" produced by administration of pancreatic glucagon requires the hepatic branch of the vagus for its expression, insofar as selective hepatic vagotomy eliminates the suppressive effects of glucagon. Selective pharmacological blockade of the vagal efferents with atropine methylnitrate did not interfere with glucagon's effect on feeding. Their results are consistent with a model that would involve vagal afferents relaying information about the metabolic status of the liver (or the duodenal field apparently innervated by the hepatic branch) to the CNS mechanisms of feeding.

Another line of evidence implicates the vagal hepatic branch (efferents, afferents, or both) in the suppressive effect of fructose on insulin-induced eating. It was first shown that fructose, a metabolic fuel readily utilized by the liver but not the brain, prevented the feeding response to insulin injections (Stricker et al., 1977). Then Friedman and Granneman found that the selective hepatic vagotomy abolished this suppressive effect of fructose but did not affect the insulin-induced eating response itself (Friedman, 1980; Friedman and Granneman, 1983; Granneman and Friedman, 1984), suggesting that either afferent information about the effects of fructose or efferent information necessary for hepatic metabolism of the hexose were compromised (Friedman and Granneman, 1983).

Novin and his colleagues (1973; 1977) have also implicated the vagal innervation of the liver in food intake. These workers have shown that hepatic portal infusions of 2-deoxy-D-glucose potentiate feeding in intact but not in vagotomized animals.

Many other experiments have suggested different roles for hepatic mechanisms in feeding behavior, but on the basis of available information, the relation of these mechanisms to the hepatic branch of the vagus is particularly tenuous. Russek (1981; see also the particularly useful comments following Russek's review) has provided a good summary of these observations. Even those cases with a clear neural basis might influence feeding through the sympathetic innervations of the liver.

2) The celiac and accessory celiac branches. These two branches are most easily considered together. In the case of the rat, the celiac branch arises as the first division of the dorsal (or right) vagal trunk on the esophagus. The accessory celiac separates from the ventral (or left) vagal trunk immediately distal to the hepatic branch separation (Fig. 1). The accessory celiac courses around the esophagus and comes to occupy a position quite close to the celiac. The two branches exit from the esophagus to the left gastric artery and from there they course (at least in part) in tandem to the celiac ganglia. As they join the artery, they intermingle with other fiber bundles, presumably of sympathetic origin (Kuntz and Jacob, 1954; Mosimann, 1954; Prechtl and Powley, 1985a). Although more thorough analyses are required, it appears that the celiac branch projects to the right celiac ganglion and the accessory celiac projects to the left (cf. McDonald and Blewett, 1981; Prechtl and Powley, 1985a; Teitelbaum, 1932).

The afferents of the celiacs (or more precisely the celiac, since the accessory celiac has not yet been as carefully analyzed) project predominantly to the commissural portion of the NST (Norgren and Smith, 1983). They appear to intermingle with the terminal field of the hepatic vagus. The efferent cell bodies of the two branches are organized as separate homotypical longitudinal cell columns at the lateral edges of the DMN; the celiac branch arises from the lateral pole of the right DMN, while the accessory branch arises from the corresponding position in the left DMN (Fox and Powley, 1985). This strict homotypy adds face validity to the apparent bilateral symmetry of the branches at the celiac ganglia.

Much remains unknown about the functional organization of these branches. Even less is known about the celiacs than the others because they cannot be followed directly to a target organ since they relay in the celiac ganglia. Furthermore, inferences about the accessory branch must be made by analogy since no experiments have looked at it exclusively.

Practically, most of the evidence indicates that the two celiac branches innervate the intestines and the pancreas. The pattern dictates the conclusion that the celiac branches must have significant roles in the control or coordination of the enteropancreatic axis. On the afferent side, intestinal sensory receptors are probably a major component of the afferent projection into the commissural nucleus. Mei and his coworkers (Mei, 1985; Ouazzani and Mei, 1981; see also Clarke and Davison, 1978; and Paintal, 1973) have identified a series of mechano-, chemo-, thermo-, and osmoreceptor afferent activity elicited from the mucosa and submucosa of the small intestine. Since, in most of this work, sensory fibers are located in the cervical vagus or the nodose ganglia and stimuli are applied to the intestine, it is impossible to be sure that the fibers are all celiac and accessory celiac fibers. The celiac efferent innervation of the gut would suggest, however, that general conclusion. As mentioned earlier, Niijima (1983) has recently described electrophysiological evidence for hepatic afferents in the celiac branch of the vagus that travel initially from the liver to the celiac ganglia by way of a splanchnic filament.

Celiac efferents innervate both the small intestine and the pancreas. Stimulation of the celiac branch in dogs (Stavney et al., 1963) indicates that the celiac branch facilitates motility from the proximal duodenum through the ascending colon. Conversely, celiac vagotomy in humans can produce ileus of the entire midgut (Griffith, 1969). Vagally elicited insulin secretion is also heavily dependent on the celiac branches. Such insulin secretion is decreased after selective vagotomy of the celiac branch (Berthoud et al., 1983). Further, cephalic or preabsorptive insulin secretion appears to require the integrity of the celiac branches (Woods and Bernstein, 1980). A series of studies employing retrograde tracers has suggested that an extensive number of DMN neurons may project to the pancreas (Laughton and Powley, 1979; Luiten et al., 1984; Weaver, 1980), although in a recent reanalysis of these results we have found evidence that, with experimental controls to avoid diffusion and leakage of the tracer, very few cells in the DMN project directly to the pancreas (Fox and Powley, submitted). These results suggest that the DMN projections in the celiac branches may actually synapse at an extrapancreatic locus such as the celiac ganglia.

A variety of experimental analyses are consistent with the conclusion that the celiac branches may modulate food intake. A number of considerations suggest that the cephalic insulin response may play a significant role in feeding behavior (for reviews, see Louis-Sylvestre et al., 1983; Louis-Sylvestre and LeMagnen, 1980; Powley, 1977; Powley and Berthoud, in press; Sahakian, 1982). Louis-Sylvestre and co-workers (1983) have argued that the component of meal size which is determined by palatability seems to be dependent on cephalic insulin responses; further, these investigators have suggested this effect depends on the celiacs because total vagotomy, but not gastric vagotomy, eliminates the effect of palatability on meal size. Sawchenko and Gold (1981) and Sawchenko et al. (1981) have also implicated the celiac branches of the vagus in food intake. They have shown that the reversal of the hyperphagia and obesity of the ventromedial hypothalamic syndrome produced by subdiaphragmatic vagotomy (see below for a discussion of this phenomenon) requires section of the celiac branches: of various combinations of selective vagotomies they tried, those combinations that included celiac vagotomies were effective in eliminating the elements of the VMH syndrome. In fact, for a full effect, Sawchenko and Gold found that they had to cut the celiac branches plus the hepatic branch, a pattern that may be required to denervate the pancreas effectively. Further, Sawchenko and co-workers (1981) found that selective celiac vagotomies reduced the feeding produced by hypothalamic infusions of norepinephrine. This reduction of norepinephrine-induced feeding was specific but not complete (Sawchenko did not use the combination of celiac and hepatic vagotomies in this experiment).

3) The gastric branches. The fibers in the anterior and posterior vagal trunks that do not course through the hepatic, accessory celiac, or celiac continue on the esophagus to form the anterior and posterior gastric branches respectively (Fig. 1). These two branches contain over half of all the fibers in the abdominal vagus (Powley et al., 1983). Each of the gastric branches ramifies into several smaller bundles above the levels of the cardiac sphincter (Prechtl and Powley, 1985a), and these bundles then distribute to innervate the lower esophageal sphincter as well as the corpus, fundus, antrum, and pylorus. Although there seem to be significant species difference in distribution, for the rat at least, anatomical (Coil and Norgren, 1979; Dennison et al., 1981; Fox and Powley, 1985) and physiological (England et al., 1979; Pritchard et al., 1968) analyses suggest that the vagal motor projections to the stomach remain unilateral and uncrossed--i.e., the left medulla projects to the ventral surface of the stomach and the right medulla projects to the dorsal surface. Gastric afferents appear to project to the brainstem in a more bilateral manner (Dennison et al., 1981; Norgren and Smith, 1983). Although often uncritically assumed to innervate the stomach, the

"gastric" branches project much more widely in the peritoneum, possibly including the liver (Magni and Carobi, 1983), the duodenum (Guillaumie, 1933), and the pancreas (Berthoud et al., 1983; Lovgren et al., 1981). They may also constitute a significant route for the substantial afferent traffic originating in the small intestine (Mei, 1985; Paintal, 1973).

On the afferent side, the stomach--including the corpus and antrum as well as the lower esophageal and pyloric sphincters--is richly innervated with mechanoreceptors registering intragastric pressure and stretch. Since gastric distension is known to serve as a salient signal in the control of food intake (Gonzales and Deutsch, 1981; Wirth and McHugh, 1983) and has been shown effective in modulating gustatory afferent activity (Glenn and Erikson, 1976; Hellekant, 1971) as well as the masseteric reflex (Pettorossi, 1983), it is relevant that gastric mechanoreceptor activity has been shown to influence activity in both the NST (Barber and Burks, 1983; Ewart and Wingate, 1983) and the hypothalamus (e.g., Jeannigros, 1984). Gastric afferents also include a population of chemoreceptors, the best-characterized of which are affected by gastric pH (Davison, 1972b; Mei, 1985; Paintal, 1973). However, gastric vagal afferents also serve to drive gastrogastric (Grundy et al., 1981) and gastropancreatic reflexes (Schwartz et al., 1979) and might not be directly involved in controlling feeding mechanisms.

Vagal efferents innervating the stomach include fibers that influence gastric secretion (Grundy and Scratcherd, 1981; Emas, 1973; Uvnas-Wallensten et al., 1976), gastric motility and emptying (Collman et al., 1983; Grundy and Scratcherd, 1982), and receptive relaxation (Lisander, 1975). Each of these responses is clearly central to an integrated pattern of digestion, propulsion, and absorption of nutrients. As has been persuasively shown, gastric emptying of food into the duodenum is one of the key steps in digestive physiology that modifies food consumption. Through its active roles in the monitoring and phasing of gastroduodenal transfer of nutrients, the abdominal vagus may exercise one of its more direct effects on feeding (Hunt, 1980; McHugh and Moran, 1979; Wright et al., 1983).

Several functional analyses have established that the integrity of the gastric branches of the vagus is required for some specific aspects of normal feeding patterns. One of the better-defined of these feeding responses is limitation of ingestion produced by the gut peptide cholecystokinin (CCK). Smith and his colleagues (1981) have shown that curtailment of spontaneous meal size by CCK exogenously administered into the peritoneal cavity requires the afferents in the gastric branches of the vagus for its expression.

4a) All subdiaphragmatic branches--evidence based on vagotomy. The vast majority of experimental analyses on the roles of the vagus in food intake have, for a number of obvious pragmatic reasons, concentrated on manipulations of the entire abdominal vagus. While the interpretation of such experimental analyses in terms of a particular response or detector is even more problematic than in the cases just reviewed, the experimentation does suggest some general conclusions. The most general is that the subdiaphragmatic branches of the vagus seem to modify food intake through a number of specific responses that influence the partitioning of energy and energy stores.

Experiments examining the impact of vagotomy on meal-taking have documented several changes in specific parameters. The varying patterns of changes reported for different maintenance conditions and diets suggest, however, not a global effect of vagotomy on feeding, but specific alterations secondary to differences in the handling and digestion of particular macronutrients and particular compositions of food. Total food intake is typically reduced in an initial period postvagotomy, and it may remain chronically depressed (Fox et al., 1976; Mordes et al., 1979). Meal sizes are smaller on liquid diets (Fox et al., 1976; Snowdon, 1970; Snowdon and Wampler, 1974), presumably because of the vagotomized stomach's "dumping response" with liquids. Other changes in feeding patterns reported include decreased intermeal intervals with liquid diets (Snowdon and Epstein, 1970), reduced diurnal fluctuations of feeding with a solid diet (Louis-Sylvestre, 1978), and altered circadian meal patterns (Geiselman et al., 1980).

In addition, operated animals self-select different dietary constituents. Under some conditions, vagotomized animals reduce the percentage of their diet taken as protein and correspondingly increase in carbohydrate consumption (Li and Anderson, 1984). While readily consuming carbohydrates, vagotomized animals evidence a decrease in simple sugar intake (sucrose--Sclafani and Kramer, 1983), an adaptive response given the mild glucose intolerance associated with vagotomy (Louis-Sylvestre et al., 1981). In related experiments, which may require a different mechanism for their explanation, Louis-Sylvestre and co-workers (1983) have suggested that the vagus (and in particular, vagally mediated cephalic phase responses of insulin secretion) may be required for the determination of the effect of palatability on meal size. More particularly, Louis-Sylvestre found that the meal size, characteristically a function of the palatability of the diet for normal animals, was unaffected by changes of palatability for vagotomized animals.

Another, somewhat "covert" role of the vagus in diet selection is the nerve's participation in conditioning taste aversion: conditioned taste aversions to substances that are gastrointestinal irritants or peripherally acting systemic toxins require an intact vagus (presumably the afferents) to establish the alteration in food selection (Kiefer et al., 1981). In different experimental conditions, vagotomy has been shown somewhat paradoxically to produce an unconditioned stimulus (e.g., weight loss, nausea, GI distress) that will support a conditioned taste aversion (Bernstein and Goehler, 1983). This latter effect, presumably operating "centrally," may account for some of the postsurgical changes in food preferences and also underscores one of the mechanisms that may explain some of the effects of vagotomy in terms of nonspecific consequences.

Surgical elimination of the abdominal vagi also leads to another consequence for energy regulation. Subdiaphragmatic vagotomy has been observed to produce a chronic reduction in the level of body weight maintenance in rats in several experiments (Mordes et al., 1979; Powley and Opsahl, 1974; but see also Ellis and Pryse-Davies, 1967). The effect is not secondary to the classical gastric emptying problems, because pyloroplasty is ineffective in correcting the weight regulation of the vagotomized animal (Mordes et al., 1979; Fox and Powley, 1984). In a particularly systematic experiment examining the reduction in weight produced by truncal vagotomy, Mordes and co-workers (1979) found additional evidence that the reduction was not the result of gastric distension, derangement of motility, or malabsorption. The reduction, ranging from 10% to 30% depending upon the experiment, may result in part from a reduction in anabolic tone of the viscera or some more direct effect upon food intake that has not yet been identified.

A possible role for the vagus in ingestion has also been suggested by experiments examining electrically elicited feeding before and after subdiaphragmatic vagotomy: stimulus-bound feeding was dramatically attenuated or even eliminated by vagotomy (Ball, 1974; Powley et al., 1978). These observations also suggested by histological correlations that the more complete or radical the vagotomy, the more extensive the disruption of feeding behavior (Powley et al., 1978). Finally, in this case, the evidence suggests that the impact of vagotomy on feeding is a consequence of the interruption of vagal afferent traffic or of slowly developing sequelae of chronic vagal dysfunction, since atropine blockade did not reproduce the effects of surgery (Powley et al., 1978).

4b) All subdiaphragmatic branches--evidence involving exaggerations of tone. Another type of experimental preparation also provides some indication of possible roles of the vagi in food intake. Destruction of the ventromedial hypothalamic area produces an immediate increase in activity in a vagal branch innervating the pancreas (Yoshimatsu et al.,

1984), an early exaggeration in vagally mediated metabolic reflexes including insulin secretion (Berthoud and Jeanrenaud, 1979) and gastric acid secretion (Weingarten and Powley, 1980), as well as an ensuing obesity and hyperphagia characteristic of the "ventromedial hypothalamic syndrome" (for review see Powley et al., 1980). As has been argued in detail in a number of previous reviews, some (but not all) of the obesity and hyperphagia associated with ventromedial hypothalamic damage may represent a response to the exaggerated vagal tone produced by the lesion (Bray and York, 1979; Powley, 1977; Powley and Opsahl, 1976; see also Sclafani, this volume). Further, one specific consequence of this lesion-induced exaggeration of anabolism has been examined more fully. Cox and Powley have found that the well-documented metabolic advantage of the animal with hypothalamic lesions is a result of increased vagal tone, since vagotomy eliminates the excess weight gain and adiposity that develop in the ventromedial hypothalamic animal pair-fed to control levels of intake (Cox and Powley, 1981a and 1981b). Further, this metabolic edge associated with exaggerations in vagal activity apparently does not result from enhanced absorption of nutrients from the gastrointestinal tract (cf. Cox and Powley, 1981b).

Correlative data suggest that exaggerated vagal activity or high levels of vagal tone might predispose some animals to anabolism and obesity. In a prospective analysis, Berthoud (1985) found, in screening a group of rats on the basis of the amplitudes of their individual cephalic insulin response to a saccharin test stimulus, that the animals with the highest responses subsequently gained the most weight when provided with a supermarket diet. Further, Rohner-Jeanrenaud and Jeanrenaud have recently (1985) reported that an exaggerated vagal outflow may be of etiologic significance in the expression of the fatty (*fa/fa*) rat obesity, since these animals display an atropine-blockable exaggerated insulin response to a glucose challenge prior to the development of the obesity.

III. THE SYMPATHETICS AND THE CONTROL OF FEEDING

A detailed review of the sympathetic nervous system contribution to feeding is beyond the scope of this chapter; furthermore, experimental observations on the organization and functions of the sympathetics are even sparser and less complete than for the parasympathetics. A few general points and two specific sympathetic mechanisms should be briefly discussed.

First, sympathetic efferents, like vagal efferents, appear to influence feeding indirectly by affecting substrate availability and energy partitioning. (With minor modifications, the organization of the second and third mechanisms in Figure 2 for the parasympathetics can be applied to the sympathetics.) The sympathetics play significant roles in glucose production and in the mobilization of triglycerides. Thus, these pathways make fuels available from energy stores when the organism needs them for "fight or flight". Furthermore, these sympathetic pathways also promote heat production and expenditure in some instances where caloric intake is high (see below).

The role of the sympathetics can be illustrated with the example of adrenergic innervation of brown adipose tissue (BAT). As is well known, BAT adipose tissue is richly innervated by the sympathetic nervous system. This mechanism has long been known to play a role in nonshivering thermogenesis and heat production stimulated by the demands of temperature regulation. More recently established is the effect on sympathetic activity of shifts in the demands on energy expenditure. By appropriate modulation of adrenergic tone, animals reduce their expenditure during starvation and increase their expenditure during overfeeding.

In the situations just described, feeding, or more precisely, the availability of ingested calories, modifies sympathetic activity. Conversely, considerable evidence suggests that a relatively chronic decrease in sympathetic tone leads to decreased BAT thermogenesis and decreased net adipose tissue lipolysis, ultimately leading to obesity and at least relative hyperphagia to replace fuels sequestered in fat. Such effects are well illustrated in several of the rodent models of obesity (Bray and York, 1979). Similarly, some experiments have found that denervation (Bartness and Wade, 1984; Dulloo and Miller, 1984) of the accessible interscapular BAT or BAT removal (Connolly et al., 1982; Moore et la., 1985) lead to weight gain and possibly increased consumption (Moore et al., 1985), although other experimental reports on BAT denervation have been negative (e.g., Cox and Lorden, 1984).

Other postulated roles of the sympathetics in modulating patterns of ingestion are associated with the celiac ganglia and their interconnections (Fig. 1). These ganglia, which form the most extensive and most rostral of the abdominal plexuses, lie adjacent to the abdominal aorta in the levels of the celiac artery (Fig. 1). The celiac ganglia communicate with a meshwork of smaller plexuses including the left gastric (innervating the esophagus and stomach), the hepatic (innervating the liver, gall bladder, stomach, duodenum, and pancreas), and splenic (innervating the spleen, pancreas, and stomach), the duodenal (innervating the duodenum and pancreas), and the superior mesenteric (innervating the pancreas, small intestine, and colon) plexuses--e.g., Kuntz and Jacobs (1954). The extensive ramifications of the system hint at its complexities and overlaps. The major preganglionic input to the celiac ganglia and their subsidiary plexuses are splanchnic nerves. Afferents and efferents from the intramural plexuses of the bowel also project to the celiac ganglia.

To date, the results of experiments that have transected, ablated, or otherwise interfered with the abdominal sympathetic system are too few and fragmentary to justify any broad conclusions. In certain experimental situations, splanchnicectomy (Opsahl, 1977), chemical sympathectomy (Powley et al., 1983), and unilateral abdominal sympathectomies (Bray et al., 1981) have been shown to affect body weight regulation and adipose tissue stores. In other experimental cases, however, chemical sympathectomy (Storlien et al., 1983; Tordoff et al., 1984) or celiac ganglionectomy (Granneman and Friedman, 1984) appear to have little impact on specific feeding responses measured or on energy regulation in either intact animals or, indeed, animals with hypothalamically induced feeding disturbances (Powley et al., 1983; Storlien et al., 1983; Tordoff et al., 1984).

A complication in interpreting the effects of ganglionectomies of the type just described is that the vagus nerve also projects through the celiac ganglia. The celiac branch of the vagus and apparently the accessory celiac branch project through celiac ganglia on their course to the pancreas, small intestine, and other tissue. The full implications of this organization are not clear, but the observations do suggest a caveat: celiac ganglionectomies are not just sympathectomies; rather, they are also selective vagotomies. Such a situation obviously must qualify our conclusions about the "sympathetics". In addition to the qualification that any effect of ganglionectomy might reflect the parasympathetic rather than the sympathetic manipulation, there is also a corollary. Parodoxically, ganglionectomies may underestimate some of the roles of the sympathetics because removing both divisions of the reciprocally organized ANS may, at least in some situations, produce less dysfunction than removing one division and leaving the other unopposed.

A final set of examples of the role of the sympathetics in the control of food intake raises much the same issue. A substantial series of experimental denervation strategies have investigated the role of the innervation of the liver in feeding. The vagal fibers innervating the liver have already been discussed. Although some modest effects have been reported in various feeding tests (e.g., Russek, 1981), a number of carefully done experiments employing complete and radical liver denervation and bypass strategies have failed to show much impact on feeding behavior (Bellinger et al., 1984; Bellinger and Williams, 1981; Koopmans, 1984; Louis-Sylvestre et al., 1980). Indeed, it can be argued that complete liver denervation has less effect on feeding than simple hepatic vagotomies. The distortions resulting from disabling one limb of the ANS may again be more disturbing than a "balanced" reduction in autonomic controls.

IV. EXPERIMENTAL PRIORITIES

Given the ample evidence that the ANS participates in the control of feeding, it is appropriate to assess current experimental priorities. More general demonstrations that the vagus or the sympathetics can affect feeding behavior will add little to our present understanding. Rather, more exacting experiments that define the pathways involved, characterize the participating physiological response, and clarify the relationship of that autonomic response to feeding consequences are needed. The vagus can be used to illustrate these points.

Our increased understanding of vagal anatomy has shown that the circuitry relevant to the vagus and feeding is immensely complex. Although much work on the vagus and feeding has assumed a one-to-one isomorphism between vagal branches and structures, these assumptions now appear simplistic and need to be reconsidered. For example, the hepatic branch of the vagus does not innervate just the liver, nor does the liver receive vagal inputs only from the hepatic branch, thus making "hepatic vagotomies" complicated. On another level, there are a number of reasons why one laboratory's vagotomy may differ from another's. The exact level of the cut may vary, and the branches spared may differ. Resections are quite different from transections. Because of the interdigitation of the sympathetics among the vagal branches, many procedures may affect some sympathetic innervation fields as well; the reverse, where sympathectomies compromise the parasympathetics--for example, the vagal projections through the celiac ganglia--is also true. Both regeneration (Fox and Powley, 1984) and physiological reorganization (Campfield et al., 1984) appear to occur after vagotomy. Therefore, caution must be exercised in interpreting the results of vagotomy experiments, especially those conducted after long postsurgical survival times. Many of the ambiguities and controversies in the literature on vagotomy may stem from the absence of accepted tests for completeness of vagotomy. With the exception of the gastric branches, most vagal branches presently do not have specific, selective, and convincing tests of completeness (cf. Louis-Sylvestre, 1983).

Functional assessments of the organization of the vagal afferents also pose significant problems. Some of these are reviewed by Mei (1985), Norgren (1983), and Davison (1972a). One major problem is relating altered afferent traffic to the specific reflexes or regulatory functions served. With only a few conspicuous exceptions, it is unclear whether identified vagal afferents participate in vago-vagal adjustments, in other metabolic reflexes, in ingestive reflexes, or in combinations of these response networks. Requirements of electrophysiological controls often make it difficult to do more than record in the nodose ganglion or NST while stimulating a distal sensitive site in an anesthetized preparation. Determining the branches over which the studied fibers travel, their destinations within the NST, their interactions with descending CNS systems, and the subsequent second-, third-, and fourth-order neurons is a major challenge for this work.

The relationship between particular autonomic responses and altered ingestive behavior needs to be specified much more completely. At one extreme, in much of the preliminary work accomplished so far, the particular afferent relay, efferent pathway, or reflex mechanism responsible for a behavioral effect is simply unknown. We can only conclude that *some* vagal response may underlie the effect under study. In other cases, with only slightly greater resolution, we are able to postulate that the candidate response must be afferent or efferent by differential effects of surgical and pharmacological blockades. In still other cases, it is possible to measure a particular response (e.g., afferent distension information, cephalic phase of gastric acid secretion, or intestinal chemoreceptor-insulin secretion reflex) and correlate it with behavioral effects. Clearly, to fully understand the relationships between autonomic physiology and behavior, far more complete identification and definition of the physiological mechanisms are required. This, in turn, would make feasible more definitive conclusions in tests for completeness of vagotomy.

Finally, in a similar vein, much more work is needed on the specification of the ingestive response systems under study. Fundamentally, many of the experimental analyses of the ANS and feeding are correlational analyses. Given the situation, the power of the analyses is limited by the accuracy and completeness of the description. The general argument that because vagotomy depresses both vagal responses and feeding behavior, the former must account for the latter, is not particularly compelling. Global correlations of that sort cannot rule out the possibility that the effects of vagal manipulations on feeding are secondary to alterations in other physiological responses of the animal. The case is more compelling if the physiological response and the ingestive response are highly correlated in terms of latency, time course, response profiles over a number of foodstuffs, and experimental challenges. The stronger and tighter the correlation, the better the analysis.

V. SUMMARY AND PERSPECTIVE

A balanced view on the important roles of the vagus and other autonomic nerves in the control of feeding should perhaps stem from the argument considered at the beginning of this review: The parasympathetic and sympathetic limbs of the autonomic nervous system are the neural link between the brain and the entire gut, including its enteric nervous system. These parasympathetic and sympathetic paths both carry afferent feedback to the CNS and return the efferent directions to the viscera. Other than humoral and endocrine links, the autonomics provide the sole communication between brain and gut. To recognize that the gut is important in feeding and that it requires linkages to the brain is to recognize *ipso facto* that the vagus and its sympathetic counterparts are important in feeding.

The argument carries extra weight because the large number of experiments that have attempted to relate autonomic function to feeding have almost certainly underestimated the extent of this involvement. For example, the changes we see in feeding behavior after a branch of the vagus nerve is cut do not describe the exact participation of a particular mechanism in feeding behavior. Rather, as in any lesion experiment, we observe what the rest of the system can do in the absence of the one branch. In many cases, redundancies and other types of compensatory adjustments are presumably able to redress partially or even fully the losses and distortions. Such compensations would seem to be the rule in other physiological systems as multiply determined (or overdetermined) as the one that determines feeding.

However, even with the proviso that presently available evidence must also certainly underestimate the role of the ANS in feeding, a number of roles have already been established. The vagus, including both the mechanoreceptor and chemoreceptor afferents as well as the efferents originating in the NA, plays a critical role in the execution of swallowing. Further, both the vagus (through its largely anabolic responses) and the sympathetics (through their predominantly catabolic mechanisms) operate to control the partitioning of energy stores in the body. As a consequence, the two divisions of the autonomic nervous system effect considerable control over the balance of energy stores and, ultimately, the energy fluxes and their correlated signals that determine feeding. Finally, vagal and sympathetic afferents are also responsible for reporting those signals to the CNS.

ACKNOWLEDGMENTS

Research described from the author's laboratory was supported by NIH grants AM27627 and AM20542.

REFERENCES

Adachi, A. 1981. Electrophysiological study of hepatic vagal projection to the medulla. *Neurosci. Lett.* 24: 19-23.

Adachi, A. 1984a. Projection of the hepatic vagal nerve in the medulla oblongata. *J. Auton. Nerv. Syst.* 10: 287-293.

Adachi, A. 1984b. Thermosensitive and osmoreceptive afferent fibers in the hepatic branch of the vagus nerve. *J. Auton. Nerv. Syst.* 10: 269-273.

Adachi, A., N. Shimizu, Y. Oomura, and M. Kobashi. 1984. Convergence of hepatoportal glucose-sensitive afferent signals to glucose-sensitive units within the nucleus of the solitary tract. *Neurosci. Lett.* 46: 215-218.

Ball, G.G. 1974. Vagotomy: effect on electrically elicited eating and self-stimulation in the lateral hypothalamus. *Science* 184: 484-485.

Barber, W.D., and T.F. Burks. Brainstem response to phasic gastric distension. *Am. J. Physiol.* 245: G242-G248.

Bartness, T.J., and G.N. Wade. 1984. Effects of interscapular brown adipose tissue denervation on body weight and energy metabolism in ovariectomized and estradiol-treated rats. *Behav. Neurosci.* 98: 674-685.

Bellinger, L.L., V.E. Mendel, F.E. Williams, and T.W. Castonguay. 1984. The effect of liver denervation on meal patterns, body weight and body composition of rats. *Physiol. Behav.* 33: 661-667.

Bellinger, L.L., and F.E. Williams. 1981. The effects of liver denervation on food and water intake in the rat. *Physiol. Behav.* 26: 663-671.

Bernstein, I.L., and L.E. Goehler. 1983. Vagotomy produces learned food aversions in the rat. *Behav. Neurosci.* 97: 585-594.

Berthoud, H.-R. 1985. Cephalic phase insulin response as a predictor of body weight gain and obesity induced by a palatable cafeteria diet. *J. Obes. Weight Reg.* 4. In press.

Berthoud, H.-R., and B. Jeanrenaud. 1979. Acute hyperinsulinemia and its reversal by vagotomy after lesions of the ventromedial hypothalamus in anesthetized rats. *Endocrinology* 105: 146-151.

Berthoud, H.-R., A. Niijima, J.-F. Sauter, and B. Jeanrenaud. 1983. Evidence for a role of the gastric, coeliac and hepatic branches in vagally stimulated insulin secretion in the rat. *J. Auton. Nerv. Syst.* 7: 97-110.

Bray, G.S., S. Inoue, and Y. Nishizawa. 1981. Hypothalamic obesity. The autonomic hypothesis and the LH. *Diabetologia* 20: 900-902.

Bray, G.A., and D.A. York. 1979. Hypothalamic and genetic obesity in experimental animals: an autonomic and endocrine hypothesis. *Physiol. Rev.* 59: 719-809.

Campfield, L.A., F.J. Smith, and R.E. Eskinazi. 1984. Glucose responsiveness and acetylcholine sensitivity of pancreatic ß-cells after vagotomy. *Am. J. Physiol.* 246: R985-R993.

Cassella, R.R., A.L. Brown, G.P. Sayre, and F.H. Ellis, Jr. 1964. Achalasia of the esophagus: pathologic and etiologic considerations. *Ann. Surg.* 160: 474.

Clarke, G.D., and J.S. Davison. 1978. Mucosal receptors in the gastric antrum and small intestine of the rat with afferent fibres in the cervical vagus. *J. Physiol.* 284: 55-67.

Clifford, D.H., P.F.T. Barboza, and J.G. Pirsch. 1980. The motor nuclei of the vagus nerve in cats with and without congenital achalasia of the oesophagus. *Br. Vet. J.* 136: 74-83.

Clifford, D.H., J.G. Pirsch, and M.L. Mauldin. 1973. Comparison of motor nuclei of the vagus nerve in dogs with and without esophageal achalasia. *Proc. Soc. Exp. Biol. Med.* 142: 878-882.

Coil, J.D., and R. Norgren. 1979. Cells of origin of motor axons in the subdiaphragmatic vagus of the rat. *J. Auton. Nerv. Syst.* 1: 203-210.

Collman, P.I., D. Grundy, and T. Scratcherd. 1983. Vagal influences on the jejunal 'minute rhythm' in the anesthetized ferret. *J. Physiol.* 345: 65-74.

Connolly, E., R.D. Morrisey, and J.A. Carnie. 1982. The effect of interscapular brown adipose tissue removal on body-weight and cold response in the mouse. *Br. J. Nutr.* 47: 653-658.

Contreras, R.J., and T. Kosten. 1981. Changes in salt intake after abdominal vagotomy: evidence for hepatic sodium receptors. *Physiol. Behav.* 26: 575-582.

Cox, J.E., and J.F. Lorden. 1984. Brown fat denervation does not alter dietary obesity. *Soc. Neurosci. Abstr.* 10: 532.

Cox, J.E., and T.L. Powley. 1981a. Intragastric pair feeding fails to prevent VMH obesity or hyperinsulinemia. *Am. J. Physiol.* 240: E566-E572.

Cox, J.E., and T.L. Powley. 1981b. Prior vagotomy blocks VMH obesity in pair-fed rats. *Am. J. Physiol.* 240: E573-E583.

Dennison, S.J., B.L. O'Connor, M.H. Aprison, V.E. Merritt, and D.L. Felten. 1981. Viscerotopic localization of preganglionic parasympathetic cell bodies of origin of the anterior and posterior subdiaphragmatic vagus nerves. *J. Comp. Neurol.* 197: 259-269.

Donald, D.E. 1952. Esophageal dysfunction in the rat after vagotomy. *Surg.* 31, 251.

Davis, P.J., and B.S. Nail. 1984. On the location and size of laryngeal motoneurons in the cat and rabbit. *J. Comp. Neurol.* 230: 13-32.

Davison, J.S. 1972a. The electrophysiology of gastrointestinal chemoreceptors. *Digestion* 7: 312-317.

Davison, J.S. 1972b. Response of single vagal afferent fibres to mechanical and chemical stimulation of the gastric and duodenal mucosa in cats. *Q. J. Exp. Physiol.* 57: 405-416.

Dubner, R., B.J. Sessle, and A.T. Storey. 1978. *The Neural Basis of Oral and Facial Function.* Plenum Press, New York.

Dulloo, A.G., and D.S. Miller. 1984. Energy balance following sympathetic denervation of brown adipose tissue. *Can. J. Physiol. Pharmacol.* 62: 235-240.

Ellis, H., and J. Pryse-Davies. 1967. Vagotomy in the rat: a study of its effects on stomach and small intestine. *Br. J. Exp. Pathol.* 48: 135.

Emas, S. 1973. Vagal influences on gastric acid secretion. *Scand. J. Gastroent.* 8: 1-4.

England, A.S., P.C. Marks, and G. Paxinos. Brain hemisections induce asymmetric gastric ulceration. *Physiol. Behav.* 23: 513-517.

Etemadi, A.A. 1961. The dorsal motor nucleus of the vagus. *Acta Anat.* 47: 328-332.

Ewart, W.R., and D.L. Wingate. 1983. Central representation and opioid modulation of gastric mechanoreceptor activity in the rat. *Am. J. Physiol.* 244: G27-G32.

Foley, J.O., and F.S. Dubois. 1937. Quantitative studies of the vagus nerve in the cat. I. The ratio of sensory to motor fibers. *J. Comp. Neurol*. 67: 49-67.

Fox, E.A., and T.L. Powley. 1984. Regeneration may mediate the sparing of VMH obesity observed with prior vagotomy. *Am. J. Physiol*. 247: R308-R317.

Fox, E.A., and T.L. Powley. 1985. Longitudinal columnar organization within the dorsal motor nucleus represents separate branches of the abdominal vagus. *Brain Res*. In press.

Fox, E.A., and T.L. Powley. 1985. Tracer diffusion has exaggerated CNS maps of direct preganglionic innervation of pancreas. Under editorial review.

Fox, K.A., S.C. Kipp, and D.A. VanderWeele. 1976. Dietary self-selection following subdiaphragmatic vagotomy in the white rat. *Am. J. Physiol*. 231: 1790-1793.

Friedman, M.I. 1980. Hepatic-cerebral interactions in insulin-induced eating and gastric acid secretion. *Brain Res. Bull*. 5, Suppl. 4: 63-68.

Friedman, M.I., and J. Granneman. 1983. Food intake and peripheral factors after recovery from insulin-induced hypoglycemia. *Am. J. Physiol*. 244: R374-R382.

Friedman, M.I., and P.E. Sawchenko. 1984. Evidence for hepatic involvement in control of ad libitum food intake in rats. *Am. J. Physiol*. 247: R106-R113.

Fryscak, T., W. Zenker, and D. Kantner. 1984. Afferent and efferent innervation of the rat esophagus. *Anat. Embryol*. 170: 63-70.

Gabella, G. 1976. *Structure of the Autonomic Nervous System*. John Wiley & Sons, Inc., New York.

Gacek, R.R. 1975. Localization of laryngeal motor neuron in the kitten. *Laryngoscope* 85: 1841-1861.

Geary, N., and G.P. Smith. 1983. Selective hepatic vagotomy blocks pancreatic glucagon's satiety effect. *Physiol. Behav*. 31: 391-394.

Geiselman, P., J.R. Martin, D.A. VanderWeele, and D. Novin. 1980. Multivariate analysis of meal patterning in intact and vagotomized rabbits. *J. Comp. Physiol. Psychol*. 94: 388-399.

Glenn, J., and R. Erickson. 1976. Gastric modulation of gustatory afferent activity. *Physiol. Behav*. 16: 561-568.

Gonzalez, M.R., and J.A. Deutsch. 1981. Vagotomy abolishes cues of satiety produced by gastric distension. *Science* 212: 1283-1284.

Granneman, J., and M.I. Friedman. 1980. Hepatic modulation of insulin-induced gastric acid secretion and EMG activity in rats. *Am. J. Physiol*. 238: R346-R352.

Granneman, J., and M.I. Friedman. 1984. Effect of hepatic vagotomy and/or coeliac ganglionectomy on the delayed eating response to insulin and 2DG injection in rats. *Physiol. Behav.* 33: 495-497.

Griffith, C.A. 1969. Significant functions of the hepatic and celiac vagi. *Am. J. Surg.* 118: 251-259.

Grijalva, C.V., and E. Lindholm. 1982. The role of the autonomic nervous system in hypothalamic feeding syndromes. *Appetite* 3: 111-124.

Grill, H.J., and R. Norgren. The taste reactivity test. I. Mimetic responses to gustatory stimuli in neurologically normal rats. *Brain Res.* 143: 263-279.

Grundy, D., A.A. Salih, and T. Scratcherd. 1981. Modulation of vagal efferent fibre discharge by mechanoreceptors in the stomach, duodenum and colon of the ferret. *J. Physiol.* 319: 43-52.

Grundy, D., and T. Scratcherd. 1982. Effect of stimulation of the vagus nerve in burst on gastric acid secretion and motility in the anesthetized ferret. *J. Physiol.* 333: 451-461.

Guillaumie, M. 1933. *Recherches éxpérimentales sur le role du nerf vague dans le fonctionnement éxocrine du pancréas.* Maurice Bremer et fils, Paris.

Hellekant, G. 1971. The effect of stomach distension on the efferent activity in the chorda tympani nerve of the rat. *Acta Physiol. Scand.* 83: 527-531.

Hermann, G.E., N.J. Kohlerman, and R.C. Rogers. 1983. Hepatic-vagal and gustatory interactions in the brainstem of the rat. *J. Auton. Nerv. Syst.* 9: 477-495.

Higgs, B., F.W.L. Kerr, F.H. Ellis, Jr. 1965. The experimental production of esophageal achalasia by electrolytic lesions in the medulla. *J. Thorac. Cardiovasc. Surg.* 50: 613-625.

Holstege, G., G. Graveland, C. Bijker-Biemond, and I. Schuddeboom. 1983. Location of motoneurons innervating soft palate, pharynx, and upper esophagus. Anatomical evidence for a possible swallowing center in the pontine reticular formation. *Brain Behav. Evol.* 23: 47-62.

Hudson, L.C., and J.F. Cummings. 1985. The origins of innervation of the esophagus of the dog. *Brain Res.* 326: 125-136.

Hunt, J.N. 1980. A possible relation between the regulation of gastric emptying and food intake. *Am. J. Physiol.*, 239: G1-G4.

Jackson, R.G. 1948. Anatomic study of vagus nerve with technique of transabdominal selective gastric resection. *Arch. Surg.* 57: 333.

Jean, A. 1984. Control of the central swallowing program by inputs from the peripheral receptors. A review. *J. Auton. Nerv. Syst.* 10: 225-233.
Jeanningros, R. 1984. Modulation of lateral hypothalamic single unit activity by gastric and intestinal distension. *J. Auton. Nerv. Syst.* 11: 1-11.
Jones, R.L. 1937. Cell fibre ratios in the vagus nerve. *J. Comp. Neurol.* 67: 469-482.
Kahrilas, P.J., and R.C. Rogers. 1984. Rat brainstem neurons responsive to changes in portal blood sodium concentration. *Am. J. Physiol.* 247: R792-R799.
Kiefer, S.W., K.W. Rusiniak, J. Garcia, and J.D. Coil. 1981. Vagotomy facilitates extinction of conditioned taste aversions in rats. *J. Comp. Physiol. Psychol.* 95: 114-122.
Koopmans, H.S. 1984. Hepatic control of food intake. *Appetite* 5: 127-131.
Kuntz, A., and M.W. Jacobs. 1955. Components of periarterial extensions of celiac and mesenteric plexuses. *Anat. Rec.* 123: 509-520.
Laughton, W., and T.L. Powley. 1979. Four central nervous system sites project to the pancreas. *Soc. Neurosci. Abstr.* 5: 46.
Lautt, W.W. 1983. Afferent and efferent neural roles in liver function. *Prog. Neurobiol.* 21: 323-348.
Lee, K.C. 1985. Reflex suppression and initiation of gastric contractions by electrical stimulation of the hepatic vagus nerve. *Neurosci. Lett.* 53: 57-62.
Leslie, R.A., D.G. Gwyn, and D.A. Hopkins. 1982. The central distribution of the cervical vagus nerve and gastric afferent and efferent projections in the rat. *Brain Res. Bull.* 8: 37-43.
Li, E.T.S., and G.H. Anderson. 1984. A role for vagus nerve in regulation of protein and carbohydrate intake. *Am. J. Physiol.* 247: E815-E821.
Lisander, B. 1975. The hypothalamus and vagally mediated gastric relaxation. *Acta Physiol. Scand.* 93: 1-9.
Louis-Sylvestre, J. 1978. Feeding and metabolic patterns in rats with truncular vagotomy or with transplanted ß-cells. *Am. J. Physiol.* 235: E119-E125.
Louis-Sylvestre, J. 1983. Validation of tests and completeness of vagotomy in rats. *J. Auton. Nerv. Syst.* 9: 301-314.
Louis-Sylvestre, J., I. Giachetti, and J. Le Magnen. 1981. A non-invasive test for completeness of vagotomy in the pancreas. *Physiol. Behav.* 26: 1125-1127.
Louis-Sylvestre, J., I. Giachetti, and J. Le Magnen. Vagotomy abolishes the differential palatability of food. *Appetite* 4: 295-299.

Louis-Sylvestre, J., and J. Le Magnen. 1980. Palatability and preabsorptive insulin release. *Neurosci. Biobehav. Rev.* 4, Suppl. 1: 43-46.

Louis-Sylvestre, J., J.M. Servant, R. Molimard, and J. Le Magnen. 1980. Effect of liver denervation on feeding pattern of rats. *Am. J. Physiol.* 239: R66-R70.

Lovgren, N.A., J. Poulsen, and T.W. Schwartz. 1981. Impaired pancreatic innervation after selective gastric vagotomy. *Scand. J. Gastroentol.* 16: 811-816.

Lu, Y.L., and H. Sakai. 1984. Cytoarchitectural study on the dorsal motor nucleus of the rat vagus. *Okajimas Folia Anat. Jpn.* 61: 221-234.

Luiten, P.G.M., G.J. ter Horst, S.J. Koopmans, M. Rietberg, and A.B. Steffens. 1984. Preganglionic innervation of the pancreas islet cells in the rat. *J. Auton. Nerv. Syst.* 10: 27-42.

Magni, F., and C. Carobi. 1983. The afferent and preganglionic parasympathetic innervation of the rat liver, demonstrated by the retrograde transport of horseradish peroxidase. *J. Auton. Nerv. Syst.* 8: 237-260.

Matsuo, Y., and A. Seki. 1978. The coordination of gastro-intestinal hormones and the autonomic nerves. *Am. J. Gastroenterol.* 69: 21-50.

McCrea, E.D. 1925. The abdominal distribution of the vagus. *J. Anat.* 59: 18-40.

McDonald, D.M., and R.W. Blewett. 1981. Location and size of carotid body-like organs (paraganglia) revealed in rats by the permeability of blood vessels to Evans blue dye. *J. Neurocytol.* 10: 607-643.

McHugh, P.R., and T.H. Moran. 1979. Calories and gastric emptying: a regulatory capacity with implications for feeding. *Am. J. Physiol.* 236: R254-R260.

Mei, N. 1985. Intestinal chemosensitivity. *Physiol. Rev.* 65: 211-237.

Miller, A.J. 1982. Deglutition. *Physiol. Rev.* 62: 129-184.

Mitchell, G.A.F., and R. Warwick. 1955. The dorsal vagal nucleus. *Acta Anat.* 25: 371-395

Mohiuddin, A. 1953. Vagal preganglionic fibres to the alimentary canal. *J. Comp. Neurol.* 99: 289-319.

Moore, B.J., T. Inokuchi, J.S. Stern, and B.A. Horwitz. 1985. Brown adipose tissue lipectomy leads to increased fat deposition in Osborne-Mendel rats. *Am. J. Physiol.* 248: R231-R235.

Mordes, J.P., M.E. Lozy, M.G. Herrera, and W. Silen. 1979. Effects of vagotomy with and without pyloroplasty on weight and food intake in rats. *Am. J. Physiol.* 236: R61-R66.

Mosimann, W. 1954. Systematisation des ramifications du nerf vague dans le plexus solaire chéz le rat blanc. *Revue Suisse de Zoologie* 61: 323-334.
Niijima, A. 1980. Glucose-sensitive afferent nerve fibers in the liver and regulation of blood glucose. *Brain Res. Bull.* 5: 175-179.
Niijima, A. 1982. Glucose-sensitive afferent nerve fibres in the hepatic branch of the vagus nerve in the guinea-pig. *J. Physiol.* 332: 315-323.
Niijima, A. 1983. Electrophysiological study on nervous pathway from splanchnic nerve to vagus nerve in rat. *Am. J. Physiol.* 244: R888-R890.
Niijima, A. 1983. Suppression of afferent activity of the hepatic vagus nerve by anomers of D-glucose. *Am. J. Physiol.* 244: R611-R614.
Norgren, R. 1983. Central neural mechanisms of taste. In *Handbook of Physiology: Neurophysiology*, Vol. III, *Sensory Processes*, 2nd ed., ed. I. Darien-Smith, 1087-1128. American Physiological Society, Bethesda, Maryland.
Norgren, R. and G.P. Smith. 1983. The central distribution of vagal subdiaphragmatic branches in the rat. *Soc. Neurosci. Abstr.* 9: 611.
Novin, D., and D. VanderWeele. 1977. Visceral involvement in feeding: there is more to regulation than the hypothalamus. In *Progress in Psychobiology and Physiological Psychology*, ed. J. Sprague and A. Epstein, 193-241. Academic Press, New York.
Novin, D., D. VanderWeele, and M. Rezek. 1973. Infusion of 2-Deoxy-D-glucose into the hepatic-portal system causes eating: evidence for peripheral glucoreceptors. *Science* 181: 858-860.
Opsahl, C.A. 1977. Sympathetic nervous system involvement in the lateral hypothalamic lesion syndrome. *Am. J. Physiol.* 232: R128-R136.
Orlando, R.C., and E.M. Bozymski. 1973. Postvagotomy dysphagia. *Am. J. Surg.* 126: 682-687.
Ouazzani, T.El. 1984. Thermoreceptors in the digestive tract and their role. *J. Auton. Nerv. Syst.* 10: 246-254.
Ouazzani, T.El. 1982. Electrophysiological properties and role of the vagal thermoreceptors of lower esophagus and stomach of cat. *Gastroenterology* 83: 995-1001.
Ouazzani, T.El., and N. Mei. 1981. Acido- ét glucorécepteurs vagaux de la région gastro-duodénale. *Exp. Brain Res.* 42: 442-452.
Paintal, A.S. 1973. Vagal sensory receptors and their reflex effects. *Physiol. Rev.* 53: 159-227.

Pettorossi, V.E. 1983. Modulation of the masseteric reflex by gastric vagal afferents. *Arch. Ital. Biol.* 121: 67-79.

Powley, T.L. 1977. The ventromedial hypothalamic syndrome, satiety, and a cephalic phase hypothesis. *Psychol. Rev.* 84: 89-126.

Powley, T.L., and H.-R. Berthoud. 1985. Diet and cephalic phase insulin responses. *Am. J. Clin. Nutr.* In press.

Powley, T.L., B.A. MacFarlane, M.S. Markell, and C.A. Opsahl. 1978. Different effects of vagotomy and atropine on hypothalamic stimulation-induced feeding. *Behav. Biol.* 23: 306-325.

Powley, T.L., and C.A. Opsahl. 1974. Ventromedial hypothalamic obesity abolished by subdiaphragmatic vagotomy. *Am. J. Physiol.* 226: 25-33.

Powley, T.L., and C.A. Opsahl. 1976. Autonomic components of the hypothalamic feeding syndromes. In *Hunger: Basic Mechanisms and Clinical Implications*, ed. D. Novin, W. Wyrwicka, and G. Bray, 313-326. Raven Press, New York.

Powley, T.L., C.A. Opsahl, J.E. Cox, and H.P. Weingarten. 1980. The role of the hypothalamus in energy homeostasis. In *Handbook of the Hypothalamus*, Vol. 3--Part A, *Behavioral Studies of the Hypothalamus*, ed. P.J. Morgane and J. Panksepp, 211-298. Marcel Dekker, Inc., New York.

Powley, T.L., J.C. Prechtl, E.A. Fox, and H.-R. Berthoud. 1983. Anatomical considerations for surgery of the rat abdominal vagus: distribution, paraganglia and regeneration. *J. Auton. Nerv. Syst.* 9: 79-97.

Powley, T.L., M.C. Walgren, and W.B. Laughton. 1983. Effects of guanethidine sympathectomy on ventromedial hypothalamic obesity. *Am. J. Physiol.* 245: R408-R420.

Prechtl, J.C., and T.L. Powley. 1985a. Organization and distribution of the rat subdiaphragmatic vagus and associated paraganglia. *J. Comp. Neurol.* 235: 182-197.

Prechtl, J.C., and T.L. Powley. 1985b. An EM analysis of the fiber composition of the hepatic branch of the vagus. *Soc. Neurosci. Abstr.* In press.

Pritchard, G.R., C.A. Griffith, and H.N. Harkins. 1968. A physiological demonstration of the anatomic distribution of the vagal system to the stomach. *Surg. Gynecol. Obstet.* 126: 791-798.

Rogers, R.C., and G.E. Hermann. 1983. Central connections of the hepatic branch of the vagus nerve: a horseradish peroxidase histochemical study. *J. Auton. Nerv. Syst.* 7: 165-174.

Rogers, R.C. and D. Novin. 1980. The neurology of hepatic osmoregulation and its implications for studies of ingestive behavior. *Brain Res. Bull.* 5, Suppl. 4: 189-194.

Rohner-Jeanrenaud, F., and B. Jeanrenaud. 1985. Involvement of the cholinergic system in insulin and glucagon oversecretion of genetic preobesity. *Endocrinology* 116: 830-834.

Russek, M. 1981. Current status of the hepatostatic theory of food intake control. *Appetite* 2: 137-143

Sahakian, B.J. 1982. The interaction of psychological and metabolic factors in the control of eating and obesity. *Scand. J. Psychol.*, Suppl. 1: 48-60.

Sakaguchi, T. 1982. Alterations in gastric acid secretion following hepatic portal injections of D-glucose and its anomers. *J. Auton. Nerv. Syst.* 5: 337-344.

Sakaguchi, T., and Y. Miyaoka. 1981. Reflex motility of the stomach evoked by electrical stimulation of the hepatic nerve. *Experientia* 37: 150-151.

Sakaguchi, T., and K. Yamaguchi. 1979a. The effect of electrical stimulation of the hepatic branch of the vagus nerve on the secretion of gastric acid in the rat. *Neurosci. Lett.* 13: 25-28.

Sakaguchi, T., and K. Yamaguchi. 1979b. Effects of electrical stimulation of the hepatic vagus nerve on the plasma insulin concentration in the rat. *Brain Res.* 164: 314-316.

Satachell, P.M. 1984. Canine oesophageal mechanoreceptors. *J. Physiol.* 346: 287-300.

Sawchenko, P.E., and M.I. Friedman. 1979. Sensory functions of the liver--a review. *Am. J. Physiol.* 236: R5-R20.

Sawchenko, P.E., and R.M. Gold. 1981. Effects of gastric vs. complete subdiaphragmatic vagotomy on hypothalamic hyperphagia and obesity. *Physiol. Behav.* 26: 281-292.

Sawchenko, P.E., R.M. Gold, and J. Alexander. 1981. Effects of selective vagotomies on knife cut-induced hypothalamic obesity: differential results on lab chow vs. high-fat diets. *Physiol. Behav.* 26: 293-300.

Sawchenko, P.E., R.M. Gold, and S.F. Leibowitz. 1981. Evidence for vagal involvement in the eating elicited by adrenergic stimulation of the paraventricular nucleus. *Brain Res.* 225: 249-269.

Schwartz, T.W., U. Grotzinger, I.-M. Schoon, and L. Olbe. 1979. Vagovagal stimulation of pancreatic-polypeptide secretion by graded distension of the gastric fundus and antrum in man. *Digestion* 19: 307-314

Sclafani, A., and T.H. Kramer. 1983. Dietary selection in vagotomized rats. *J. Auton. Nerv. Syst.* 9: 247-258.

Sessle, B.J., and A.G. Hannam, eds. 1976. *Mastication and Swallowing: Biological and Clinical Correlates*. University of Toronto Press, Toronto.

Shapiro, R.E., and R.R. Miselis. 1985. The central organization of the vagus nerve innervating the stomach of the rat. *J. Comp. Neurol.* In press.

Smith, G.P., C. Jerome, B.J. Cushin, R. Eterno, and K.J. Simansky. 1981. Abdominal vagotomy blocks the satiety effect of cholecystokinin in the rat. *Science* 213: 1036-1037.

Snowdon, C.T. 1970. Gastrointestinal sensory and motor control of food intake. *J. Comp. Physiol. Psychol.* 71: 68-76.

Snowdon, C.T., and R.S. Wampler. 1974. Effects of lateral hypothalamic lesions and vagotomy on meal patterns in rats. *J. Comp. Physiol. Psychol.* 87: 399-409

Stavney, L.S., T. Kato, C.A. Griffith, L.M. Nyhus, and H.N. Harkins. 1963. A physiological study of motility changes following selective gastric vagotomy. *J. Surg. Res.* 3: 390-394.

Stellar, E. 1954. The physiology of motivation. *Psychol. Rev.* 61: 5-22.

Storlien, L.H., W.P. Bellingham, and G.A. Smythe. 1983. Effect of guanethidine sympathectomy on intake and body weight of intact and LHA-lesioned rats. *Physiol. Behav.* 31: 401-404.

Stricker, E.M., N. Rowland, C.F. Saller, and M.I. Friedman. 1977. Homeostasis during hypoglycemia: central control of adrenal secretion and peripheral control of feeding. *Science* 196: 79-81.

Teitelbaum, H.A. 1932. The nature of the thoracic and abdominal distribution of the vagus nerves. *Anat. Rec.* 55: 297-317.

Tordoff, M.G., C.V. Grijalva, D. Novins, L.L. Butcher, J.H. Walsh, F.X. Pi-Sunyer, and D.A. VanderWeele. 1984. Influence of sympathectomy on the lateral hypothalamic lesion syndrome. *Behav. Neurosci.* 98: 1039-1059.

Tordoff, M.G., J. Hopfenbeck, and D. Novin. 1982. Hepatic vagotomy (partial hepatic denervation) does not alter ingestive responses to metabolic challenges. *Physiol. Behav.* 28: 417-424.

Uvnäs-Wallensten, K., B. Uvnäs, and G. Nilsson. 1976. Quantitative aspects of the vagal control of gastrin release in cats. *Acta Physiol. Scand.* 96: 19-28.

Weaver, F.R.C. 1980. Localization of parasympathetic preganglionic cell bodies innervating the pancreas within the vagal nucleus and nucleus ambiguus of the rat brainstem: evidence of dual innervation based on the retrograde axonal transport of horseradish peroxidase. *J. Auton. Nerv. Syst.* 2: 61-69.

Weingarten, H.P., and T.L. Powley. 1980. Ventromedial hypothalamic lesions elevate basal and cephalic phase gastric acid output. *Am. J. Physiol.* 239: G221-G229

Wirth, J.B., and P.R. McHugh. 1983. Gastric distension and short-term satiety in the rhesus monkey. *Am. J. Physiol.* 245: R174-R180.

Woods, S.C., and I.L. Bernstein. 1980. Cephalic insulin response as a test for completeness of vagotomy to the pancreas. *Physiol. Behav.* 24: 485-488.

Wright, R.A., S. Krinksy, C. Fleeman, J. Trujillo, and E. Teague. 1983. Gastric emptying and obesity. *Gastroenterol.* 84: 747-751.

Yoshimatsu, H., A. Niijima, Y. Oomura, K. Yamabe, and T. Katafuchi. 1984. Effects of hypothalamic lesions on pancreatic autonomic nerve activity in the rat. *Brain Res.* 303: 147-152.

Chapter 4

CAUDAL BRAINSTEM CONTRIBUTIONS TO THE INTEGRATED NEURAL CONTROL OF ENERGY HOMEOSTASIS

Harvey J. Grill

I. INTRODUCTION

Life-maintaining processes such as active transport and biosynthesis require energy, and ingested food is the ultimate source of this energy. Once swallowed, food is rapidly transformed by digestion into metabolic fuels that are immediately utilized or else stored for subsequent use. The ability to liberate and then utilize stored fuels such as fats sustains life when food is scarce, but there are limitations on how long metabolism during fasting can go on. New supplies of metabolic fuels are required, and feeding behavior is the exclusive means of procuring this new energy. The central nervous system (CNS) orchestrates the behavioral, autonomic, and endocrine responses that maintain relatively constant internal energy resources; this process has been called energy homeostasis and is based on the concepts of Claude Bernard and Walter Cannon. An important question for behavioral neuroscientists is, what neural mechanisms are involved in energy homeostasis? Particularly, what types of signals acting on what portions of the nervous system initiate and maintain feeding behavior?

A. Concepts Available to Us Ten Years Ago: Isomorphism between Psychological Concepts and Hypothalamic Unit Activity

For over 30 years, research on this question has focused on the hypothalamus, a focus derived from data gathered by various experimental techniques including focal electrical and chemical stimulation and lesions, neurological case studies, and electrophysiological recording from hypothalamic neurons. These data were interpreted to mean that one population of hypothalamic cells was responsive to signals of metabolic deficit, and once activated, would effect behavioral and autonomic responses resulting in increased levels of metabolic fuels, decreased activation of these hypothalamic neurons, and increased activation of another population of hypothalamic neurons whose activity was deemed the neural basis of satiety. The parsimony of controlling feeding behavior through the activation and inhibition of two functionally distinct populations of hypothalamic neurons made the hypothalamic hypothesis attractive. However, this hypothesis fostered the impression that simply monitoring the activity of small localized populations of neurons would provide an adequate explanation for complex psychological processes like hunger and satiety. As Powley has noted (1977), such an assumption reduced the need to explain how hypothalamic neural activity was integrated with activity in other neural systems to produce behaviors collectively categorized as hunger or satiety. Within the past 10 years the limitations of the hypothalamic or single integrator hypothesis have emerged. The demonstration that neural systems outside the hypothalamus (Gold, 1978; Stricker and Zigmond, 1976) and even outside the brain (Russek, 1971; Sawchenko and Friedman, 1979) contributed to energy homeostasis challenged the hypothalamus as *the* site of detection of metabolic fuels and control of feeding behavior. Hypothalamic participation with other neural systems in a more broadly based, integrated control of energy homeostasis remains unquestioned (Rolls et al., 1976).

As mentioned above, the maintenance of adequate metabolic fuel levels is central to the concept of energy homeostasis. Since there are several metabolic fuels, and different organ systems utilize different fuels, the question arises: Which fuel or fuels are monitored and at what sites? Glucose is the principal metabolic fuel of the CNS. The CNS locus of the hypothalamic hypothesis called special attention to glucose utilization as *the* monitored signal for energy homeostasis. Until very recently, the contribution to energy homeostasis of fuels other than glucose, monitored at sites outside the CNS, was simply not entertained.

B. Concepts Emerging within the Past Ten Years: Toward a More Integrated System

1. Metabolic Afferent Input from the Viscera

Three significant papers appearing in *Psychological Review* in 1976 and 1977 exemplified the contribution of new ideas to reshaping experimentation in this area of research. Friedman and Stricker (1976) reminded the field that the gastrointestinal (GI) tract is the first organ system to receive metabolic fuels (putative signals) during a meal. While acknowledging that the CNS is critical for initiating compensatory behavioral and autonomic responses, these authors also suggested that the detection of metabolic fuels, their metabolites, or digestive hormones could take place in the periphery as well as within the CNS. Peripherally detected signals would be conveyed to the CNS via the afferent portion of the autonomic nervous system. The importance of autonomic afferent systems projecting to the CNS at the caudal brainstem level was thus highlighted. The shift of attention to peripherally utilized metabolic fuels expanded the list of putative signals to include a variety of metabolic fuels in addition to glucose. For example, Friedman, Stricker and colleagues (Sawchenko and Friedman, 1979) demonstrated that an hepatic portal infusion of fructose, a fuel not normally used by the CNS, was effective in terminating insulin-induced compensatory feeding, gastric acid secretion, and GI motility. These effects of fructose were blocked by sectioning the hepatic branch of the vagus nerve (note that the vagus is a mixed nerve).

2. CNS Efferent Output to the Viscera

Since most animals eat discrete meals ending before the absorption of ingested fuels is complete, the role of preabsorptive or cephalic responses in metabolic homeostasis was warranted (see below for a discussion of the role of conditioning). While Pavlov was the first to describe these responses, Powley's provocative 1977 review used the neurology of these responses to develop ideas giving a mechanistic rather than functional explanation to the obesity that follows forebrain (ventromedial hypothalamic or VMH) lesions. Following lesions of the VMH, the amplitudes of parasympathetically controlled digestive responses, such as basal insulin secretion, were shown to rise rapidly (Berthoud and Jeanrenaud, 1979). In addition, the amplitudes of cephalic autonomic-endocrine reflexes, such as preabsorptive insulin and gastric acid secretions, were also elevated in VMH-lesioned rats (Louis-Sylvestre, 1976). The preabsorptive reflex focus allowed the obesity to be interpreted as the metabolic consequence of elevated levels of digestive hormones favoring fat storage.

Traditionally, VMH obesity was viewed as a direct result of the overeating that accompanied the loss of hypothalamic "satiety neurons". This notion, however, could not explain why the VMH-lesioned rat whose food ration was yoked to that of a non-lesioned littermate still gained more weight than did the littermate. Whether the elevated amplitudes of preabsorptive digestive reflexes provide the best explanation for VMH obesity is currently being debated; evidence qualifying this hypothesis is accumulating (King and Frohman, 1982). Whatever the outcome of this debate, however, the cephalic phase hypothesis has helped to focus attention on the role of the CNS in the control of peripheral metabolism. Other experimental approaches support the importance of CNS efferent control of peripheral metabolism in energy homeostasis. For example, Shimazu (1981) has shown that electrical stimulation of the hypothalamus produces changes in liver enzyme activity.

3. Focus on Autonomic Afferent and Efferent Activity

Study of autonomic anatomy and physiology has long been neglected. An interest in specifying the central and peripheral anatomy of the vagal and splanchnic systems has emerged from the autonomic contribution to energy homeostasis just discussed. For example, anatomical and physiological investigations of the dorsal motor nucleus of the vagus and the nucleus of the solitary tract have highlighted an organotopic organization for these caudal brainstem nuclei such that the connections of individual digestive organs are segregated within these nuclei (e.g., Kahlia and Mesulum, 1980; Powley and Laughton, 1981; Coil and Norgren, 1979; Shapiro and Miselis, 1983).

4. Integration of Taste and Peripherally-detected Postingestive Feedback Influences the Duration of Feeding Behavior

In their 1977 review, Davis and Levine developed a mathematical model for the integration of taste and osmotic feedback from the digestive tract that accurately predicted the amount of a sweet-tasting liquid food that rats would consume. They again emphasized the importance of autonomic afferent input to the caudal brainstem in the control of feeding behavior, and stated that taste alone does not predict ingestive consummatory behavior. This paper stimulated interest in determining the sites of gustatory and visceral afferent integration whose output directly influenced behavior.

5. The Role of the Caudal Brainstem in Energy Homeostasis

The parsimony of the hypothalamic hypothesis directed analysis away from the caudal brainstem despite the provocative earlier work of Claude Bernard (1849), which showed that lesions near the fourth ventricle, in the pons, produced hyperglycemia. As the single integrator or hypothalamic model gave way to a broader systems analysis, however, the caudal brainstem and its important afferent and efferent connections to the viscera have begun to be investigated by several laboratories.

The approach that Norgren and I have taken to the problem of the neural organization of energy homeostasis derives from Hughlings Jackson's notion of a hierarchical organization of function within the nervous system and from more recent formulations of this theme (Mountcastle, 1978). The hierarchical approach views function (e.g., energy homeostasis) as a property of the activity of the nervous system as a whole, and not the activity of any single neural locus. While subfunction (e.g., salivation, swallowing, vagally-mediated insulin release, or discriminative responses to taste stimuli) can be effected by restricted parts of the nervous system, it is hypothesized that in the intact nervous system these restricted parts are systematically influenced by the activity of other neural circuits. The hierarchical perspective is helpful in explaining why a lesion of a restricted part of the system (e.g., the lateral hypothalamic area) rarely destroys the system's function (e.g., feeding behavior) completely. Recovery of function after focal lesions can be viewed as evidence for the adaptive capacity of hierarchically organized systems to achieve a behavioral goal, albeit slowly and with some error, using the remaining neural apparatus (Mountcastle, 1978). The goal of our initial experiments, begun in 1974, was to first describe what the simplest behavioral units of ingestion and rejection were and then, having defined what a sufficient neural substrate for these simple responses was, to examine the neural requirements of progressively more complex aspects of the neural control of feeding behavior.

II. THE CAUDAL BRAINSTEM IS SUFFICIENT FOR THE PRODUCTION OF DISCRIMINATIVE RESPONSES TO TASTE

The final or consummatory acts of feeding behavior, ingestion or rejection, involve highly stereotyped responses guided principally by taste stimuli. We view these ingestion and rejection oral behavior patterns as the basic functional units of feeding behavior. The basic sensory input (gustatory and somatosensory) and motor output (chewing, salivating, and swallowing) of discriminative responses to taste might be organized within the caudal brainstem itself. Sherrington (Miller and Sherrington, 1916) described reflex components of ingestion and rejection in acute decerebrate cats. It was not clear from these data or that of others (Macht, 1951) whether the same discriminative taste stimuli eliciting ingestion and rejection responses in the intact cat were effective in the decerebrate cat. Therefore we developed a chronic, supracollicular

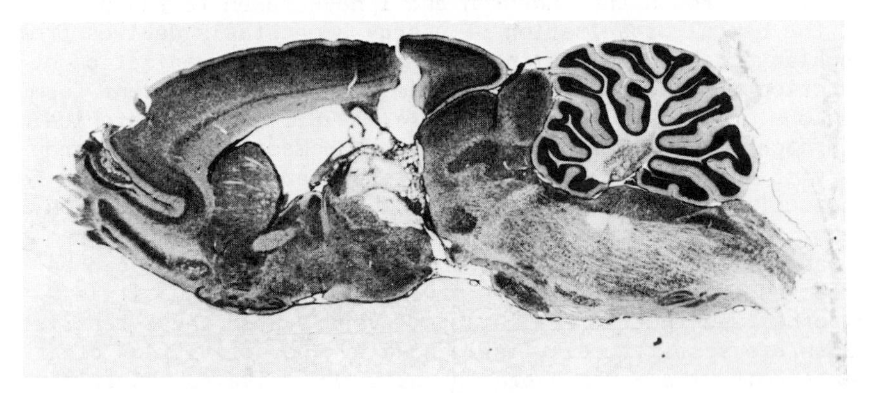

Fig. 1. Sagittal section from the brain of a representative chronic decerebrate rat stained with cresyl violet; survival time, 37 days. The supracollicular plane of section is highly similar for each rat. The tissue posterior to the transection appears normal in the light microscope. A cavity filling the space normally occupied by portions of the thalamus and hippocampus is present anterior to the transections of rats surviving 30 days or more.

decerebrate rat preparation to examine the sufficiency of the isolated caudal brainstem for producing discriminative responses to taste. Histological and general behavioral data from the chronic decerebrate rat are shown in Figures 1 and 2.

The chronic decerebrate rat never initiates a spontaneous meal. To circumvent this lack of spontaneous feeding, tastes were presented directly into the rat's mouth via oral cannulae anchored to the skull (Fig. 3). In an apparatus like the one shown in Figure 4, responses elicited by remotely presenting taste stimuli into the mouths of freely moving rats were videotaped and subsequently analyzed by single frame. A lexicon of response topographies, called taste reactivity responses, was described first in intact rats (Grill and Norgren, 1978a). In

Fig. 2. Chronic decerebrate rats exhibit no spontaneous activity other than grooming but often overreact with well-coordinated movements to seemingly inappropriate stimuli. Tail pinch facilitates a brisk, well-coordinated sequence of cage climbing (a-e). Decerebrate rats maintain their fur; face washing (f), grooming of the flanks (g and h), anal (i), and genital grooming involve complicated postures which are executed in a coordinated fashion by these rats.

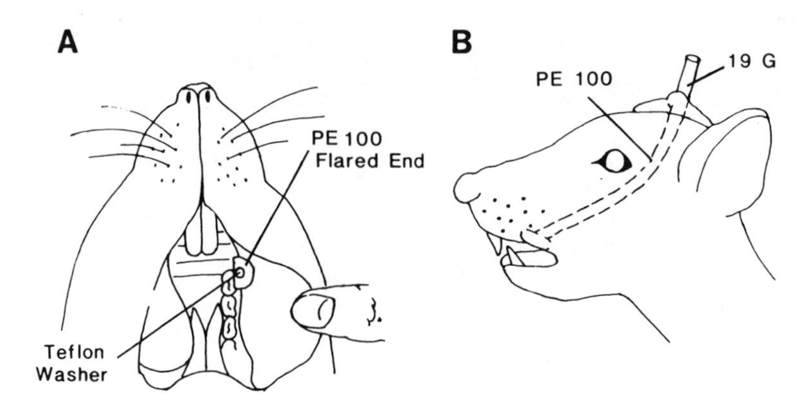

Fig. 3. Diagram of the intraoral catheter. The intraoral end is placed just rostral to the first maxillary molar. The tubing is led out subcutaneously to the skull and secured to a short piece of 19-gauge (19 G) stainless steel tubing with dental acrylic. (A) Ventral view. (B) Lateral view.

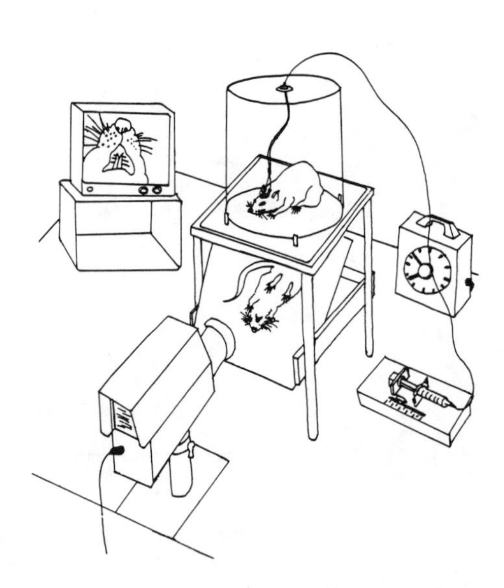

Fig. 4. Apparatus for videotaping taste reactivity responses to taste stimuli injected into the mouth via chronic intraoral catheters. Videotaping is done via a mirror located beneath the plexiglass floor in the upper right-hand corner of the figure.

Fig. 5. Taste-elicited fixed action patterns. Ingestive responses, top, are elicited by oral infusions of glucose, sucrose, and isotonic sodium chloride, and include rhythmic mouth movements, tongue protrusions, lateral tongue protrusions and paw licking. Aversive responses, bottom, are elicited by infusions of quinine, caffeine and sucrose octaacetate solutions and include gapes, chin rubs, head shakes, paw wipes, forelimb flailing, and locomotion (not shown). From Berridge, K., H. Grill, and R. Norgren. 1981. Relation of consummatory responses and preabsorptive insulin release to palatability and learned taste aversions. *Journal of Comparative and Physiological Psychology* 95: 363-382. Copyright 1981 by the American Psychological Assn. Adapted by permission of the author.

response to sucrose, an *ingestion response sequence* of rhythmic mouth movements, rhythmic tongue protrusions, and larger lateral tongue protrusions accompanied swallowing, as shown in Figure 5. Movement was restricted to the oral region in response to sucrose; rats did not move about or in any way increase their activity in response to sucrose. Taste reactivity to quinine differed strikingly. Figure 5 shows that intact rats gape, chin rub, head shake, face wash, forelimb shake, and paw rub in response to orally delivered quinine. Quinine produced a *rejection response sequence* that included locomotion, rearing, and rejection of the orally-presented taste.

In response to a variety of 50-microliter orally-delivered taste stimuli, chronic decerebrate rats demonstrated the same ingestion and rejection behavioral sequences as intact rats did (Fig. 5) (Grill and Norgren, 1978b). Furthermore, the threshold of individual taste reactivity response components of decerebrates was identical to that of intact rats. For example, the gape response threshold was 0.03 mM quinine for both chronic decerebrate and pair-fed control rats. These data can be generalized by comparison with observations on the taste-elicited responses of anencephalic and intact human newborns (Steiner, 1977). The isolated caudal brainstem is therefore sufficient for the production of discriminative responses in response to taste stimulation.

A given taste does not always elicit the same discriminative response, be it ingestion or rejection. In other words, knowledge of the input (the taste stimulus) is insufficient to predict the output (the response that follows). Rather, the neural mechanisms effecting ingestion and rejection responses are affected by the integration of the peripheral gustatory signals with neural signals arising from the orginism's present physiological state and past experience to produce responses appropriate to the prevailing physiological and environmental conditions. For example, signals arising from the digestive tract, such as the osmotic signals described by Davis and Levine (1977), can act directly or indirectly upon the caudal brainstem circuitry effecting ingestive consummatory behavior to alter the response elicited by a particular taste. Examples of neural control signals are the neural correlates of food and sodium depletion and repletion, as well as those of taste-illness association. These neural control systems impart greater complexity to behaviors guided by taste. In classical Jacksonian neurology, a neural control system is located rostral to the circuitry effecting the most basic behavioral responses. A distributed systems neurology allows for control mechanisms to be located within the same neural level as the circuitry that produces the basic behavioral responses (Mountcastle, 1978). The approach of my laboratory was to manipulate a given physiological or associative condition known to alter the discriminative responses to taste or intake volume of the intact rat in order to determine the sufficient neural substrate for the integration of that particular neural control signal. A series of experiments was performed to examine whether the caudal brainstem, in the absence of neural participation of the forebrain, was sufficient to integrate neural control signals with taste afferent input so as to alter ingestive consummatory responses.

A. Food Deprivation and Insulin-induced Hypoglycemia Control the Ingestive Consummatory Behavior of Chronic Decerebrate Rats.

The effect of simple food deprivation on food consumption was tested first. To measure the food intake of the aphagic decerebrate rat we took advantage of the taste reactivity methodology just described. Instead of presenting a variety of taste stimuli to the oral cavity in 50 μl aliquots, we continuously infused a single taste stimulus. Intake of orally-infused sucrose (0.03 M; 0.6 ml/min) was examined under two conditions: sated (1 hour following a tube-fed meal) and deprived (24 hours after the same meal). Sucrose infusion

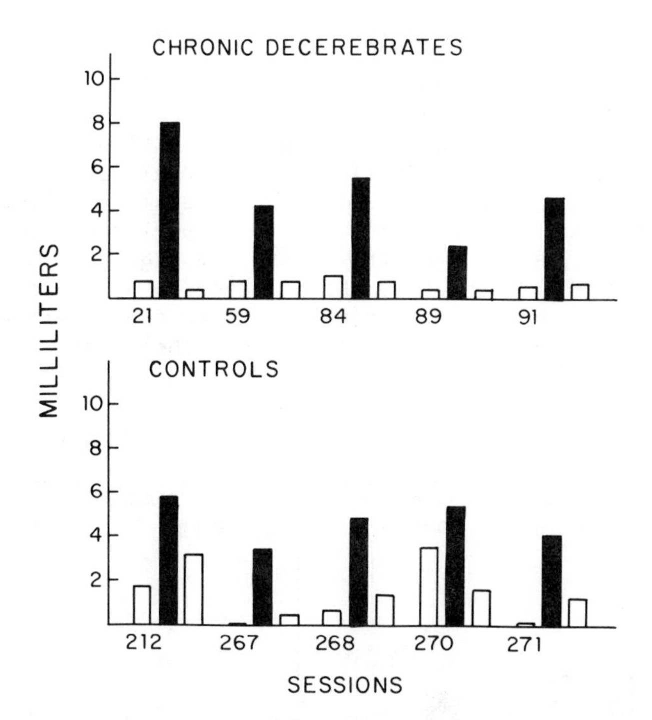

Fig. 6. (Above) 0.03 M sucrose intraoral intakes of chronic decerebrate and rats for just-fed (open) and deprived (black = 24-hr-deprived) conditions. (Below) Intakes of pair-fed control rats for just-fed (open) and 24-hr-deprived (black) conditions. Rat code numbers are depicted below the abcissa.

stopped as soon as a drop of fluid was observed to drip from the rat's mouth (rejection) and started again 30 seconds later. Termination of intake was defined as two consecutive instances of sucrose rejection. As shown in Figure 6, food-deprived chronic decerebrates, like control rats, ingested 2 to 3 times the volume of sucrose they had consumed in the sated condition (Grill and Norgren, 1978c; Grill, 1980). To be certain that the effects of food deprivation were specific to sucrose (a potent food cue) and did not reflect a general facilitation of fluid intake, the effects of deprivation on another taste stimulus, water, were examined. Neither food-deprived control nor decerebrate rats increased their water intake above their sated levels.

Numerous physiological consequences of food deprivation and repletion could be integrated with food taste signals to alter the ingestive behavior of chronic decerebrate rats. They include alterations in blood-borne levels of metabolic fuels and digestive hormones, and the state of the digestive tract. To limit this list of candidate signals, an additional set of experiments was performed. The effect of insulin (5 U/kg or 10 U/kg) or saline control injections on the oral sucrose intake of sated decerebrate and control rats was examined. Insulin treatment produced comparable levels of hypoglycemia in decerebrate and pair-fed control rats. As Figure 7 shows, Flynn and I (1983) found that chronic decerebrate rats made hypoglycemic by the insulin injection, like their pair-fed controls, increased their sucrose but not their water intakes.

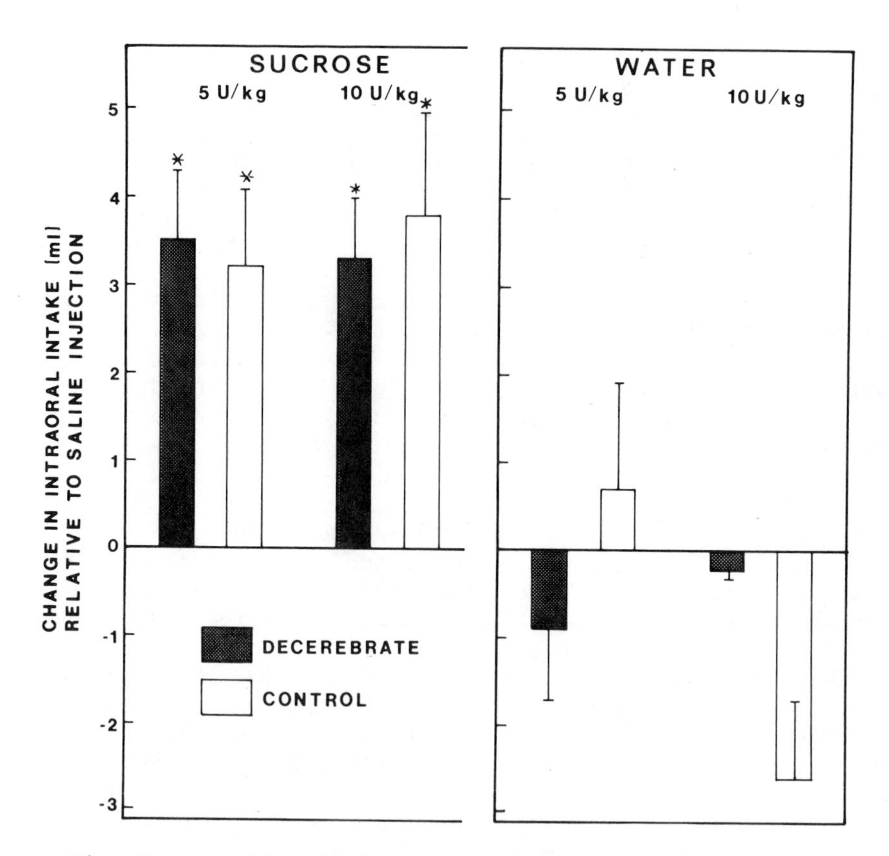

Fig. 7. Insulin-elicited change in sucrose and water intake relative to saline control injection.

As just stated, the neural correlates of food deprivation and repletion controlling the ingestive consummatory behavior of decerebrate rats include the absorption of metabolic fuels, osmotic alterations, the mechanical state of the gastrointestinal tract, and gastrointestinal hormone secretion. Experiments on lateral hypothalamic lesioned (LHX) rats are another approach to defining what correlates of a tube-fed meal might reduce the sucrose intake of chronic decerebrate rats. 50 μl presentations of sucrose elicit ingestive taste reactivity in intact or decerebrate rats regardless of whether these rats are tested after food deprivation or tube feeding. In both preparations, repeated presentations of sucrose (continuous infusions) result in a more rapid cessation of intake in the tube-fed or sated condition. The taste reactivity responses of aphagic LHX rats in the food-deprived state were identical to those of decerebrate and intact rats. In contrast, Fluharty and I (1980) have shown that the tube-fed LHX rat rejects all orally delivered tastes, even concentrated sucrose. We took advantage of the LHX rat's dramatic sensitivity to tube feeding to highlight the potential correlates of repletion for decerebrate rats. The tube feeding induced suppression of sucrose intake, and ingestion taste reactivity responses in the LHX rat were _not_ duplicated by inserting the feeding tube without delivering diet, delivering a non-absorbable substance (mannitol) of an equivalent osmolarity and volume, or delivering the water content of the tube-fed meal. Figure 8 shows that equivolemic and equiosmotic glucose was the only intubated substance duplicating the dramatic effects of a tube-fed meal on the taste reactivity and intra-oral intake of LHX rats. Parallel experiments with decerebrate rats will follow up this suggested role of calorically related feedback in altering intake and taste reactivity. In a recent experiment we have shown that decerebrates, like controls, are sensitive to elevated levels of cholecystokinin (CCK), a digestive hormone released in the course of normal digestion. When treated with CCK, 24-hour-food-deprived chronic decerebrate rats, like their pair-fed controls, reduced both their intake of intraoral sucrose and the number of ingestive taste reactivity responses that accompany sucrose ingestion (Grill et al., 1983).

B. The Caudal Brainstem Is Sufficient to Produce Autonomic Responses of Energy Homeostasis

Hunger has been operationally defined as the increase in food intake that correlates with a magnitude of food deprivation (Silverstone, 1976). As noted above, the control of feeding exerted by the correlates of deprivation was thought to

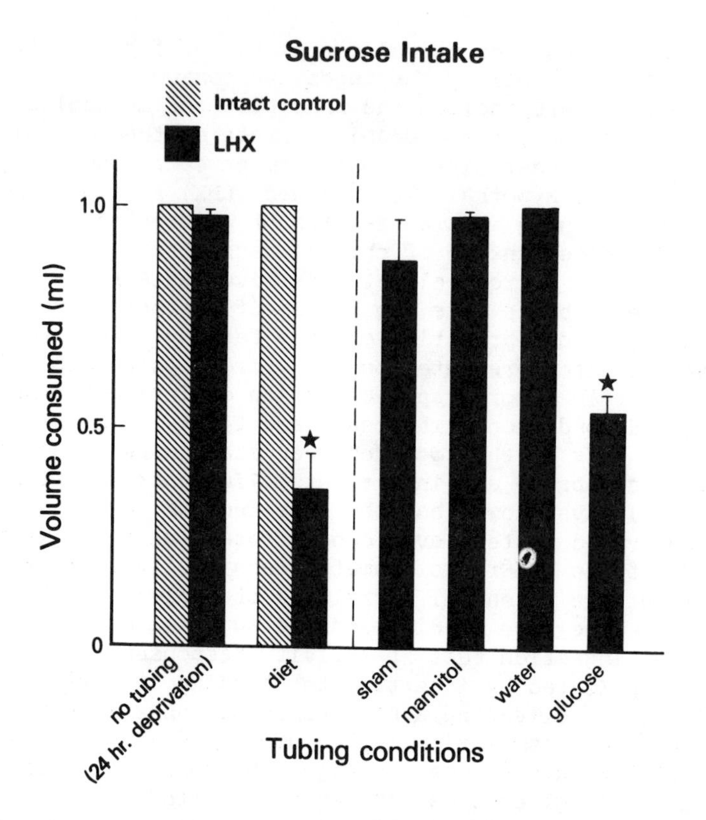

Fig. 8. The effect of six randomly presented intragastric intubation conditions on intraoral sucrose intake. The conditions are 12 ml of liquid diet, 0.2 M mannitol, 1.4 M glucose, 8 ml of distilled water (the portion of 12 ml diet that is water), passing the tube but not delivering any liquid (sham), or passing neither the tube nor any liquid (24-hr-deprived). Ninety minutes after each intubation condition, 1 ml of 1.0 M sucrose (1 ml/min) was intraorally infused into the mouths of lateral hypothalamically lesioned and yoked control rats. The volume of the sucrose ingested is shown as the amplitude of each histogram. Analysis of variance reveals that sucrose intake following diet and glucose tubed conditions is significantly lower (* = $P > 0.01$) than other conditions for the lesioned rat.

require the participation of the forebrain, particularly the hypothalamus. The data just described, however, made a preliminary case for the sufficiency of the caudal brainstem in mediating aspects of deprivation-derived control.

This case for caudal brainstem control of compensatory behavioral responses in energy homeostasis gained support from demonstrations of intact sympathetic and parasympathetic-endocrine responses in chronic decerebrate rats. Figure 9 shows that in the absence of food, chronic decerebrate rats compensate for decreases in glucose utilization with sympatho-adrenal hyperglycemia (DiRocco and Grill, 1979). Orally infused glucose elicited a parasympathetically mediated, pre-absorptive insulin secretion in chronic decerebrate rats and pair-fed controls (Grill and Berridge, 1981).

C. The Caudal Brainstem Is a Site of Metabolic Interoceptors

Our experiments with chronic decerebrate rats show that caudal brainstem mechanisms are sufficient for the integration of metabolic and gustatory signals and the efferent control of both behavioral and autonomic compensatory responses. The location of the metabolic interoceptors triggering these responses in decerebrate rats, however, is not specified by these experiments. These receptor cells may reside within the caudal brainstem itself, the periphery, or both.

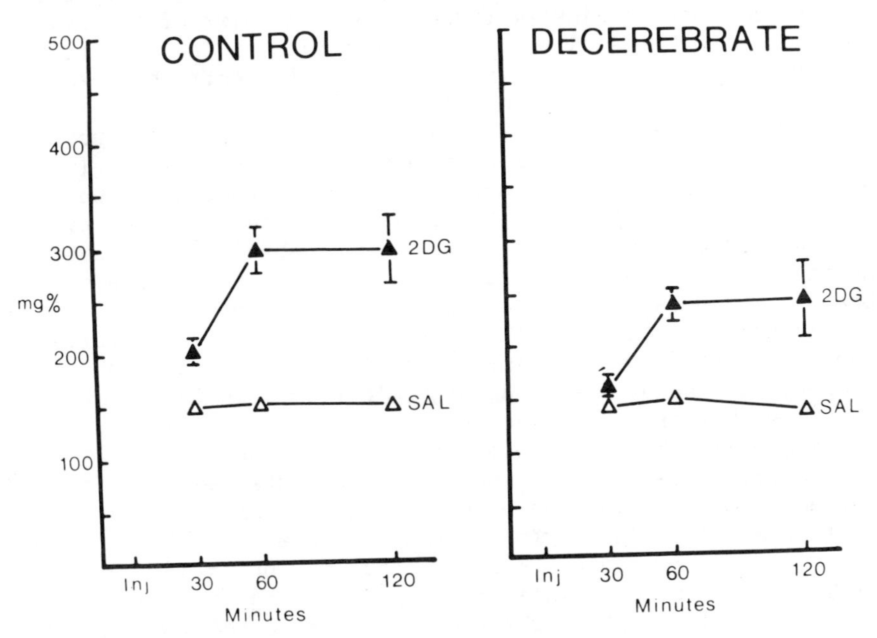

Fig. 9. Plasma glucose concentrations of decerebrate and control rats 30, 60 and 120 min after intraperitoneal injection (Inj) of 2DG and physiological saline.

Glucose is the primary substrate for CNS metabolism. When this substrate is insufficent for the metabolic needs of CNS neurons, compensatory behavioral and autonomic responses are elicited to elevate circulating glucose levels (Himsworth, 1970; Smith and Epstein, 1969). While CNS efferent neurons provide the final path for the production of these compensatory responses, it remained to be shown that CNS metabolic interoceptors participated in triggering either response. The existence of CNS interoceptors was supported by the ability of lateral intracerebroventricular (i.c.v.) injections of phlorizin, a glucose transport inhibitor (Betz et al., 1975), to stimulate feeding; plasma glucose levels were not measured in the study (Glick and Mayer, 1968). Subsequently, inhibition of brain glycolysis by lateral i.c.v. injections of 2-deoxy-D-glucose (2DG) (Wick et al., 1957) was shown to evoke both behavioral and autonomic compensatory responses (Berthoud and Mogenson, 1977; Coimbra et al., 1979; Miselis and Epstein, 1975).

Due to the normal caudalward movement of cerebrospinal fluid, lateral i.c.v. injections make the glucodynamic drug available to the entire ventricular system (Bradbury, 1979); peripheral detection is eliminated in these experiments by the ventricular application of small drug dosages or appropriate control experiments. Despite the contact of the glucodynamic drug with caudal brainstem tissue, the Zeitgeist of the hypothalamic hypothesis directed interpretation of the elicited compensatory responses to the existence of electrophysiologically identified hypothalamic interoceptors (Chhina et al., 1971; Desiraju et al., 1968; Oomura, 1976). The existence of hypothalamic and other forebrain neurons responsive to glucose, however, demonstrated neither that these neurons were the exclusive site of interoception, nor that they were functional in the efferent control of compensatory behavioral and autonomic responses in the intact rat. Furthermore, in the decerebrate rat, these cells have been neurally eliminated!

It has recently been shown that metabolically sensitive neurons in the caudal brainstem provide an afferent limb for compensatory behavioral and autonomic responses in intact rats. R. Ritter and colleagues demonstrated that injections of 5-thioglucose (5TG), an antimetabolite of glucose, restricted to the fourth ventricle by means of a cerebral aqueduct plug, produced both feeding and hyperglycemia compensatory responses (Ritter et al., 1981). In contrast, lateral ventricular injections of 5TG restricted to the forebrain ventricles by a cerebral aqueduct plug stimulated neither feeding nor hyperglycemia (Ritter et al., 1981). These data suggested that the effects of lateral i.c.v. phlorizin on feeding (Glick and Mayer, 1968) might also be mediated by caudal brainstem mechanisms. To investigate this possibility and to extend the investigation of caudal brainstem interoception initiated by the important study of Ritter et al., we have applied both phlorizin and 5TG to the fourth ventricle of intact rats (Flynn and Grill, 1984).

Rats previously implanted with cannulae aimed at the fourth ventricle were given, in a random order, 4th i.c.v. injections of saline (N = 15), 150 μg 5TG (N = 9), 210 μg 5TG (N = 9), 6.5 μg phlorizin (3.0 mM, N = 8) and 13.0 μg phlorizin (6.0 mM, n = 15), and food intake was measured. The range of phlorizin doses tested was restricted by the drug's limited solubility. All injections were of a 5.0 μl volume. Food intake was measured at 3, 6, and 24 hours after injection.

To verify that any effect of phlorizin on food intake was mediated by its central actions and not by a systemic action of the drug, food intake was next measured in response to systemic phlorizin injection. Rats (N = 8) were injected i.p. with 13.0 μg phlorizin or saline (0.15 M), and food intake was measured 3 and 6 hours later.

After the food intake experiments were completed, plasma glucose and insulin concentrations were monitored in response to the 4th i.c.v. injections in the absence of food. The rats' food was removed prior to tail blood collection and returned after the collection of the last sample. Rats were injected on alternate days with 4th i.c.v. 5TG (150 μg, 210 μg), phlorizin (13 μg), or saline. Tail blood samples (250 μl) were collected prior to the 4th i.c.v. injection, and 30 minutes, 1 hour, 2 hours, 3 hours, and 6 hours after the injection. Glucose concentrations were determined with a Beckman glucose oxidase analyzer, and a radioimmunoassay for insulin was performed on 50 μl plasma samples by the Diabetes Center at the University of Pennsylvania.

Histological analyses revealed that the cannulae were positioned on the midline, just anterior to the primary cerebellar fissure. Ink (5 μl) injected prior to sacrifice was restricted to the fourth ventricle in all cases; there were no traces of ink rostral to the recess of the inferior colliculus. As shown in Figure 10, cumulative food intake 3 hours and 6 hours following 4th i.c.v. 13.0 μg phlorizin, 150 μg 5TG, and 210 μg 5TG was significantly greater than that consumed following 4th i.c.v. saline. Phlorizin- and 5TG-treated rats ate similar amounts of food. In contrast, systemically injected phlorizin (13.0 μg) did not affect food intake.

Figure 11 reveals that fourth ventricular injections of 5TG (210 μg) elicited hyperglycemia. Within 30 minutes of 4th i.c.v. injection of 210 μg 5TG, plasma glucose values were 81 ± 25.2 mg% higher than preinjection values and 54.5 ± 20.2 mg% higher than saline injection values. In contrast to the sympathetic arousing effect of 5TG, 4th i.c.v. phlorizin (13 μg) injection did not elicit sympathoadrenal hyperglycemia.

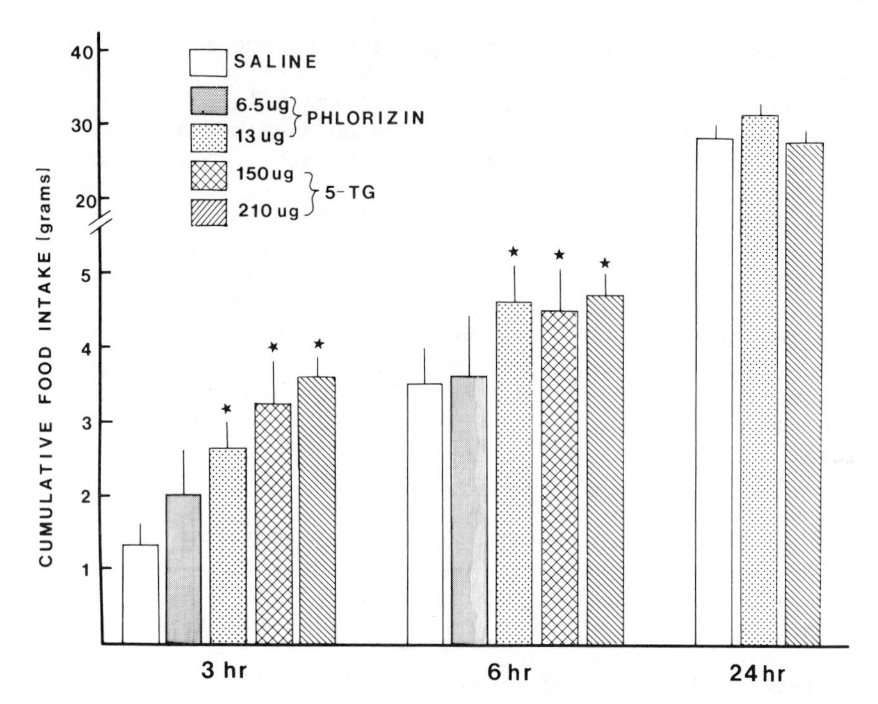

Fig. 10. Cumulative 3 hr, 6 hr, and 24 hr food intake (mean ± standard error of the mean) following 4th i.c.v. saline, 6.5 μg phlorizin, 13.0 μg phlorizin, 150 μg 5TG, and 210 μg 5TG. (Asterisk indicates significantly different from saline.)

Fourth ventricular injections of phlorizin did not elicit sympathoadrenal hyperglycemia. The inability of phlorizin to stimulate hyperglycemia cannot be attributed to cannula placement or patency since injections of 5TG through the same cannula elicited hyperglycemia. In addition, the absence of hyperglycemia following 4th i.c.v. injections of phlorizin cannot be attributed to a concomitant stimulation of insulin secretion since the results from our insulin assays do not support this outcome; insulin secretion was not altered by the i.c.v. injection of phlorizin. Therefore, these data indicate that phlorizin stimulates a caudal brainstem system, controlling behavioral but not autonomic compensatory responses. Similarly, S. Ritter and Strang (1982) reported that fourth ventricular injection of subtoxic alloxan doses stimulates feeding but not sympathoadrenal hyperglycemia in rats. Thus, the actions of phlorizin and subtoxic alloxan appear specific for that mechanism controlling compensatory ingestive behavior.

The neural systems controlling compensatory ingestive and autonomic responses differ on other measures. First, in response to the metabolic challenge provided by 2DG, the neural mechanism mediating sympathoadrenal hyperglycemia develops prior to that of compensatory feeding in the rat (Houpt and Epstein, 1973). Second, the neural system mediating feeding in response to systemic 2DG is more susceptible to the effects of a 4th i.c.v. toxic dose of alloxan than is that system mediating sympathoadrenal hyperglycemia (Ritter et al., 1982). Last, compensatory behavioral and autonomic responses are dissociable by intravenous infusions of different metabolic fuels (Stricker

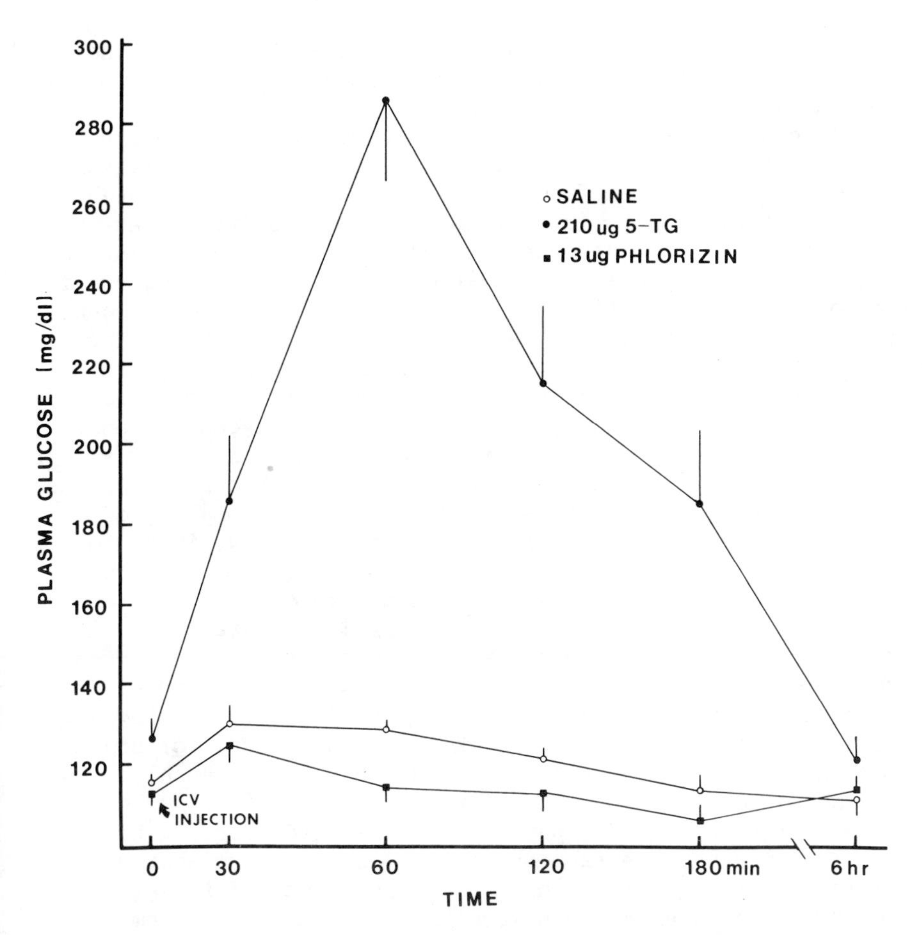

Fig. 11. Plasma glucose concentrations following 4th i.c.v. saline, 13.0 μg phlorizin, and 210 μg 5TG.

et al., 1977). It would thus appear that different neural systems or metabolic events mediate compensatory behavioral and autonomic responses.

Our findings using 4th i.c.v. injections of 5TG support the conclusion of Ritter et al. (1981) that fourth ventricularly applied 5TG, restricted to the caudal brainstem by a cerebral aqueduct plug, stimulates feeding and hyperglycemia. We also demonstrated that 4th i.c.v. injections of phlorizin that were confined to the caudal brainstem stimulated feeding. Therefore, feeding stimulated by lateral i.c.v. phlorizin cannot be exclusively attributed to phlorizin's action on forebrain neurons (Glick and Mayer, 1968). Our data, taken together with the identification by Ritter et al. (1981) of caudal brainstem metabolic receptors, argue that lateral i.c.v. phlorizin is acting on caudal brainstem mechanisms to stimulate feeding. These fourth ventricular injection studies demonstrate that the detection of metabolically relevant signals takes place within the caudal brainstem of intact rats.

A related series of experiments from our laboratory has demonstrated that caudal brainstem mechanisms, in isolation of the forebrain, are sufficient for the integration of metabolic and gustatory signals in the control of ingestive and autonomic responses. Chronic decerebrate and intact rats increase their ingestion of intraorally delivered sucrose, but not water, in response to food deprivation (Grill and Norgren, 1978c) and insulin-induced hypoglycemia (Flynn and Grill, 1983). The caudal brainstem interoceptive mechanisms revealed by 4th i.c.v. injections in intact rats may trigger the compensatory ingestive behavior and sympathoadrenal hyperglycemia seen in the the decerebrate rat. However, before we can conclude that caudal brainstem mechanisms are alone sufficient for both the detection and integration signals controlling compensatory responses, we need to demonstrate that 4th i.c.v. injection of 5TG stimulates sucrose intake in chronic decerebrate rats.

As noted above, chronic decerebrate rats never spontaneously feed. These rats die without oral or gastric infusion of nutrients. Obviously then, the case made for certain neural controls on feeding behavior contained within the caudal brainstem must be viewed within the context of a larger integrated nervous system. From a hierarchical perspective, forebrain mechanisms can act directly on caudal brainstem neural circuits to inhibit or facilitate particular responses that would regulate energy input and expenditure. As noted above, a hierarchical perspective assumes some degree of redundancy and would therefore presume that some mechanisms and receptors are found within more than one level of the neuraxis. From a functional perspective, the forebrain may be required to initiate action (appetitive behavior) and mediate between competing behaviors prior to the decision to feed.

D. The Role of Conditioning in Energy Homeostasis: Broadening the Strict Homeostatic Model

An implicit assumption in the majority of research on the neural control of feeding, one that survives even the reevaluation of the hypothalamic hypothesis, is that a *deficit signal* is the antecedent condition for feeding. This deficit signal, or strict homeostatic model, enables a fair degree of prediction for the initiation of feeding behavior of laboratory animals maintained on a continuously available, constant quality food in a standard single rat cage. This assumption, however, often fails in prediction when applied outside the standard laboratory setting. The strict homeostatic model is inadequate in explaining the initiation of human feeding behavior as well as the feeding of laboratory animals maintained on varied ("supermarket") diets or in enriched environments (Sclafani and Springer, 1976).

This lack of adequate explanation must be addressed. As discussed above, attempts at identifying the antecedent internal change(s) that initiates or terminates feeding behavior have largely focused on glucose utilization while numerous other possibilities have not yet been thoroughly examined (e.g., Sawchenko and Friedman, 1979). Different approaches to this problem have attracted attention. Let us recall that animals eat discrete meals which end *before* fuel absorption is complete. If fuel absorption is not complete by the time the meal has terminated, what, then, signals deficit reduction to the CNS? The role of conditioning in predicting satiety (deficit reduction) has been examined in several thoughtful reviews (e.g., Stunkard, 1975; Booth, 1977).

Rats and humans are capable of adjusting food intake in response to adulteration of their diet with nonnutritive bulk and maintain their body weight in the face of this challenge. As Stunkard (1975) notes, these adjustments of meal size to varying caloric densities appear to represent the result of Pavlovian conditioning. The sight, smell, taste, and texture of food can serve as conditioned stimuli for ending a meal. After meal termination and nutrient absorption, signals deriving from absorption in the GI tract are conveyed to the nervous system where they can serve as unconditioned stimuli. The preabsorptive or cephalic responses that rapidly follow cephalic stimulation may also serve as unconditioned stimuli in this process. When food quality is comparable for successive meals, the conditioned stimuli (e.g., tastes) that were associated with the metabolic consequences of the previous meal act to terminate feeding of the next meal, following the consumption of approximately the same amount of food. For example, Booth (1977) has shown that the ingestion or avoidance of a flavor paired with either dilute or concentrated starch solutions can be predicted by the changing metabolic state of an animal as it eats a meal. The flavor paired with concentrated starch (presumably signalling rapidly available carbohydrate fuel) is ingested when presented early in the meal. In contrast, the same flavor is avoided when presented late in the meal. When either meal quality or meal interval vary, however, the strength of the unconditioned stimuli signals either excess or deficient consumption. Conditioned stimuli lead to increases or decreases in meal size by their capacity to predict (anticipate) the metabolic consequences of a particular food. In Booth's (1977) view, satiety as well as feeding responses are conditioned "aversions" (quotation marks mine) and preferences dependent on nutritional state. The role of conditioning in meal initiation has not attracted as much attention as satiety but is certain to in the future (see however, the recent work of Weingarten, 1982).

The inadequacy of the strict homeostatic model is clearly described in a paper by Toates (1979) and commentaries that follow it. The paper details the limited (laboratory) explanation of the traditional homeostatic model of feeding control and suggests that a broadening of these ideas would yield a more useful and complete model. The author stresses the importance of incorporating the results of psychological and ethological studies into the more physiological, deficit signal oriented framework of homeostasis. Specifically, the results of research on cues associated with feeding, time elapsed since last feeding, food availability, and competing need states have clearly shown that these factors also regulate feeding and require further study (e.g., McFarland, 1974; deCastro, 1980).

III. CONCLUSIONS

A systematic perspective on the neural control of food intake has superseded the hypothalamic hypothesis. This perspective includes a vigorous neural and humoral analysis of the GI tract, a focus on the contribution of the caudal brainstem itself and through its connections to the GI tract via the autonomic nervous system, an appreciation of the powerful changes in peripheral metabolism that follow CNS damage or stimulation, a reevaluation of the strict homeostatic model of feeding to include data from ethological and conditioning studies, an awareness of a role for a wide range of preabsorptive responses that accompany contact with food, and a more hierarchically organized, less localized approach to the neural coordination of these activities.

The recent attention paid to the GI tract has already revealed that several types of stimuli acting on a variety of receptor sites within the digestive organs are all logical candidates for satiety-inducing signals. An integrated systems approach would see these stimuli acting in concert or synergistically to insure efficient digestion, as well as to initiate satiety. It is already clear that there is no single organ nor single type of visceral afferent stimulus that can account for all of the complexities of meal termination in the intact organism. Further analysis is required of the temporal organization or pattern of the different types of afferent signals generated as each successive organ is stimulated by the flow of nutrients within the digestive tract. It has also been difficult to parcel out the degree of integration accomplished at different neural levels, such as the visceral ganglia and the caudal brainstem nuclei and ganglia, due to a paucity of information on visceral afferent anatomy and physiology. We hope that in ten years we will find more of these questions answered than not. I am encouraged by our progress within the past decade, many examples of which are described in this book.

ACKNOWLEDGMENTS

My thanks to my collaborators Ralph Norgren, Kent Berridge, Bill Flynn, and Richard DiRocco. I am grateful for Eva Kosar's and Gary Schwartz's critical reading of the manuscript. This work was been supported by AM 21397.

REFERENCES

Bernard, C. 1849. Chiens rendu diabetique. *Comptes Rendus Soc. Biol.* 1: 60.

Berridge, K., H.J. Grill, and R. Norgren. 1981. Relation of consummatory responses and preabsorptive insulin release to palatability and learned taste aversions. *J. Comp. Physiol. Psych.* 95: 363-382.

Berthoud, H.R., and B. Jeanrenaud. 1979. Acute hyperinsulinemia and its reversal by vagotomy after lesions of the ventromedial hypothalamus in anesthetized rats. *Endocrinology* 105: 146-151.

Berthoud, H.R., and G.J.Mogenson. 1977. Ingestive behavior after intracerebral and intracerebroventricular infusions of glucose and 2-deoxy-D-glucose. *Am. J. Physiol.* 223: R127-R133.

Betz, A.L., R. Drewes, and D.D. Gilboe. 1975. Inhibition of glucose transport into brain by phlorizin, phloretin and glucose analogues. *Biochem. Biophys. Acta* 406: 505-515.

Booth, D.A. 1977. Satiety and appetite are conditioned reactions. *Psychosom. Med.* 39: 76-81.

Bradbury, M. 1979. *The Concept of a Blood-Brain Barrier*. John Wiley & Sons, New York.

Chhina, G.S., B.K. Anand, and P.S. Rao. 1971. Effect of glucose on hypothalamic feeding centers in deafferented animals. *Am. J. Physiol.* 221: 662-667.

Coil, J., and R. Norgren. 1979. Cells of origin of motor axons in the subdiaphragmatic vagus of the rat. *J. Auton. Nerv. Syst.* 1: 203-210.

Coimbra, C.C., J.I. Gross, and R.H. Migliorini. 1979. Intraventricular 2-deoxyglucose, insulin and free fatty acid mobilization. *Am. J. Physiol.* 207: E317-E329.

Davis, J.D., and M.W. Levine. 1977. A model for the control of ingestion. *Psychol. Rev.* 84: 379-412.

deCastro, J.M. 1980. Feeding behavior: Establishing causation in a systems environment. *Brain Res. Bull.* 5: 89-95.

Desiraju, T., M.G. Benerjee, and B.K. Anand. 1968. Activity of single neurons in the hypothalamic feeding centers: Effect of 2-deoxy-D-glucose. *Physiol. Behav.* 3: 757-760.

DiRocco, R.J., and H.J. Grill. 1979. The forebrain is not essential for sympathoadrenal hyperglycemic response to glucoprivation. *Science* 204: 1112-1114.

Fluharty, S.J., and H.J. Grill. 1980. Taste reactivity of lateral hypothalamic lesioned rats: effects of deprivation and tube feeding. *Soc. Neurosci. Abstr.* 6: 28.

Flynn, F.W., and H.J. Grill. 1983. Insulin elicits ingestion in decerebrate rats. *Science* 221: 188-190.

Flynn, F.W., and H.J. Grill. 1984. Fourth ventricular phlorizin dissociates feeding from sympathoadrenal hyperglycemia in rats. *Brain Res.* In press.

Flynn, F.W., H.J. Grill, and D. Rooney. 1983. Fourth ventricular phlorizin injection stimulates feeding but not hyperglycemia. *Soc. Neurosci. Abstr.* 9: 190.

Friedman, M.I., and E.M. Stricker. 1976. The physiological psychology of hunger: A physiological perspective. *Psychol. Rev.* 83: 409-431.

Glick, Z., and J. Mayer. 1968. Hyperphagia caused by cerebral ventricular infusion of phlorizin. *Nature* 219: 1374.

Gold, R.M. 1978. Hypothalamic obesity: The myth of the ventromedial nucleus. *Science* 182: 488-490.

Grill, H.J. 1980. Production and regulation of ingestive consummatory behavior in the chronic decerebrate rat. *Brain Res. Bull.* 5: 79-87.

Grill, H.J. and K.C. Berridge. 1981. Chronic decerebrate rats demonstrate preabsorptive insulin secretion and hyperinsulinemia. *Soc. Neurosci. Abstr.* 7: 29.

Grill, H.J., and R. Norgren. 1978a. The taste reactivity test. I. Mimetic responses to gustatory stimuli in neurologically normal rats. *Brain Res.* 143: 263-279.

Grill, H.J., and R. Norgren. 1978b. The taste reactivity test. II. Mimetic responses to gustatory stimuli in chronic thalamic and chronic decerebrate rats. *Brain Res.* 143: 281-297.

Grill, H.J., and R. Norgren. 1978c. Chronically decerebrate rats demonstrate satiation but not bait-shyness. *Science* 201: 267-269.

Grill, H.J., D. Ganster, and G.P. Smith. 1983. CCK-8 decreases sucrose intake in chronic decerebrate rats. *Soc. Neurosci. Abstr.* 9: 903.

Himsworth, R.L. 1970. Hypothalamic control of adrenalin secretion in response to insufficient glucose. *J. Physiol.* 206: 411-417.

Houpt, K.A., and A.N. Epstein. 1973. Ontogeny of controls of food intake in the rat: GI fill and glucoprivation. *Am. J. Physiol.* 225: 58-66.

Kalia, M., and M.M. Mesulum. 1980. Brain stem projections of sensory and motor components of the vagus complex in the cat: II. Laryngeal, tracheobronchial, pulmonary, cardiac, and gastrointestinal branches. *J. Comp. Neurol.* 193: 467-508.

King, B.M., and L.A. Frohman. 1982. The role of vagally mediated hyperinsulinemia in hypothalamic obesity. *Neurosci. Biobehav. Rev.* 6: 205-214.

Louis-Sylvestre, J. 1976. Preabsorptive insulin release and hypoglycemia in rats. *Am. J. Physiol.* 203: 56-60.

Macht, M.B. 1951. Subcortial localization of certain "taste" responses in the cat. *Fed. Proc.* 10: 88.

McFarland, D. 1974. Experimental investigation of motivational state. In *Motivational Control Systems Analysis*, ed. D. McFarland, 251-282. Academic Press, New York.

Miller, F.R., and C.S. Sherrington. 1916. Some observations on the buccopharyngeal stage of reflex deglutination in the cat. *Q. J. Exp. Physiol.* 9: 147-186.

Miselis, R.R. and A.N. Epstein. 1975. Feeding induced by

intracerebroventricular 2-deoxy-D-glucose in the rat. *Am. J. Physiol.* 229: 1438-1447.
Mountcastle, V.B. 1978. An organizing principle for cerebral function: the unit module and the distribution system. In *The Mindful Brain*, ed. V.B. Mountcastle and G.M. Edelman. MIT Press, Cambridge.
Oomura, Y. 1976. Significance of glucose, insulin and free fatty acid on the hypothalamic feeding and satiety neurons. In *Hunger: Basic Mechanisms and Clinical Implications*, ed. D. Novin, W. Wyriwicka, and G.A. Bray. Raven Press, New York.
Powley, T.L. 1977. The ventromedial hypothalamic syndrome, satiety, and cephalic phase hypothesis. *Psychol. Rev.* 84: 89-126.
Powley, T.L., and W. Laughton. 1981. Neural pathways involved in the hypothalamic integration of autonomic responses. *Diabetologica* 20: 378-387.
Ritter, R.C., P.G. Slusser, and S. Stone, S. 1981. Glucoreceptors controlling feeding and blood glucose: Location in the hindbrain. *Science* 213: 451-453.
Ritter, S., J.M. Murane, and E.E. Landenheim. 1982. Glucoprivic feeding is impaired by lateral or fourth ventricular alloxan injection. *Am. J. Physiol.* 243: R312-R317.
Ritter S. and M. Strang. 1982. Fourth ventricular alloxan injection causes feeding but not hyperglycemia in rats. *Brain Res.* 249: 198-201.
Rolls, E.T., M.J. Burton, and F. Mora. 1976. Hypothalamic neuronal responses associated with the sight of food. *Brain Res.* 111: 53-66.
Russek, M. 1971. Hepatic receptors and the neurophysiological mechanisms controlling feeding behavior. In *Neuroscience Research*, Vol. 4, ed. S. Ehrenpreis and D.C. Solnitzky, 214-282. Academic Press, New York.
Sawchenko, P.W., and M.I. Friedman. 1979. Sensory functions of the liver--a review. *Am. J. Physiol.* 236: R5-R20.
Scalfani, A., and D. Springer. 1976. Dietary obesity in adult rats: similarities to hypothalamic and human obesity syndromes. *Physiol. Behav.* 17: 461-471.
Shapiro, R.E., and R.R. Miselis. 1983. Organization of gastric efferent and afferent projections within the dorsal medulla oblongata in rat. *Anat. Rec.* 205: 182A.
Shimazu, T. 1981. Central nervous system regulation of liver and adipose tissue metabolism. *Diabetologica* 20: 343-356.
Silverstone, J.T. 1976. The CNS and feeding: Group report. In *Dahlem Workshop on Appetite and Food Intake*. Abakon Verlagsgesellschaft, Berlin.
Smith, G.P., and A.N. Epstein. 1969. Increased feeding in

response to decreased glucose utilization in the rat and monkey. *Am. J. Physiol.* 217: 1083-1087.

Steiner, J.E. 1977. Facial expressions of the neonate infant indicating the hedonics of food-related chemical stimuli. In *Taste and Development: The Genesis of Sweet Preference*, ed. J.M. Weiffenbach. U.S. Dept. of Health, Education and Welfare, Bethesda, Maryland.

Stricker, E.M., N. Rowland, C. Saller, and M.I. Friedman. 1977. Homeostasis during hypoglycemia: Central control of adrenal secretion and peripheral control of feeding. *Science* 196: 79-81.

Stricker, E.M. and M.J. Zigmond. 1976. Recovery of function following damage to central catecholamine-containing neurons: A neurochemical mode for the lateral hypothalamic syndrome. In *Progress in Psychobiology and Physiological Psychology*, ed. J.M. Sprague and A.N. Epstein. Academic Press, New York.

Stunkard, A. 1975. Satiety is a conditioned reflex. *Psychosom. Med.* 37: 383-387.

Toates, F.M. 1979. Homeostasis and drinking. *Behav. Brain Sci.* 2: 95-139.

Weingarten, H.P. 1983. Conditioned cues elicit feeding in sated rats: a role for learning in meal initiation. *Science* 220: 431-433.

Wick, A.N., D.R. Dury, H.I. Nakada, and J.B. Wolfe. 1957. Localization of the primary metabolic block produced by 2-deoxyglucose. *J. Biol. Chem.* 224: 963-969.

Chapter 5

DORSOMEDIAL HINDBRAIN PARTICIPATION IN CONTROL OF FOOD INTAKE

Robert C. Ritter and Gaylen L. Edwards

I. INTRODUCTION

Nearly fifty years ago the advent of stereotaxic surgery led rapidly to the realization that the ventromedial and lateral hypothalamus are important for the normal control of food intake (Brobeck et al., 1943; Anand and Brobeck, 1951). The behavioral changes associated with lesions or stimulation of these two forebrain structures are so dramatic that the first forty years of experimental investigation of brain controls of food intake were devoted almost exclusively to the hypothalamus. More recently, the wealth of neuroanatomical information concerning connectivity of the hindbrain and forebrain have fostered the view that controls of feeding may be integrated in systems which are networked along the entire rostrocaudal extent of the neuroaxis rather than being centered in the hypothalamus. Furthermore, recognition of visceral afferent participation in ingestive controls (Snowdon, 1970), together with the study of oral reflexes and consumption by decerebrate animals (Grill and Norgren, 1978a,b), has sparked interest in participation of the hindbrain neurons in the control of food intake.

Two regions of the hindbrain are of particular interest relative to ingestive controls. First, the rostrolateral portion of the nucleus of the solitary tract (NST) has been examined extensively by anatomists and sensory physiologists involved in the study of gustation (for example see Norgren, 1984). This portion of the NST receives most of the terminals of primary taste afferents from cranial nerves VII and IX (Hamilton and Norgren, 1984). Although traditionally it has not been considered to be a site for integration of taste with other behaviorally relevant sensations, some recent electrophysiological and behavioral data suggest that changes in responsiveness in the rostrolateral NST may be associated with changes in food intake (Chang and Scott, 1984, but see also Yaxley et al., 1985). Furthermore, metabolic changes influencing food intake cause changes in the electrical response of second-order gustatory units of this region (Giza and Scott, 1983). A second hindbrain area that may participate in the control of food intake encompasses the caudal subnuclei of the NST, the dorsal vagal motor nuclei, and the area postrema (AP). Because both the caudal NST and the AP receive vagal sensory terminals (Kalia and Sullivan, 1982; Shapiro and Miselis, 1985) and because the dorsal motor nucleus is a major source of vagal efferents (Contreras et al., 1982), these structures are often referred to collectively as the vagal complex. Electrophysiological data indicate that vagal sensory fibers which carry chemo- (Davison, 1972; Ewart and Wingate, 1983b; Mei, 1978; Rehbould et al., 1985) and mechanoreceptive (Harding and Leek, 1973; Ewart and Wingate, 1983a) information from the abdominal cavity synapse in the NST at this level. In addition, the AP is a probable receptor site for blood-borne hormones (Harding, et al., 1981; van Houten et al., 1979; Zarbin et al., 1985; Zarbin et al., 1983) and metabolites (Adachi and Kobashi, 1985). These features make the caudal NST and AP attractive candidates for participation in the control of feeding behavior. In fact, a growing number of studies of animals in which the AP and caudal portions of the NST have been ablated suggest that these structures may play an important part in the control of ingestion. This review will concentrate on recent evidence implicating the AP, the caudal NST, or both in the control of food intake.

II. HISTORICAL AND ANATOMICAL OVERVIEW

The first anatomical description of the AP was published by Retzious in 1898 (see Wilson, 1906), while Cajal (1909) is generally credited with the first detailed description of the NST. Subsequent to these early accounts there has been a burgeoning of neuroanatomical data concerning these structures. While it is not the intent of this article to extensively review the neuroanatomy of the AP and adjacent NST, a brief nonexhaustive overview of the connectivity of these areas in the rat seems appropriate.

The AP resides at the caudal end of the fourth cerebral ventricle (Fig. 1). It is a prototypical circumventricular organ in that it is perfused via fenestrated capillaries permeable both to large molecular weight substances and highly polar substances that do not readily enter other brain areas (Broadwell and Brightman, 1976; Murabe et al., 1981; Tervo et al., 1978). The AP contains a variety of unusual glial cells as well as a chemically heterogeneous population of small neurons (Morest, 1960; Clemente and van Breeman, 1955; Spacek and Parazek, 1969; Klara and Brizzee, 1977). The structure is bordered laterally and ventrally by the nucleus of the solitary tract (Fig. 1), with which it has extensive reciprocal neuronal connections (Morest, 1967; Norgren, 1978; Shapiro and Miselis, 1985; van der Kooy, 1983). The AP is reported to communicate with the NST via a portal vascular connection (Roth and Yamamoto, 1968). Although vagal sensory neurons generally terminate most heavily in the NST, the AP itself appears to be the site of termination for a significant number of vagal afferent fibers innervating both thoracic and abdominal viscera (for example see Kalia and Mesulam, 1980; Kalia and Sullivan, 1982). The AP has also been reported to receive afferent terminals from the trigeminal nerve (Jacquin et al., 1982).

In addition to its close and extensive connections with the NST, the AP connects with other hindbrain nuclei. The lateral parabrachial nucleus appears to be the recipient of the densest projection of efferents from the AP (van der Kooy and Koda, 1983; Shapiro and Miselis, 1985). The AP in turn also receives a terminal projection from the lateral parabrachial nucleus

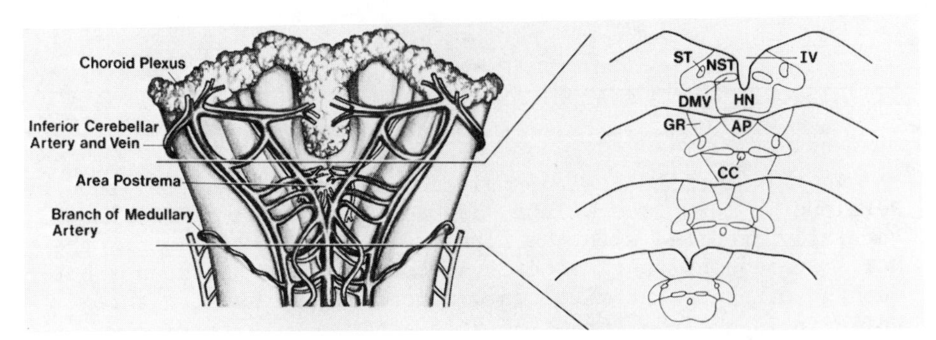

Some Connections of Rat Area Postrema

AFFERENTS

- Vagus Nerve
- Trigeminal Nerve
- N. of Solitary Tract
- Lateral Parabrachial N.
- Dorsal Tegmentum
    - Pericentral N.
    - Lateral N.
- Hypothalamus
    - Dorsomedial N.
    - Perifornical N.
    - Paraventricular N.

EFFERENTS

- N. of Solitary Tract
- Dorsal Motor N. of Vagus
- Ambiguus Complex
- Lateral Parabrachial N.

Some Connections of the Caudal Nucleus of the Solitary Tract

AFFERENTS

- Vagus Nerve
- Trigeminal Nerve
- N. of Solitary Tract
    (Intranuclear)
- Ambiguus Complex
- Lateral Parabrachia N.
- Cerebellum
- Spinal Cord
- Hypothalamus
    - Paraventricular N.
    - Lateral Area
    - Dorsomedial N.
    - Arcuate N.
    - Septum
- Bed N. of Stria Terminalis
- Amygdala
- Insular Cortex

EFFERENTS

- N. of Solitary Tract
    (Intranuclear)
- Area Postrema
- Ambiguus Complex
- Lateral Parabrachial N.
- Locus Coeruleus
- Raphe N.
    (Magnus and Pallidus)
- Midbrain Central Grey
- Hypothalamus
    - Paraventricular N.
    - Lateral Area
    - Dorsomedial N.
- Bed N. of Stria Terminalis
- Amygdala

Fig. 1. Anatomy of dorsal hindbrain at the level of the obex. The left panel illustrates the topography of the hindbrain surface. The cerebellum and more rostral brainstem have been removed to better visualize the obex. Note the extensive blood supply to the AP. The right panel diagrammatically illustrates the cross-sectional anatomy of the obex region as coronal sections taken at approximately 200μm intervals through the region bracketed on the left panel. In both panels the rostral end of the diagram is at the top of the page. Listed below are some of the structures connected with the AP and caudal NST. Abbreviations: AP = area postrema, CC = central canal, DMV = dorsal motor nucleus of the vagus, GR = nucleus gracilus, HN = hypoglossal nucleus, IV = fourth ventricle, NST = nucleus of the solitary tract, ST = solitary tract.

(van der Kooy and Koda, 1983; Shapiro and Miselis, 1985). At least one report indicates that the AP projects to the pericentral and lateral dorsal tegmental nuclei (van der Kooy and Koda, 1983). These projections, however, have not been confirmed by other investigators (Shapiro and Miselis, 1985). The AP also receives direct afferent projections from the dorsomedial (Hosoya and Matsushita, 1981a; Buijs, 1978) and lateral hypothalamus (Hosoya and Matsushita, 1981b) and from the lateral parvicellular portion of the paraventricular hypothalamic nucleus (Swansen and Kuypers, 1980).

As previously mentioned, the AP is bordered laterally, ventrally, and caudally by the NST. The NST begins just caudal and ventral to the AP but extends approximately 3 mm rostral to the AP as bilateral lenticular columns that merge at the level of the AP to form an open V-shaped structure. The NST of the rat has been divided into six subnuclei (Kalia and Sullivan, 1982) on the basis of cytoarchitectural criteria. The portions of the NST lying immediately adjacent to the AP, and which are likely to be damaged by lesions directed toward the AP, include the commissural, medial, and caudal subnuclei (Fig. 1 and 2). At the level of the AP the NST receives heavy afferent input from the vagus nerve (Beckstead and Norgren, 1982; Kalia and Sullivan, 1982). There are also projections from the glossopharyngeal nerve (Contreras et al., 1982) as well as some input from the trigeminal (Jacquin et al., 1982). The NST has extensive connections with the cranial nerve motor nuclei (III, VII, IX, X, and XII), with especially heavy projections to the dorsal motor nucleus of the vagus (Morest, 1967; Norgren, 1978; Rogers et al., 1980) and the ambiguus complex (Norgren, 1978; Ricardo and Koh, 1978; van der Kooy and Koda, 1983; Shapiro and Miselis, 1985). Projections from the NST to the raphe have been reported (Ross et al., 1981). There are extensive reciprocal connections between the caudal portions of the NST and the pontine parabrachial nuclei (Saper and Loewy, 1980; Ricardo and Koh, 1978; Norgren 1978; van der Kooy and Koda, 1983; Shapiro and Miselis, 1985), and the NST has been reported to have both afferent and efferent connections with the cerebellum (Ross et al., 1981; Somana and Walberg, 1979). There are extensive interconnections between the NST and several hypothalamic nuclei, including but not restricted to the paraventricular, dorsomedial, and lateral hypothalamic nuclei (for example see Ricardo and Koh, 1978). Other forebrain connections of the caudal portions of the NST include the bed nucleus of the stria terminalis and the amygdala, both of which connect reciprocally with the NST. Finally, there are afferents to the NST from the insular cortex (van der Kooy et al., 1982).

The close physical association of the AP and mNST, coupled with their intimate neural and vascular connections, has not

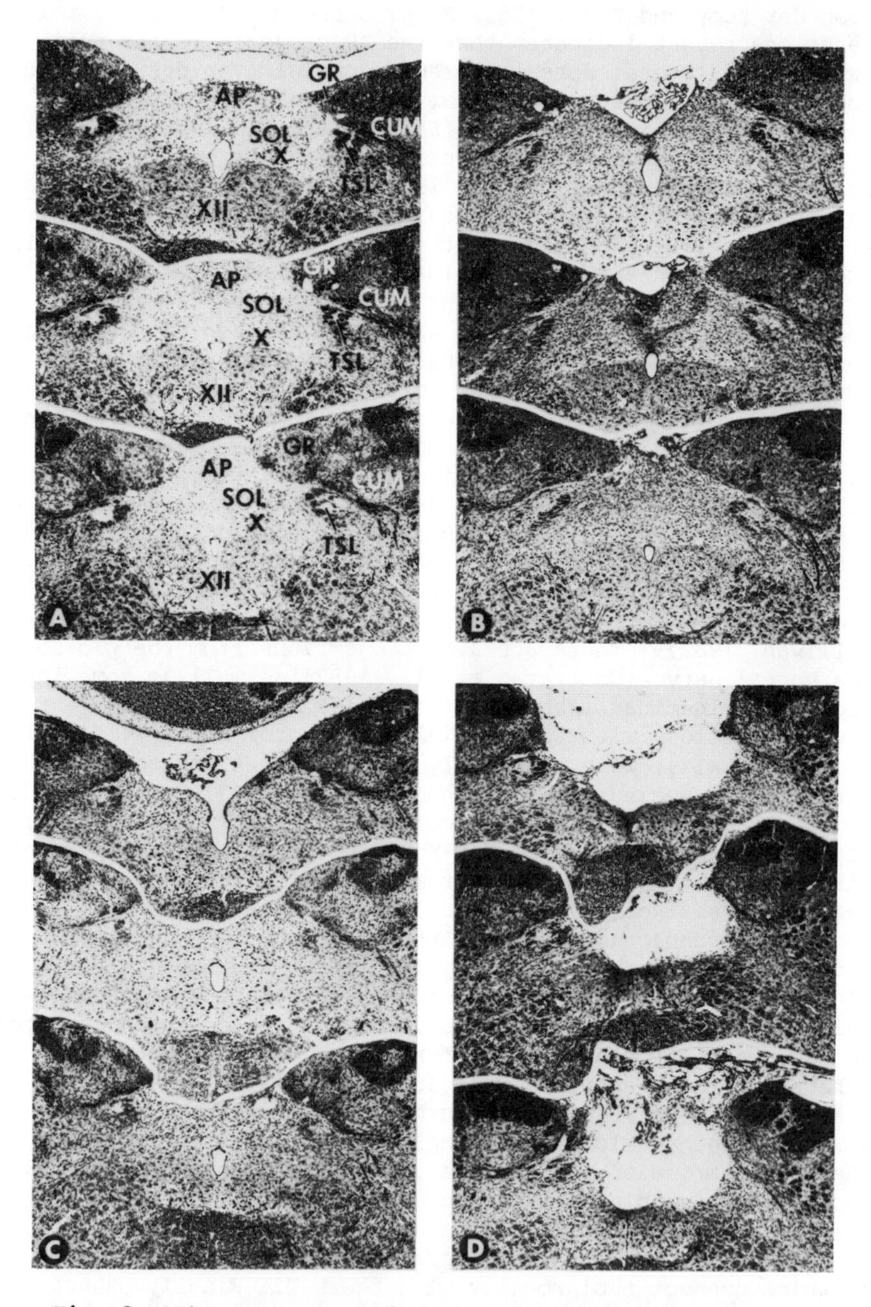

Fig. 2. Photomontages of AP lesions demonstrating variations in lesion size at three different anterior posterior levels. Panel A shows sections from a nonlesioned control rat.

permitted lesions of one structure without probable damage to the other. All of the behavioral phenomena discussed here are associated with total destruction of the AP. Nevertheless, it should be understood that when the term "AP lesion" is used, some damage to the adjacent mNST also is assumed. Figure 2 shows the range of tissue damage encountered following thermal lesions directed at the AP.

III. INGESTIVE PHENOMENA ASSOCIATED WITH LESIONS OF THE AREA POSTREMA

A. Attenuation of Learned Taste and Food Aversions

Examinations of the effect of AP lesions upon drug-induced emesis were among the first studies of AP function. The work of Borison and co-workers demonstrated that destruction of the AP abolished or attenuated the emetic response to some drugs, such as cardiac glycosides, while leaving emesis in response to other stimuli intact (for review see Borison, 1974). Because AP lesions reduced the emetic response to specific drugs and not to other emetic stimuli, Borison proposed that AP lesions destroyed a chemosensory organ and did not interfere with efferent aspects of the emetic reflexes. He consequently dubbed the AP a "chemoreceptor trigger zone" for vomiting in response to systemic intoxication. Drug-induced emesis appears to be a sign of malaise in animals. Therefore, the possibility that the AP's proposed chemoreceptive role might serve to mediate learned avoidance of foods has been investigated by a growing number of individuals. Berger et al. (1973) appear to have been the first to demonstrate that AP lesions impair the formation of conditioned taste aversions (CTAs). They found

---

Panel B shows a very small lesion with little or no damage to the NST. Panel C shows a typical large lesion, and panel D shows a very large lesion, which destroys much of the NST and DMV and damages the hypoglossal nucleus. Abbreviations: AP = area postrema, CUM = medial cuneate nucleus, GR = nucleus gracilis, SOL = solitary nucleus, TSL = solitary tract, X = dorsal motor nucleus of the vagus nerve, XII = hypoglossal nerve nucleus. Reprinted with permission from *Physiology and Behavior* 29. Edwards, G.L., and R.C. Ritter. Area postrema lesions increase drinking to angiotensin and extracellular dehydration.

that scopalamine methylnitrate, an antimuscarinic agent, which does not pass the blood-brain barrier, served as an unconditioned stimulus for taste aversions in intact rats but not in AP-lesioned rats. Subsequently, others have demonstrated that AP lesions also abolish taste aversions conditioned with other toxicants, such as LiCl (Ritter et al., 1980) or paraquat (Dey et al., 1984). The AP lesion does not impair the acquisition of aversions when amphetamine, a drug that crosses the blood-brain barrier, is used as the unconditioned stimulus (Berger et al., 1973; Ritter et al., 1980). Therefore, it is clear that the AP lesion does not impair the animal's ability to detect the conditioned stimulus (taste). Furthermore, the lesion does not impair formation of aversions conditioned by intragastric copper sulfate infusion (Dey et al., 1984), a stimulus that appears to act via vagal afferents (Coil et al., 1978). Therefore, AP-lesion-induced impairment of taste aversion conditioning is not caused by damage to the central terminals of vagal sensory neurons. It is interesting that AP lesions do attenuate the formation of taste aversions in response to exposure to ionizing radiation (Ossenkopp, 1983). Since radiation-induced CTAs are formed even when the head itself is shielded from radiation exposure (Garcia and Kimeldorf, 1960; Smith et al., 1981), it appears that the unconditioned stimulus for radiation-induced taste aversions may be an endogenously produced chemical that reaches the AP via the blood. Some support for this proposal is provided by a report claiming that when one partner of a parabiotic pair of rats is irradiated, the non-irradiated partner may also acquire a taste aversion (Hunt et al., 1965).

Another interesting condition under which the AP is implicated in the acquisition of a food aversion conditioned by an endogenously produced humoral factor has recently been reported by Bernstein et al. (1985). These investigators have demonstrated that anorexia and loss of weight, which occur during the growth of an implanted tumor, are abolished in AP-lesioned rats. Furthermore, Bernstein et al. have previously demonstrated that rats develop conditioned food aversions following tumor implantation (Bernstein and Sigmundi, 1980), and this tumor-induced food aversion is apparently prevented by lesioning the AP. Since the implanted tumors used in these experiments were not highly invasive and had not metastasized, it seems likely that AP lesions attenuate tumor-induced ingestive effects by interfering with the detection of a blood-borne tumor product. The apparent participation of the AP in suppression of food intake by a growing implanted tumor raises the suspicion that the AP may be involved in control of food intake exerted by products secreted by non-neoplastic tissue under physiological conditions. This suspicion is fed by the apparent presence in the AP and/or adjacent mNST of high affinity binding sites for such blood-borne substances as insulin (van Houten, 1979), bombesin (Zarbin et al., 1985), and cholecystokinin (Zarbin et al., 1983), all of which have been implicated in the control of food intake.

B. Alterations of Body Weight

Lesions that destroy the AP typically cause acute loss of body weight. This effect of AP lesions was first reported by Carlisle and Reynolds (1961), but has subsequently been observed by several other investigators (Edwards and Ritter, 1981; Hyde and Miselis, 1983; Kenney et al., 1984). We have found post-lesion weight loss to occur gradually over a period of two to three weeks. By this time the body weights of AP-lesioned rats are generally 12-25% below those of sham-operated controls. By two to three weeks postsurgery, the weight of the lesioned rats stabilizes and they begin to gain weight at the same rate as sham-operated controls. Nevertheless, most lesioned rats do not reattain body weights comparable to those of controls. This general pattern has now been reported by several laboratories active in the investigation of hindbrain participation in the control of food intake (Edwards and Ritter, 1981; Ritter and Edwards, 1984; Hyde and Miselis, 1983; Kenney et al., 1984).

The cause of chronically reduced body weight in the AP-lesioned rat is still unknown. Several reports suggest, however, that reduced body weight of the AP-lesioned rat is due to loss of body fat. For example, Kott et al. (1985) have reported that the epididymal fat pads of AP-lesioned rats are markedly reduced in size. By extrapolation of their epididymal fat pad data, they suggest that reduced body weight may be due entirely to reduced body fat in AP-lesioned rats. Hyde and Miselis (1984) have also reported reduced body fat content in AP-lesioned rats. Results from our work in collaboration with Dr. Roy Martin at the University of Georgia are in general agreement with Hyde and Miselis and Kott et al. In this work we found that AP-lesioned (30 days post-lesioning), lean littermates of Zucker fatty rats exhibited body weight reduction of 13.0% relative to sham-lesioned controls. Carcasses of the lesioned lean rats also contained significantly less fat than those of sham-lesioned controls. Lean control rats exhibited 15.6% of their body weight as fat, while in lesioned rats only 11.2% of body weight was fat. Carcass contents of protein, water and ash were also reduced, but these reductions did not reach statistical significance. Although AP-lesioned fatty rats did lose weight after surgery, there were no significant differences in body weight, carcass fat, or percentage carcass composition by 30 days postsurgery. These data support the hypothesis that weight reduction in AP-lesioned rats is primarily due to loss of body fat. It is possible that this change in composition is caused by severe intake reduction during a period when the animals are normally adding fat. We found, however, that carcass fat content of food-restricted, lean, sham-lesioned rats, which were returned to ad libitum food intake 14 days postsurgery, did not differ from non-restricted rats. Furthermore, we have previously reported that pair-fed control rats rapidly regain lost weight when free access to food is reinstituted. AP-lesioned rats, on the other hand, tend to defend their reduced body weights. Thus it would be premature to rule out a metabolic or endocrine explanation for reduced body weight in this lesion preparation. A report that AP lesions abolish the increased weight gain associated with hypothalamic knife cuts (Hyde and Miselis, 1983) is therefore most interesting. Although no carcass analyses were reported with this study, one would assume that prevention of weight gain in hypothalamic knife cut rats was due specifically to reduced adiposity. Findings such as these may be compatible with reduced insulin secretion in AP-lesioned rats. However, no clear evidence of reduced insulin secretion in this preparation has been reported thus far. In fact, one recent preliminary report indicates that rats with AP lesions exhibit elevated postprandial insulin concentrations (Lacour, et al., 1985).

C. Impaired Glucoprivic Feeding

Several reports indicate that the hindbrain contains the glucoreceptors which mediate feeding in response to antimetabolic glucose analogues (Ritter et al., 1981; Flynn and Grill, 1983). (For review see the chapter by Sue Ritter in this volume.) Because of the AP's intimate association with the blood vascular system and its neural connections with vagal sensory fibers and the hypothalamus, the AP or adjacent NST is a reasonable site for the location of glucoreceptors that mediate feeding behavior. Consequently, several investigators have examined the glucoprivic control of food intake in AP-lesioned rats (Hyde and Miselis, 1983; Contreras et al., 1982; Bird et al., 1983; Nonavinakere and Ritter, 1984).

Contreras and co-workers (1982) have reported virtual abolition of feeding in response to systemic injection of 2-deoxy-D-glucose (2DG) following AP lesions. This report is contrary to the observation of Hyde and Miselis (1983) which indicates that feeding in response to 2DG is unimpaired by AP lesions. Nonavinakere and Ritter (1984) have systematically analyzed feeding in response to both systemic 2DG and fourth ventricular infusion of 5-thio-D-glucose (5TG). The results of these studies indicate that AP lesions do not totally abolish feeding in response to systemic or cerebral glucoprivation. Sixty to 75% of lesioned rats continue to eat in response to subcutaneous 2DG or fourth ventricular 5TG. The responses are not normal, however. AP-lesioned rats display a significant delay in the onset of glucoprivic feeding (Fig. 3) after subcutaneous 2DG, and in some groups of rats the absolute magnitude of the response is also diminished. The lesions examined in the studies of Nonavinakere and Ritter (1984) were very large. The entire AP and at least 63% of the adjacent NST were destroyed. Therefore, at least some of the receptors that mediate feeding in response to glucoprivation are probably located outside the AP and adjacent NST. Nevertheless, there is mounting electrophysiological evidence for glucoreceptive neurons in the NST (Mizuno et al., 1983) and the AP (Adachi and Kobashi, 1985). In view of the fact that AP lesions impair glucoprivic feeding, it remains possible, and even likely, that the AP or adjacent NST are locations for some of the glucoreceptors responsible for mediating glucoprivic feeding.

D. Alteration of Water and Sodium Intake

Several reports now indicate that lesions of the AP and adjacent NST cause increased water and sodium intake. We have previously reported that rats with AP lesions drink nearly twice as much water as control rats in response to subcutaneously injected angiotensin II or in response to extracellular dehydration or ß-adrenergically induced hypotension (Edwards and Ritter, 1982). Hyde and Miselis (1984) have partially replicated these findings. However, their lesioned rats also drank more than controls in response to cellular dehydration and had elevated 24-hour water intakes.

It also appears that rats with AP lesions exhibit an exaggerated sodium appetite. Contreras et al. (1983) reported that this increase in the consumption of sodium is not associated with increased sodium excretion. Others (Hyde and Miselis, 1984), however, have found that AP-lesioned rats excrete relatively more sodium/gram of body weight than do intact rats. Since this increased sodium loss may be associated with increased water excretion as well, both increases in drinking and increased sodium appetite may be secondary to heightened excretion.

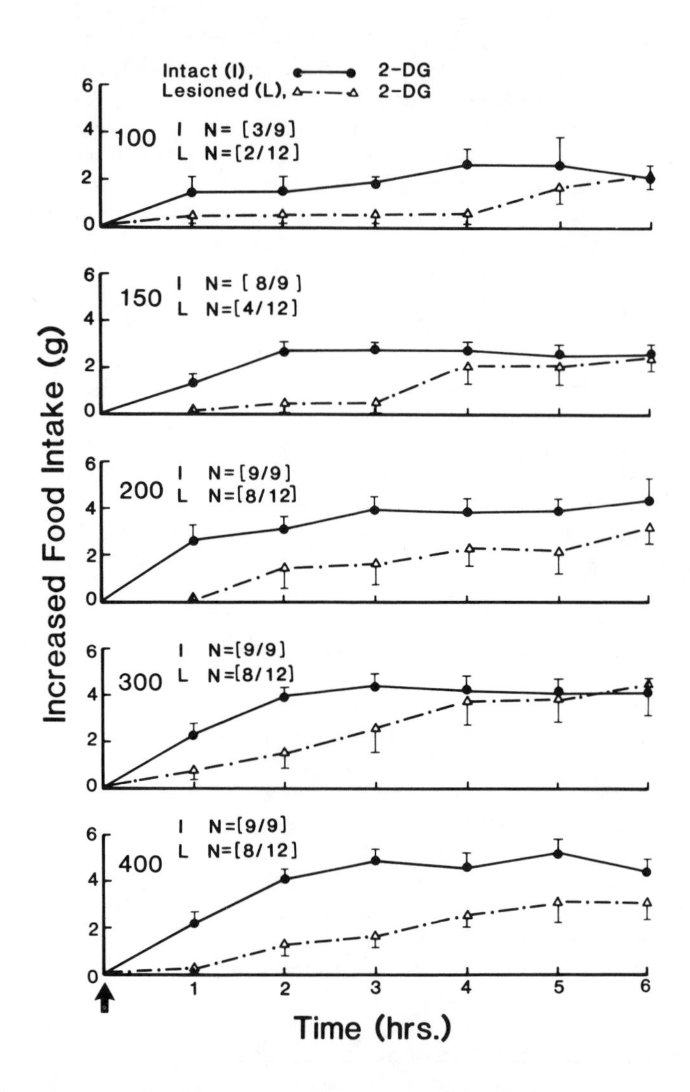

Fig. 3. Cumulative increase in food intake by AP-lesioned and intact rats following subcutaneous injections of 100, 150, 200, 300, or 400 mg/kg 2-deoxy-D-glucose (2DG). Arrow indicates time of injection. Numbers within parentheses are proportions of rats that increased their food intake in response to 2DG. Although 6-hour intakes by lesioned and intact rats are similar, lesioned rats show significant delays in the onset of feeding compared to intact rats. Also notice that not all lesioned rats ate in response to 2DG.

E. Altered Sensitivity to the Ingestive Effects of Hormones

Because it is endowed with fenestrated capillaries, the AP is in more intimate contact with blood-borne substances than most other brain regions. Therefore, the notion that the AP may serve as a receptor site for circulating hormones and metabolites is plausible, and the possibility that some endocrine effects on feeding behavior may be mediated by receptors in or near the AP is appealing. It is therefore interesting that high affinity binding for several hormones and peptides thought to participate in the control of ingestion is present at high concentrations in the AP or the adjacent NST. For example, van Houten et al. (1979) have demonstrated that the NST immediately adjacent to the AP is densely labeled by radioiodinated insulin and that the labeling is saturable. Zarbin et al. have shown that the AP and NST exhibit high affinity specific binding for bombesin (Zarbin et al., 1983) and CCK (Zarbin et al., 1985). However, it appears that AP lesions fail to attenuate the well-described suppression of food intake that occurs following systemic administration of either bombesin or CCK (Edwards and Ritter, 1981; Edwards et al., 1985; Edwards and Ritter, 1986).

The role of the AP and the NST in mediating suppression of feeding by CCK deserves special consideration because of conflicting reports in the literature concerning the sensitivity of AP-lesioned rats to CCK (Van der Kooy, 1984). Smith et al. (1981) have demonstrated that suppression of food intake by CCK is abolished following section of the gastric branches of the vagus nerve. The gastric vagal afferents terminate in the NST, especially subjacent to and anterior to the AP (Shapiro and Miselis, 1985). Therefore, it is not unreasonable to expect that lesions which destroy the AP and extensively damage the NST will impair suppression of feeding by CCK. Nevertheless, we have repeatedly found (Edwards and Ritter, 1981; Edwards et al., 1984; Edwards and Ritter, 1986) that lesions which destroy the AP and moderately damage the adjacent NST do not even attenuate CCK-induced suppression of food intake. A recent report by van der Kooy (1984) suggesting that lesion of the AP impaired CCK-induced suppression stimulated us to systematically examine the role of the AP and the NST in this phenomenon (Edwards et al., 1985, 1986). We found that lesions that destroy the AP and damage the subjacent NST do not impair suppression of feeding by CCK (Fig. 4). Lesions limited to the NST just rostral to the AP are also without apparent effect. However, in rats in which we made lesions of the AP, subjacent NST, and the NST rostral to the AP, suppression of feeding by CCK was markedly attenuated (Fig. 4). This study in itself does not prove that circulating CCK does not alter ingestion

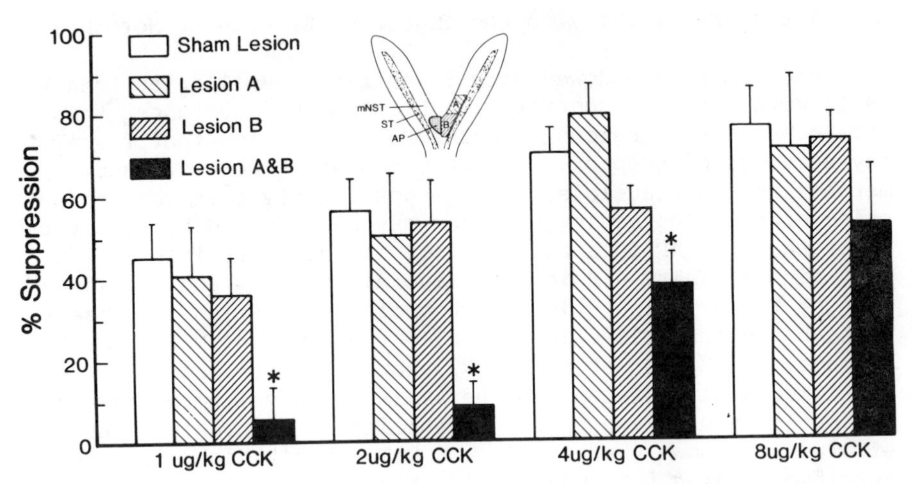

Fig. 4. Attenuation of CCK-induced suppression of food intake by lesions that extensively damage afferent projections of the gastric branches of the vagus nerve. Inset shows placement of lesions on a schematic diagram of the AP and NST. Lesion A damaged the medial nucleus of the solitary tract (mNST) rostral to the AP. Lesion B destroyed the AP and extensively damaged the adjacent portions of the NST. Lesion A&B was a combination of both lesions. This lesion destroyed most of the NST receiving vagal sensory terminals from the stomach. Note that only lesion A&B significantly attenuated CCK-induced suppression of food intake. Lesion B, which destroyed the AP, did not attenuate suppression of food intake by CCK. From Edwards et al. (1986). *Am. J. Physiol.*, in press.

via some action on the AP, but it does show that the AP is not necessary for the known effect of CCK on feeding behavior. Furthermore, our results are consistent with the interpretation that abdominal vagal afferents mediate CCK's effects on feeding and that brainstem lesions which attenuate these effects do so by destroying central terminals of vagal sensory neurons.

F. Increased Intake of Palatable Foods and Solutions

Foods that are avidly consumed because of their positive orosensory qualities are generally regarded as highly palatable foods. One of the most striking ingestive changes that occurs following AP lesions is an increase in the lesioned animal's ingestive response to palatable foods when such foods are offered over a short period of time (30 minutes to 6 hours).

For example, when lesioned and sham-lesioned rats, maintained on a standard laboratory rodent diet, are presented with a sweet, vanilla-flavored, milk-based food (instant breakfast), the lesioned rats consume nearly twice as much as sham-lesioned rats over a 30-minute period (Edwards and Ritter, 1981). This increase in intake begins within the first 5 days post-lesioning and persists for as long as we have maintained the animals (10 months). Increased intake of palatable substances is not limited to liquid foods. It is also induced by exposure of rats to palatable solid foods, such as cookies. Therefore, this effect is not a manifestation of altered drinking behavior, which has been reported by our laboratory and others (Edwards and Ritter, 1982; Hyde and Miselis, 1984). Finally, the overconsumption of palatable food is not accompanied by any increase in intake of the maintenance diet. In fact, after recovery from surgery, intakes of laboratory diet by lesioned rats and sham-lesioned rats are not statistically different either under ad libitum conditions or during the compensatory ingestive response to food deprivation (Edwards and Ritter, 1981; Ritter and Edwards, 1984). Therefore, the overingestion of some foods by AP-lesioned rats appears to be a specific response to orosensory qualities of the food. In addition, when AP-lesioned rats were offered an ascending concentration

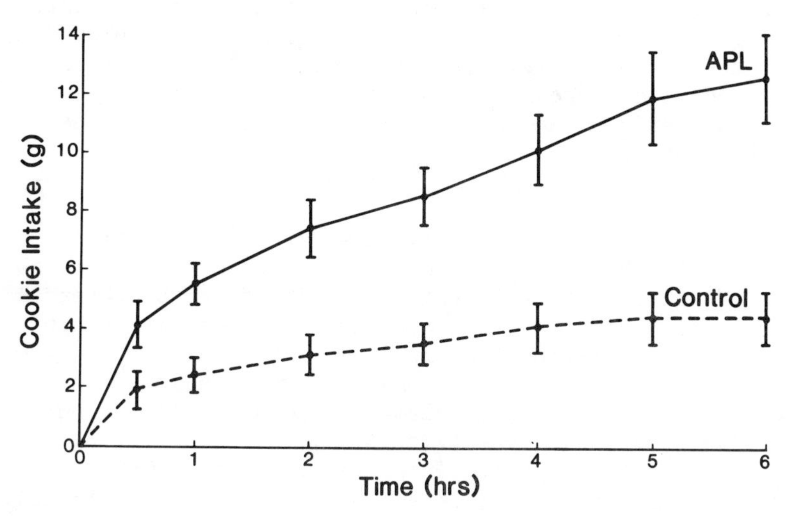

Fig. 5. Intake of cookies by AP-lesioned rats (APL) and sham-operated control rats. The intake by lesioned rats was significantly greater than that of controls beginning 30 min after presentation of the cookies (Edwards and Ritter, 1981).

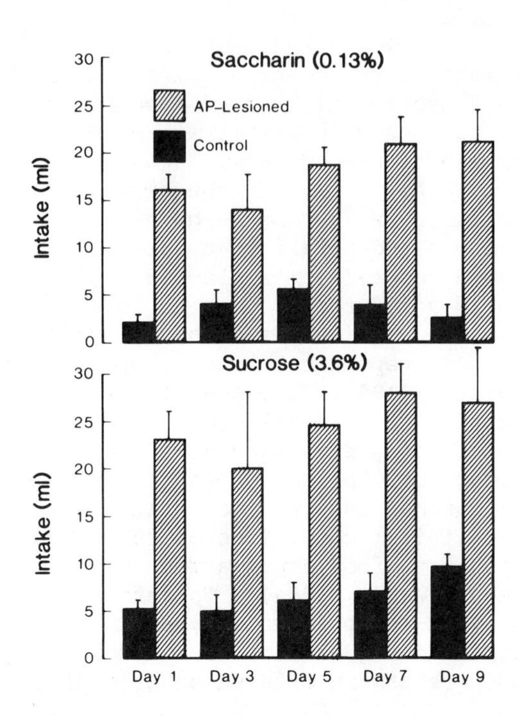

Fig. 6. Sixty-minute intake of preferred concentrations of saccharin and sucrose by AP-lesioned and sham-lesioned control rats. Intakes of lesioned rats were significantly greater than those of control rats on every test day. Data displayed is for odd-numbered test days. On even-numbered days the rats were presented with a fresh tube of water for 1 hour. Intakes of water for lesioned and control rats were indistinguishable (Ritter, 1984).

series of sucrose or saccharin in short (30-60 minute) one-bottle tests they exhibited preference thresholds comparable to those of intact rats (Ritter, 1984). In fact, at threshold concentrations of the sweeteners the lesioned rats exhibit consumptions indistinguishable from those of intact controls. At suprathreshold concentrations, however, lesioned rats consume significantly larger volumes of both nutritive and non-nutritive sweet solutions (Fig. 6). It is interesting that AP-lesioned rats exhibit quinine rejection thresholds identical to those of intact rats (Ritter, unpublished observations). These data suggest that the AP lesion does not impair taste sensitivity, but only responsiveness at tastant concentrations above threshold. In addition, these results suggest that the apparent lesion-induced increase in responsiveness does not extend to aversive tastes, i.e., bitter. Of course, the preference threshold cannot be construed as a measure of the animal's absolute ability to detect a tastant in solution (detection threshold). The detection threshold is usually well below the preference threshold. Nevertheless, the preference threshold is an indication of the minimum concentration that elicits ingestion and would be expected to be diminished if the detection threshold were diminished.

AP lesion-induced overingestion of highly palatable substances is not limited to sweets. We have offered control and AP-lesioned rats fat (solid vegetable shortening) in a 60-minute ingestion test. Under this condition AP-lesioned rats consume significantly more fat than control animals. Kenney et al. (1984), however, have found that when AP-lesioned rats are given continual access to fat in a caloric self-selection paradigm, they take fewer calories as fat than do sham-operated rats during a period from 4 to 13 days postsurgery. They point out that the postsurgical reduction of caloric consumption by lesioned rats is due entirely to reduction of fat intake and suggest that perhaps the lesioned rats have developed an aversion to fat. The possibility of a lesion-related aversion to fat is tantalizing and potentially important. However, Kenney et al. also find that fat intake is no longer significantly depressed between 14 and 23 days postsurgery. Therefore, it is possible that the failure of lesioned rats to select a larger percentage of their total caloric intake as fat is a phenomenon limited to the acute post-lesion period, whereas overingestion of highly palatable foods appears to persist for the life of the lesioned animal.

The mechanism by which AP lesions cause increased ingestion of palatable foods is not yet understood. However, several alternative explanations have been suggested. Hyde and Miselis (1983) have speculated that the AP lesion may constitute a sort of central sensory vagotomy. Consequently, the overingestion of palatable foods by lesioned rats might be due to loss of vagally mediated gastrointestinal satiety cues. It is difficult to explain, however, why interference with visceral afferent input would cause overingestion of some foods and not others. This fact as well as additional experimental evidence (see Section IV,A) argue strongly against this hypothesis. An alternative hypothesis is that the AP or adjacent mNST is directly involved in modulating the activity of orosensory modalities either within the hindbrain or via connections with the forebrain. Damage to these structures may then remove a net inhibitory influence on the response to some orosensory quality of the diet.

There are several potential, and non-mutually exclusive, mechanisms through which AP lesions may alter orosensory responsiveness. One intriguing hypothesis has been offered by Kenny (personal communication). She suggests that the AP lesion serves as the unconditioned stimulus for the development of a conditioned taste aversion to the animal's familiar diet. The overresponse to other novel foods is then viewed as a preference shift. In support of this hypothesis, Kenney and co-workers (Kott and Kenney, 1984) find that rats fed sweetened milk diet while they are recovering from the lesion subsequently exhibit a transient overingestion of pelleted rat diet. Rats that recover from the lesion while consuming pelleted diet subsequently overconsume sweetened milk. The possibility that the AP lesion causes altered response to taste as a result of a learned taste aversion is interesting but not entirely satisfying. First, intact rats that are maintained on a familiar diet (milk diet, for instance) during the recovery period also appear to display an increased intake of a novel diet (e.g., pelleted rodent diet) when it is subsequently introduced. The response, however, is not nearly so dramatic as that observed in lesioned rats. Therefore, this effect may reflect the potential role of novelty as a determinant of palatability. Furthermore, when lesioned rats that are fully recovered from the lesioning procedure have continuous, as opposed to episodic, access to a novel palatable diet, their overingestion moderates and in fact they do not consume more than controls (Tomoyasu and Kenney, 1985). However, when a new palatable diet is introduced they again exhibit overingestion. Therefore, if overingestion is the result of a learned food aversion, the process must be one of repeated acquisition of aversion in response to sequentially presented foods.

Another possible explanation for increased orosensory responsiveness in AP-lesioned rats is that the intact AP or adjacent NST modulates ingestive responses to oral sensation either via connections with hindbrain gustatory neurons or through connections with forebrain structures. AP lesions would abolish such a modulatory influence, allowing exaggerated responses to foods preferred by intact rats. Although results that directly demonstrate a role for the AP or adjacent NST in the modulation of taste responsiveness have not yet appeared, some recent electrophysiological evidence may support this proposition. Herman and Rogers (1985) have recently shown that some third-order gustatory neurons in the parabrachial area are also activated by stimulation of the caudal nucleus of the solitary tract adjacent to the AP. This finding indicates that neurons, perhaps including AP efferents to the parabrachial area, which are usually destroyed by the AP lesion, can influence the discharge of taste neurons.

IV. NEURAL SUBSTRATES OF AP LESION-INDUCED ALTERATION OF INGESTIVE BEHAVIOR

A. Role of General Visceral Afferents

As previously discussed, the AP and medial NST possess connections that are anatomically diverse and neurochemically heterogeneous. Therefore, we should not be surprised when lesions impinging on these structures produce alterations in a variety of behavioral and physiological phenomena. It is probable, however, that individual sequels of AP lesions are related to disruption of distinct neuronal networks. Consequently, interventions that disrupt specific neural connections of the AP and adjacent NST might produce, individually, behavioral alterations observed as part of the aggregate following AP lesions.

One of the most reasonable ideas concerning AP lesion-induced behavioral changes is that some lesion-related changes in behavior are due to interference with vagal function. Both the AP and the adjacent NST are innervated by vagal afferent fibers (Kalia and Sullivan, 1982; Beckstead and Norgren, 1982; Shapiro and Miselis, 1985a), some of which are of abdominal origin. In addition, the dendritic field of the dorsal vagal motor nucleus extends to the ventral border of the AP (Shapiro and Miselis, 1985b). Therefore, AP lesions may also interfere with vagal efferent function.

We have studied lesioned and sham-lesioned rats with and without total subdiaphragmatic vagotomies to examine the possibilities that damage to general visceral afferent fibers or to vagal motor function can account for overingestion of palatable foods by AP-lesioned rats. We have also examined the feeding of lesioned and control rats with open gastric fistulas (sham feeding) to further assess the role of impaired gastrointestinal feedback in lesion-induced overingestion (Edwards and Ritter, 1986). The results of our experiments do not support the hypothesis that lesion-induced overingestion is causally related to destruction of primary visceral afferent or efferent neurons. The results of our vagotomy experiments can be summarized as follows. First, unlesioned rats with total subdiaphragmatic vagotomies do not overingest highly palatable foods. This finding suggests that indiscriminate destruction of abdominal vagal afferents or efferents is not responsible for AP lesion-induced overingestion of palatable foods. Furthermore, AP-lesioned rats with vagotomies eat significantly more palatable food than either vagotomized or non-vagotomized sham-lesioned rats. If deranged vagal efferent function were the cause of the altered feeding behavior of AP-lesioned rats, one would expect vagotomy to abolish the overingestion. It does not. Finally, if overingestion by AP-lesioned rats was due to attenuation of vagal or non-vagal gastrointestinal sensory cues, then actually removing such cues in intact and lesioned rats should cause both groups of animals to respond similarly to a palatable solution. During sham feeding, ingesta do not accumulate in the stomach, and gastric emptying is minimal or absent. Therefore, sham feeding should approximate a condition in which GI feedback is greatly reduced. AP-lesioned rats consume more of a palatable sucrose solution than do sham-lesioned controls even when both groups are sham fed. In fact, opening of the gastric fistulas appears to potentiate ingestion in lesioned rats more than it does in sham-lesioned control rats. These results indicate that increased feeding due to removal of neural feedback from the gastrointestinal tract is additive with the increased intake produced by AP lesions. Consequently, it is not likely that overingestion of palatable foods by lesioned rats is the direct result of impaired visceral afferent function. Rather, these data are consistent with the hypothesis that overingestion of some foods by lesioned rats is caused by an alteration in orosensory responsiveness.

The role of general visceral afferent function in other behavioral manifestations of the AP lesion has not been systematically investigated. However, abdominal vagotomy is not associated with increased drinking to thirst challenges and is in fact associated with decreased drinking in response to some hydrational challenges (Kraly et al., 1975; Kraly, 1979). Therefore, there are no data to suggest that interference with abdominal vagal sensory neurons is involved in increased drinking by AP-lesioned rats. Likewise, section of the abdominal vagus or its branches does not cause increased sodium appetite. In fact, hepatic vagotomy is reported to decrease sodium appetite (Contreras and Kosten, 1981). Therefore, removal of abdominal vagal afferents does not mimic the effect of AP lesions on sodium appetite. Of course, none of the experiments mentioned rule out participation of thoracic vagal afferents in AP lesion-induced behavioral changes. Nevertheless, the available data argue against interpreting AP lesion-induced behavioral changes as simply the result of vagal deafferentation.

B. Role of Parabrachial Connections

The area postrema and adjacent mNST possess numerous afferent and efferent connections with other hindbrain structures as well as with more rostral brain regions. It is therefore possible that individual behavioral changes, evident following AP lesions, are referable to interference with specific connections of the AP, mNST, or both. Most neuroanatomical work indicates that both the AP and adjacent mNST provide dense projections to the lateral parabrachial nucleus (PBN) and a less dense innervation to the medial parabrachial nucleus (MPN) (Van der Kooy and Koda, 1983; Shapiro and Miselis, 1985b). The lateral parabrachial projection is by far the most prominent ascending projection of the AP and mNST (Shapiro and Miselis, 1985b) and it appears that this region is a major relay for visceral afferent information headed for the forebrain (Saper and Loewy, 1980). Therefore, if the behavioral effects of AP lesions involve hindbrain-forebrain connections through the parabrachial region, then lesions of the LPN or MPN may either mimic or antagonize some behavioral aspects of AP lesions. In fact, large lesions of the LPN do produce overdrinking in response to angiotensin II (Ohman and Johnson, 1983; Ritter and Ladenheim, 1983). These large LPN lesions do not cause deficits in glucoprivic feeding (Ritter and Ladenheim, 1983). On the other hand, large MPN lesions do not cause overdrinking in response to angiotensin II but do impair glucoprivic feeding (Ritter and Ladenheim, 1983). It seems quite plausible that AP lesion-induced alterations in drinking are caused by interference with AP efferents to the PBN. The impairment of glucoprivic feeding observed after MPN lesions is more difficult to assess because so many brain lesions have been shown to interfere with glucoprivic feeding that the specificity of lesion effects on this control is always suspect.

Neither large LPN or MPN lesions consistently cause overconsumption of palatable foods. Nevertheless, recent work in our laboratory indicates that lesions of the dorsal portions of the rostral end of the parabrachial area attenuate or abolish overingestion of palatable foods by rats with AP lesions (Edwards and Ritter, unpublished). Thus it may be that some parabrachial neurons are essential for expression of overingestion by AP-lesioned rats. It would, however, be premature to conclude that the intact AP or adjacent NST acts via the parabrachial area to control food intake. Nonetheless, it is interesting that the region of the parabrachial area to which our lesions were directed has recently been shown to contain neurons that are activated by both gustatory stimulation and stimulation of the caudal NST (Herman and Rogers, 1985). Since the caudal NST contains cell bodies damaged by AP lesions, as well as efferents from the AP itself, it may be that our parabrachial lesions interfere with AP or caudal NST input to pontine gustatory neurons. Such connections with the gustatory system do support speculation that some effects of AP lesions may be due to changes in responsiveness to gustatory stimuli.

C. Relative Importance of AP versus NST Damage for Lesion-induced Overingestion of Palatable Foods

As previously mentioned, the intimate association of the AP and NST generally precludes exclusive damage to one structure without damage to the other. Consequently, we cannot be sure whether damage to the AP or to the NST accounts for the behavioral effects of the AP lesion. Nevertheless, several shreds of circumstantial evidence suggest that the overingestion of palatable foods observed after lesions of the AP and adjacent NST may require damage to the AP. First, Berthoud and Powley (1985) have recently reported results of behavioral testing of bipipyridyl mustard (BPM)-treated rats. This substance appears to be a selective neurotoxin for a few specific brain regions, including the NST adjacent to the AP. Their conventional histology from BPM-treated rats suggests that the AP is intact, but the cell bodies and neuropil of the adjacent NST have disappeared almost completely. Hence, the BPM-lesioned rat may represent a preparation in which those portions of the NST likely to be damaged by thermal lesions of the AP are destroyed, leaving the AP itself intact. Berthoud and Powley, however, found that BPM-lesioned rats did not overingest palatable foods and responded normally to glucoprivation. If, indeed, the AP and its connections are intact in the BPM-lesioned rat, then this work suggests that NST damage probably does not account for overingestion of palatable foods.

A second piece of circumstantial evidence suggesting that damage to the AP may account for the overingestion of palatable foods following lesions comes from our work with hundreds of thermally lesioned rats. We routinely test all of our lesioned animals for their response to sweet foods or solutions. In subsequent histological analyses we have encountered a significant number of rats in which the AP was not totally destroyed and consequently there was no histologically discernible damage to the NST. Nonetheless, many rats with extensive damage to but incomplete destruction of the AP consumed nearly double the amount of sweet food or solution as their respective controls. These observations, together with the results described above, suggest that the AP may be the critical tissue destroyed by lesions causing overingestion of palatable foods.

D. Overingestion and Selective Neurotoxic Damage to the AP and Adjacent NST.

As mentioned in Section IV,D, BPM, a neurotoxin which damages the NST but does not produce apparent damage to the AP, does not cause overingestion of preferred foods (Berthoud and

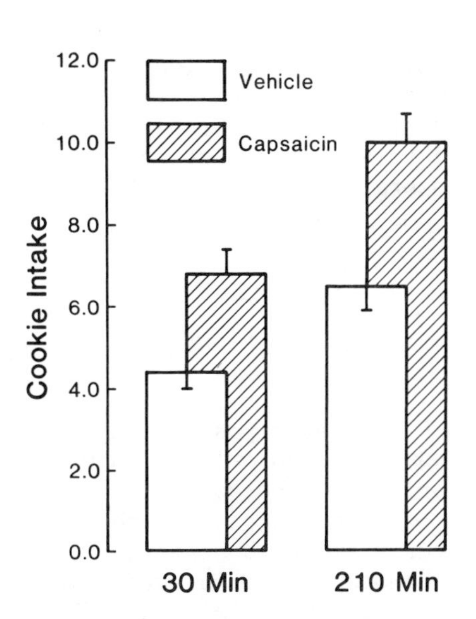

Fig. 7. Enhanced intake of cookies by rats injected with capsaicin directly into the dorsal hindbrain. Capsaicin-injected rats consumed significantly more than vehicle-injected rats at both 30 and 210 min after cookie presentation. Capsaicin-injected rats did not display any of the other alterations of ingestive behavior associated with physically produced AP lesions (South and Ritter, 1983).

Powley, 1985). Nevertheless, some other neurotoxins seem to produce behavioral effects similar to those produced by AP lesions. For example, Kanarek and Marks-Kaufman (1981) have shown that rats treated with monosodium glutamate (MSG) as neonates exhibit exaggerated preference for sucrose-containing diets as adults. MSG destroys glutamate-sensitive neurons in the several brain regions, including the AP (Olney et al., 1977; Meharg et al., 1985). Although MSG-induced damage is not limited to the AP, the behavioral sequels of this toxin are consistent with selective physical destruction of the AP. Thus, investigation of the role of excitatory amino acids in the behavioral function of the AP may prove fruitful for identifying specific neurons involved in ingestive control.

Another neurotoxin that has been reported to produce overingestion of preferred foods following direct injection into the AP and immediate vicinity is capsaicin (South and Ritter, 1983) (Fig. 7). Capsaicin treatment does not cause chronic weight loss, overdrinking, or depressed responsiveness to glucoprivation. When injected systemically in neonatal (Jancso et al., 1977; Jancso and Kiraly, 1980) or adult rats (Dinh and Ritter, 1985) capsaicin destroys small unmyelinated primary sensory neurons, including vagal sensory neurons which terminate in the AP and NST. However, Dinh and S. Ritter (S. Ritter, personal communication) have found that systemic injection of capsaicin also damages some neurons that do not appear to be primary sensory cells. It is possible that the overingestion produced by capsaicin is due to vagal afferent damage and, therefore, is not the same phenomenon observed after physical destruction of the AP. It is also possible, however, that direct application of capsaicin to the AP may selectively destroy some neurons that are not primary sensory cells. Such a selective effect of local capsaicin application may provide a useful tool to dissect the neural substrates involved in the behavioral effects of physical lesions.

V. CONCLUSION

The area postrema and immediately adjacent portions of the NST have anatomical features and connections that seem appropriate to the control of a variety of physiological activities including ingestive behavior. Putative roles for these structures in the control of food intake range from that of primary chemosensor to integrator of diverse viscerosensory cues. The neural connections through which the AP and adjacent NST may influence ingestion are still uncertain. Nevertheless, the currently available evidence suggests that lesions of these structures enhance behavioral responses to some of the sensory cues that drive ingestion. Convincing evidence concerning the mechanism(s) for such changes in responsiveness require further experimentation. Likewise, assurance that the intact AP and adjacent NST participate in the control of ingestion awaits the results of further studies.

REFERENCES

Adachi, A., and M. Kobashi. 1985. Chemosensitive neurons within the area postrema of the rat. *Neurosci. Lett.* 55: 137-140.

Anand, B.K., and J.R. Brobeck. 1951. Hypothalamic control of food intake in rats and cats. *Yale J. Biol. Med.* 24: 123-140.

Berger, B.D., C.D. Wise, and L. Stein. 1973. Area postrema damage and bait shyness. *J. Comp. Physiol. Psychol.* 81: 21-26.

Bernstein, I.I., and R.A. Sigmundi. 1980. Tumor anorexia: a learned food aversion? *Science* 209: 416-418.

Berthoud, H-R., and T.L. Powley. 1985. Altered plasma insulin and glucose after obesity-producing bipiperidyl brain lesions. *Am. J. Physiol.* 248: R46-R53.

Bird, E., C.C. Cardone, and R.J. Contreras. 1983. Area postrema lesions disrupt food intake induced by cerebroventricular infusions of 5-thioglucose in the rat. *Brain Res.* 270: 193-196.

Borison, H.L. 1974. Area postrema: chemoreceptor trigger zone for vomiting--is that all? *Life Sci.* 14: 1807-1817.

Broadwell, R.D., and M.W. Brightman. 1976. Entry of peroxidase into neurons of the central and peripheral nervous system from extracerebral and cerebral blood. *J. Comp. Neurol.* 166: 257-284.

Brobeck, J.R., J. Tepperman, and C.N.H. Long. 1943. Experimental hypothalamic hyperphagia in the albino rats. *Yale J. Biol. Med.* 15: 831-853.

Buijs, R.M. 1978. Intra and extrahypothalamic vasopressin and oxytocin pathways in the rat. *Cell Tissue Res.* 192: 423-435.

Cajal, S.R. 1909. *Histologie du systeme Nerveux de L'Homme et des Vertebres.* A. Maloine, Paris.

Carlisle, H.J., and R.W. Reynolds. 1961. Effect of amphetamine on food intake in rats with brainstem lesions. *Am. J. Physiol.* 201: 965-967.

Chang, F-C.T., and T.R. Scott. 1984. Conditioned taste aversions modify neural responses in the rat nucleus tractus solitarius. *J. Neurosci.* 4: 1850-1862.

Clemente, C.D., and V.L. van Breeman. 1955. Nerve fibers in the area postrema of cat, rabbit, guinea pig and rat. *Anat. Rec.* 123: 65-79.

Coil, J.D., R.C. Rogers, J. Garcia, and D. Novin. 1978. Conditioned taste aversions: vagal and circulatory mediation of toxic unconditioned stimulus. *Behav. Biol.* 24: 509-519.

Contreras, R.J., and T. Kosten. 1981. Changes in salt intake after abdominal vagotomy: Evidence for hepatic sodium receptors. *Physiol. Behav.* 26: 575-582.

Contreras, R.J., P.W. Stetson, and T. Kosten, T. 1981. Changes in salt intake with lesions of the area postrema and nucleus of the solitary tract. *Brain Res.* 211: 355-356.

Contreras, R.J., R.M. Beckstead, and R. Norgren. 1982. The central projections of the trigeminal, facial, glossopharyngeal and vagus nerves: An autoradiographic study in the rat. *J. Auton. Nerv. Syst.* 6: 303-322.

Contreras, R.J., E. Fox, and M.L. Drugovich. 1982. Area postrema lesions produce feeding deficits in the rat: Effects of preoperative dieting and 2-deoxy-D-glucose. *Physiol. Behav.* 29: 875-884.

Davison, J.S. 1972. Response of single vagal afferent fibers to mechanical and chemical stimulation of the gastric and duodenal mucosa in cats. *Q. J. Exp. Physiol.* 57: 405-416.

Dey, M.S., G.L. Edwards, R.I. Krieger, B.E. Renzi, and R.C. Ritter. 1982. Paraquat-induced conditioned taste aversions and weight loss: Mediation by the area postrema. *The Toxicologist* 3: 96.

Dinh, T., and S. Ritter. 1985. Capsaicin induces neuronal degeneration in the brain and spinal cord of adult rats. *Soc. Neurosci. Abstr.* 11: 349.

Edwards, G.L., and R.C. Ritter. 1981. Ablation of the area postrema causes exaggerated consumption of preferred foods in the rat. *Brain Res.* 216: 265-276.

Edwards, G.L., and R.C. Ritter. 1982. Area postrema lesions increase drinking to angiotensin and extracellular dehydration. *Physiol. Behav.* 29: 943-947.

Edwards, G.L., and R.C. Ritter. 1986. Area postrema lesions: Cause of overingestion is not altered visceral nerve function. *Am. J. Physiol.* In press.

Edwards, G.L., E.E. Ladenheim, and R.C. Ritter. 1985. Dorsal hindbrain participation in CCK-induced satiety. *Soc. Neurosci. Abstr.* 11: 37.

Edwards, G.L., E.E. Ladenheim, and R.C. Ritter. 1986. Dorsal hindbrain participation in cholecystokinin-induced satiety. *Am. J. Physiol.* In press.

Ewart, W.R., and D.L. Wingate. 1983a. Central representation and opioid modulation of gastric mechanoreceptor activity in the rat. *Am. J. Physiol.* 244: G27-G32.

Ewart, W.R., and Wingate, D.L. 1983b. Central representation of arrival of nutrient in the duodenum. *Am. J. Physiol.* 246: G750-G756.

Flynn, F.W., and H. Grill. 1983. Insulin elicits ingestion in decerebrate rats. *Science* 220: 188-189.

Garcia, J., and D.J. Kimeldorf. 1960. Some factors which influence radiation-conditioned behavior in rats. *Radiat. Res.* 12: 719-727.

Giza, B.K., and T.R. Scott. 1983. Blood glucose selectively affects taste-evoked activity in the rat nucleus tractus solitarius. *Physiol. Behav.* 31: 643-650.

Grill, H.J., and R. Norgren. 1978a. The taste reactivity test. II. Mimetic responses to gustatory stimuli in chronic thalamic and chronic decerebrate rats. *Brain Res.* 143: 281-297.

Grill, H.J., and R. Norgren. 1978b. Chronic decerebrate rats demonstrate satiation but not bait-shyness. *Science* 201: 267-269.

Hamilton, R.B., and R. Norgren. 1984. Central projections of gustatory nerves in the rat. *J. Comp. Neurol.* 222: 560-577.
Harding, J.W., L.P Stone, and J.W. Wright. 1981. The distribution of angiotensin II binding sites in rodent brain. *Brain Res.* 205: 265-274.
Harding, R., and B.F. Leek. 1973. Central projection of gastric afferent vagal inputs. *J. Physiol. (Lond.)* 228: 73-90.
Hermann, G.E., and R.C. Rogers. 1985. Convergence of vagal and gustatory afferent input within the parabrachial nucleus of the rat. *J. Auton. Nerv. Syst.* 13: 1-17.
Hoysoya, H., and M. Matsushita. 1981a. A direct projection from the hypothalamus to the area postrema in the rat, as demonstrated by the horseradish peroxidase method and autoradiographic methods. *Brain Res.* 214: 144-149.
Hoysoya, H., and M. Matsushita. 1981b. Brainstem projections from the lateral hypothalamic area in the rat, as studied with autoradiography. *Neurosci. Lett.* 24: 111-116.
Hunt, E.L., H.W. Carroll, and D.J. Kimeldorf. 1965. Humoral mediation of radiation-induced motivation in parabiont rats. *Science* 150: 1747-1748.
Hyde, T.M., and R.R. Miselis. 1983. Effects of area postrema/ caudal medial nucleus of the solitary tract lesions on food intake and body weight. *Am. J. Physiol.* 244: 577-587.
Hyde, T.M., and R.R. Miselis. 1984a. Changes in body composition following area postrema/caudal medial nucleus of the solitary tract lesions in rats. *Soc. Neurosci. Abstr.* 10: 1014.
Hyde, T.M., and R.R. Miselis. 1984b. Area postrema and adjacent nucleus of the solitary tract in water and sodium balance. *Am. J. Physiol.* 247: R173-R182.
Jacquin, M.F., K. Semba, R.W. Rhoades, and M.D. Egger. 1982. Trigeminal primary afferents project bilaterally to dorsal horn and ipsilaterally to cerebellum, reticular formation, and cuneate, solitary, supratrigeminal and vagal nuclei. *Brain Res.* 246: 285-291.
Jansco, G., and E. Kiraly. 1980. Distribution of chemosensitive primary sensory afferents in the nervous system of the rat. *J. Comp. Neurol.* 190: 781-792.
Jansco, G., E. Kiraly, and A. Jansco-Gabor. 1977. Pharmacologically induced selective degeneration of chemosensitive primary sensory neurons. *Nature* 270: 741-743.
Kalia, M., and M.-M. Mesulam. 1980. Brainstem projections of the afferent and efferent fibers of the vagus nerve in the cat: II. Laryngeal, tracheobronchial, pulmonary, cardiac and gastrointestinal branches. *J. Comp. Neurol.* 193: 523-533.

Kalia, M., and J.M. Sullivan. 1982. Brainstem projections of the sensory and motor components of the vagus nerve in the rat. *J. Comp. Neurol.* 211: 248-264.

Kanarek, R.B., and R. Marks-Kaufman. 1981. Increased carbohydrate consumption induced by neonatal administration of monosodium glutamate to rats. *Neurobehav. Toxicol. Teratol.* 3: 343-350.

Kenney, N.J., J.N. Kott, and C.L. Ganfield. 1984. Diet selection following area postrema/nucleus of the solitary tract lesions. *Physiol. Behav.* 32: 749-753.

Klara, P.M., and K.R. Brizzee. 1977. Ultrastructure of the feline area postrema. *J. Comp. Neurol.* 171: 409-432.

Kott, J.N., and N.J. Kenney. 1984. Further evaluation of diet selection in rats following area postrema ablation. *Soc. Neurosci. Abstr.* 10: 1015.

Kraly, F.S. 1979. Abdominal vagotomy inhibits osmotically induced drinking in the rat. *J. Comp. Physiol. Psychol.* 92: 999-1013.

Kraly, F.S., J. Gibbs, and G.P. Smith. 1975. Disordered drinking after abdominal vagotomy in rats. *Nature* 258: 226-228.

Meharg, S., T.T. Dinh, and S. Ritter. 1985. Neurotoxicity of systemic monosodium glutamate in the adult rat: analysis with silver stain degeneration techniques. *Soc. Neurosci. Abstr.* 11: 1195.

Mei, N. 1978. Vagal glucoreceptors in the small intestine of the cat. *J. Physiol. (Lond.)* 282: 485-506.

Mizuno, Y., Y. Oomura, K. Hattori, T. Minami, and N. Shimizu. 1983. Glucose responding neurons in the nucleus tractus solitarius of the rat. *Neurosci. Lett.* Suppl. 13: 561.

Morest, D.K. 1960. A study of the area postrema with Golgi methods. *Am. J. Anat.* 107: 291-303.

Morest, D.K. 1967. Experimental study of the projections of the nucleus of the tractus solitarius and the area postrema in the cat. *J. Comp. Neurol.* 130: 277-293.

Murabe, Y., K. Nishida, and Y. Sano. 1981. Cells capable of uptake of horseradish peroxidase in some circumventricular organs of the cat and rat. *Cell Tissue Res.* 219: 85-92.

Nonavinakere, V.K., and R.C. Ritter. 1984. Destruction of the area postrema does not abolish glucoprivic control of feeding or blood glucose. *Soc. Neurosci. Abstr.* 10: 654.

Norgren, R. 1978. Projections from the nucleus of the solitary tract in the rat. *Neurosci.* 3: 207-218.

Norgren, R. 1984. Central neural mechanisms of taste. In *Handbook of Physiology*, Section I. *The Nervous System*, Volume III. *Sensory Processes*, Part I, ed. J.M. Brookhart and V.B. Mountcastle, 1087-1128. American Physiological Society, Bethesda, Maryland.

Ohman, L.E., and A.K. Johnson. 1983. Evidence for involvement of the lateral parabrachial nucleus in the maintenance of body fluid balance. *Soc. Neurosci. Abstr.* 9: 199.
Olney, J.W., V. Rhee, and T. De Gubareff. 1977. Neurotoxic effects of glutamate on mouse area postrema. *Brain Res.* 120: 151-157.
Ossenkopp, K.-P. 1983. Taste aversions conditioned with gamma radiation: attenuation by area postrema lesions in rats. *Behav. Brain Res.* 7: 297-305.
Ricardo, J.A., and E.T. Koh. 1978. Anatomical evidence of direct projections from the nucleus of the solitary tract to the hypothalamus, amygdala and other forebrain structures in the rat. *Brain Res.* 153: 1-26.
Ritter, R.C. 1984. The role of the rat in research on the control of ingestion. *J. Anim. Sci.* 59: 1373-1380.
Ritter, R.C., and G.L. Edwards. 1984. Area postrema lesions cause overingestion of palatable foods but not calories. *Physiol. Behav.* 32: 923-927.
Ritter, R.C., and E.E. Ladenheim. 1983. Behavioral features of area postrema lesions are reproduced separately by lesions of the parabrachial nuclei. *Soc. Neurosci. Abstr.* 9: 185.
Ritter, R.C., J.J. McGlone, and K.W. Kelley. 1980. Absence of lithium-induced taste aversion after area postrema lesion. *Brain Res.* 201: 501-506.
Ritter, R.C., P.G. Slusser, and S. Stone. 1981. Brain glucoreceptor controlling feeding and blood glucose: location in the hindbrain. *Science* 213: 451-453.
Rogers, R.C., H. Kita, L.L. Butcher, and D. Novin. 1980. Afferent projections of the dorsal motor nucleus of the vagus. *Brain Res. Bull.* 5: 365-373.
Ross, C.A., D.A. Ruggerio, and D.J. Reis. 1981. Afferent projections to the cardiovascular portions of the nucleus tractus solitarius in the rat. *Brain Res.* 223: 402-408.
Roth, G.I., and W.S. Yamamoto. 1968. The microcirculation of the area postrema in the rat. *J. Comp. Neurol.* 133: 329-340.
Saper, C.B., and A.D. Loewy. 1980. Efferent connections of the parabrachial nucleus in the rat. *Brain Res.* 197: 291-317.
Shapiro, R.E., and R.R. Miselis. 1985a. The central organization of the vagus nerve innervating the stomach of the rat. *J. Comp. Neurol.* 238: 473-488.
Shapiro, R.E., and R.R. Miselis. 1985b. The central neural connections of the area postrema in the rat. *J. Comp. Neurol.* 234: 344-364.

Smith, J.C., G.R. Hollander, and A.C. Spector. 1981. Taste aversions conditioned with partial body radiation exposures. *Physiol. Behav.* 27: 903-913.

Smith, G.P., C. Jerome, B.J. Cushin, R. Eterno, and K.J. Simansky. 1981. Abdominal vagotomy blocks the satiety effects of cholecystokinin in the rat. *Science* 213: 1036-1037.

Snowdon, C.D. 1970. Gastointestinal sensory and motor control of food intake. *J. Comp. Physiol. Psychol.* 71: 68-76.

Somana, R., and F. Walberg. 1979. Cerebellar afferents from the nucleus of the solitary tract. *Neurosci. Lett.* 11: 41-47.

South, E.H., and R.C. Ritter. 1983. Overconsumption of preferred foods following capsaicin pretreatment of the area postrema and adjacent nucleus of the solitary tract. *Brain Res.* 288: 243-250.

Spacek, J., and J. Parizek. 1969. The fine structure of the area postrema of the rat. *Acta. Morphologica Acad. Sci. Hung.* 17: 17-34.

Swanson, L.W., and H.G.J.M Kuypers. 1980. The paraventricular nucleus of the hypothalamus: cytoarchitectonic subdivisions and organization of projections to the pituitary, dorsal vagal complex and spinal cord, as demonstrated by retrograde fluorescence double-labeling methods. *J. Comp. Neurol.* 194: 555-570.

Tervo, T., F. Joo, A. Palkama, and L. Salminen. 1978. Penetration barrier to sodium fluorescein and fluorescein-labeled dextrans of various molecular sizes in brain capillaries. *Experientia* 35: 252-254.

Tomoyasu, N., and N.J. Kenney. 1985. Attenuation of hypophagia following area postrema ablation by continuous diet change. *Soc Neurosci. Abstr.* 11: 345.

Van der Kooy, D. 1984. Area postrema: site where CCK acts to decrease food intake. *Brain Res.* 295: 345-347.

Van der Kooy, D., and L.Y. Koda. 1983. Organization of the projections of a circumventricular organ: the area postrema in the rat. *J. Comp. Neurol.* 219: 328-338.

Van der Kooy, D., J.F. McGinty, L.Y. Koda, C.R. Gerfen, and F.E. Bloom. 1982. Visceral cortex: a direct connection from prefrontal cortex to the solitary nucleus in the rat. *Neurosci. Lett.* 33: 123-127.

van Houten, M., B.I. Posner, B.M. Kopriwa, and J.R. Brawer. 1979. Insulin binding sites in the rat brain: *In vivo* localization to the circumventricular organs by quantitative autoradiography. *Endocrinology* 105: 666-673.

Wilson, J.T. 1906. On the anatomy of the calamus region of the human bulb; with and account of a hitherto undescribed "nucleus postremus." *J. Anat. Physiol.* 40: 357-386.

Yaxley, S., E.T. Rossl, Z.J. Sienkiewicz, and T.R. Scott. 1985. Satiety does not affect gustatory activity in the nucleus of the solitary tract of the alert monkey. *Brain Res*. 347: 85-93.

Zarbin, M.A., R.B. Innis, J.K. Wamsley, S.H. Snyder, and M.J. Kuhar. 1983. Autoradiographic localization of cholecystokinin receptors in rodent brain. *J. Neurosci*. 4: 877-906.

Zarbin, M.A., M.J. Kuhar, T.L. O'Donohue, S.S. Wolf, and Moody, T.W. 1985. Autoradiographic localization of (^{125}I Tyr^4) bombesin-binding sites in rat brain. *J. Neurosci*. 5: 1985.

Chapter 6

NEURONAL ACTIVITY RELATED TO THE CONTROL OF FEEDING

Edmund T. Rolls

I. INTRODUCTION

Since early in this century it has been known from clinical evidence that damage to the base of the brain can influence food intake and body weight. It was later demonstrated that one critical region is the ventromedial hypothalamus, where bilateral lesions in animals lead to hyperphagia and obesity (see Grossman, 1967,1973; Rolls, 1981b). In 1951 Anand and Brobeck discovered that bilateral lesions of the lateral hypothalamus can produce a reduction in feeding and body weight. In the 1950s and 1960s evidence of this type led to the view that food intake is controlled by two interacting "centers," a feeding center in the lateral hypothalamus and a satiety center in the ventromedial hypothalamus (see Stellar, 1954; Grossman, 1967, 1973).

Problems with this evidence for a dual center hypothesis of the control of food intake soon appeared. For example, lesions of the lateral hypothalamus that were effective in producing aphagia also damaged fiber pathways coursing nearby, such as the dopaminergic nigrostriatal bundle, and damage to these pathways outside the lateral hypothalamus could produce aphagia (Marshall et al., 1974). Thus, by the middle 1970s it was clear that evidence from lesion studies for a lateral hypothalamic feeding center was not straightforward, since at least part of the effect of the lesions was due to damage to the fibers of passage traveling through or near the lateral hypothalamus (Stricker and Zigmond, 1976). It was not clear by this time what, if any, role the hypothalamus played in feeding. In more recent investigations it has been possible to damage the cells in the lateral hypothalamus without damaging fibers of passage, using the neurotoxin ibotenic acid (Winn et al., 1984; Dunnett et al., 1985). With this technique, it has been shown that damage to lateral hypothalamic cells does produce a long-lasting decrease in food intake and body weight, and that this is not associated with dopamine depletion due to damage to dopamine pathways, or with the akinesia and sensorimotor deficits produced by damage to the dopamine systems (Winn et al., 1984; Dunnett et al., 1985). Thus the recent lesion evidence does suggest that the lateral hypothalamus is involved in the control of feeding and body weight.

The evidence just described implicates the hypothalamus in the control of food intake and body weight, but does not show what functions important in feeding are being performed by the hypothalamus and by other brain areas. More direct evidence on the neural processing involved in feeding, based on recordings of the activity of single neurons in the hypothalamus and other brain regions such as the amygdala, prefrontal cortex, and striatum during feeding, is described next.

II. NEURONAL ACTIVITY IN THE LATERAL HYPOTHALAMUS DURING FEEDING

We have found that there is a population of neurons in the lateral hypothalamus and substantia innominata of the monkey with responses related to feeding (see Rolls, E.T., 1981a,b; 1983). These neurons, which made up 13.6% of one sample of 764 hypothalamic neurons, responded to the taste and/or sight of food. The response of these neurons to taste was such that they responded only when certain substances, for example, glucose solution, but not water or saline, were in the mouth, and their firing rates were related to the concentration of the substance to which they were responding (Rolls, E.T., et al., 1980). The neurons responding to taste (4.3% of the sample of 764 neurons) did not respond simply in relation to mouth movements. The responses of the neurons associated with the sight of food (11.8% of the sample of 764 neurons) occurred as soon as the monkey saw food, before the food was in its mouth, and occurred only to foods and not to non-food objects (Rolls, E.T., et al., 1976, 1980). Some of these neurons (2.5% of the total sample) responded to both the sight and taste of food (Rolls, E.T., et al., 1976, 1980). The finding that there are neurons in the lateral hypothalamus of the monkey which respond to the sight of food has been confirmed by Ono et al. (1980).

A. Effect of Hunger

The neurons with feeding-related responses respond to the sight or taste of food only if the monkey is hungry (Burton et al., 1976). Thus, neuronal responses to food occuring in the hypothalamus depend on the motivational state of the animal. In the monkey, several signals reflect the motivational state and perform this modulation. One signal is gastric distension, as shown by the finding that after a monkey has fed to satiety, relief of gastric distension by drainage of ingested food through a gastric cannula leads to the almost immediate resumption of feeding (Gibbs et al., 1981). Because feeding is reinstated so rapidly, it is probably due to relief of gastric distension rather than to the altered availability of metabolites in the gastrointestinal tract. Another signal is provided by the presence of food in the duodenum and later parts of the gut, as shown by the finding that if ingested food is allowed to drain from a duodenal cannula (situated near the pylorus), then normal satiety is not shown, and the monkey feeds almost continuously (Gibbs et al., 1981). Under these conditions, food does not accumulate normally in the stomach, showing that influences of duodenal or more distal origin are required to control gastric emptying and thus allow gastric distension to play a role in satiety. In this way an "enterogastric loop" contributes to satiety, as shown by the finding that duodenal infusions of food at rates similar to those of gastric emptying reduce the rate of feeding (Gibbs et al., 1981). Other signals influencing hunger and satiety presumably reflect the metabolic state of the animal and may include glucose and insulin levels (Le Magnen, 1980; Friedman and Stricker, 1976; Woods et al., 1980).

The hypothalamus is not necessarily the first stage at which hunger modulates processing of such environmental stimuli. To investigate whether hunger modulates neuronal responses in those parts of the visual system through which visual information is likely to reach the hypothalamus, the activity of neurons in the inferior temporal visual cortex was recorded in the testing situations described above. We found that here the neuronal responses to visual stimuli were not dependent on hunger (Rolls, E.T., et al., 1977). Nor were neuronal responses in the amygdala, which connects the inferior temporal visual cortex to the hypothalamus, found to depend on hunger (Sanghera et al., 1979), although there still could be other prehypothalamic areas in which motivational modulation of processing occurs. However, in the orbitofrontal cortex, which receives inputs from the inferior temporal visual cortex and projects into the hypothalamus (see below and Russchen et al., 1985), neurons with visual responses to food are found, and these visual responses are modulated by hunger (Thorpe et al., 1983; see below). Thus for visual processing, neuronal responsiveness only at late stages of sensory processing and in the hypothalamus has been found to be modulated by hunger. The situation for taste processing in the primate is comparable, in that it is only at high levels of gustatory information processing, and in the hypothalamus, that neuronal responses to taste are modulated by hunger. This is further evidence that the hypothalamic neurons which respond to food are likely to be involved in the responsiveness of the organism to food (see also Rolls, E.T., 1982).

B. Sensory-Specific Modulation of the Responsiveness of Lateral Hypothalamic Neurons and of Appetite

During these experiments on satiety we observed that if a lateral hypothalamic neuron had ceased to respond to a food on which the monkey had been fed to satiety, then the neuron might still respond to a different food. This occurred for neurons having responses associated with the taste (see Rolls, E.T., 1981b) or sight (see Rolls, E.T., and Rolls, 1982) of food (see also Rolls et al., 1986). Corresponding to this neuronal specificity of the effects of feeding to satiety, the monkey rejected the food on which it had been fed to satiety, but accepted other foods that it had not been fed.

As a result of these neurophysiological and behavioral observations showing the specificity of satiety in the monkey, we performed experiments to determine whether, in man, satiety is similarly specific to foods eaten. We found that the pleasantness of the taste of food eaten to satiety decreased more than it did for foods which had not been eaten (Rolls, B.J., et al., 1981). One implication of this finding is that if one food is eaten to satiety, appetite reduction for other foods is often incomplete, and that in man, too, at least some of the other foods will be eaten. This has been confirmed in an experiment in which either sausages or cheese with crackers were eaten for lunch. The appetite for the food eaten decreased more than for the food not eaten, and when an unexpected second course was offered, more was eaten if a subject had not been given that food in the first course than if he had (98% vs. 40% of the first course intake eaten in the second course, $P < 0.01$, Rolls, B.J., and Rolls et al., 1981). A further implication of these findings is that if a variety of foods is available, the total amount consumed will be more than when only one food is offered repeatedly. This prediction has been confirmed in a study in which humans ate more when offered a variety of sandwich fillings or types of yogurt that differed in taste, texture, and color (Rolls, B.J., and Rowe et al., 1981). It has also been confirmed by a study in which humans were offered a relatively normal meal of four courses. Intake was significantly enhanced by the change of food at each course (Rolls, B.J., et al., 1984). Because sensory factors such as similarity of color, shape, flavor, and texture usually have a greater influence than metabolic equivalence in terms of protein, carbohydrate, and fat content on how foods interact in this type of satiety, it has been termed "sensory-specific satiety" (Rolls, E.T., and Rolls, 1977, 1982; Rolls, B.J., et al., 1982; Rolls, B.J., and Rolls et al., 1981; Rolls, B.J., and Rowe et al., 1981). It should be noted that this effect is distinct from alliesthesia, which is a change in the pleasantness of sensory inputs produced by internal signals (such as glucose in the gut) (see Cabanac, 1971; Cabanac and Fantino, 1977; Cabanac and Duclaux, 1970), whereas sensory-specific satiety is a change in the pleasantness of sensory inputs that is accounted for at least partly by the external sensory stimulation received (such as the taste of a particular food), and which, as shown above, is at least partly specific to the external sensory stimulation received.

The parallel between feeding in humans and the neurophysiology of hypothalamic neurons in monkeys has been extended by the observations that in humans, sensory-specific satiety occurs for the sight as well as for the taste of food (Rolls, B.J., et al., 1982). Further, to complement the finding that some neurons in the hypothalamus respond differently to food and to water (Rolls et al., unpublished observations), and that satiety with water can decrease the responsiveness of hypothalamic neurons which respond to water, it has been shown that in man, motivation-specific satiety can also be detected. For example, satiety with water decreases the pleasantness of the sight and taste of water, but not of food (Rolls, E.T., and Rolls et al., 1983).

The operation of sensory-specific satiety results in the enhancement of eating when a variety of foods is available. This may have been advantageous in evolution by ensuring that foods with different nutrients were consumed; today, when a wide variety of foods is readily available, it may be a factor leading to overeating and obesity in man. In a test of this in the rat, it has been found that variety itself can lead to obesity (Rolls, B.J., et al., 1983).

Advances in understanding the neurophysiological mechanisms of sensory-specific satiety are being made through analyses of information processing in the gustatory system, as described below.

In addition to the sensory-specific satiety which operates primarily during meals (see above) and in the post-meal period (Rolls, B.J., et al., 1984), there is now evidence for a long-term form of sensory-specific satiety (Rolls and de Waal, 1985). This was shown in a study in an Ethiopian refugee camp, in which it was found that refugees who had been in the camp for six months found the taste of their three regular foods less pleasant than the taste of three comparable foods which they had not been eating. The effect was a long-term form of sensory-specific satiety in that it was not found in refugees who had been in the camp and eaten the regular foods for two days (Rolls and de Waal, 1985). It is suggested that it is important to recognize the operation of long-term sensory-specific satiety in conditions such as these, for it may enhance malnutrition if the regular foods become less acceptable and so are rejected, exchanged for other less nutritionally effective foods or goods, or inadequately prepared. It may be advantageous under these circumstances to attempt to minimize the operation of long-term sensory-specific satiety by providing some variety, perhaps even with spices.

C. Effects of Learning

The responses of these hypothalamic neurons in the primate become associated with the sight of food through learning. This is shown by experiments in which the neurons come to respond to the sight of a previously neutral stimulus, such as a syringe, from which the monkey is fed orally; in which the neurons cease to respond to a stimulus if it is no longer associated with food (in extinction or passive avoidance); and in which the responses of these neurons remain associated with whichever visual stimulus is associated with food reward in a visual discrimination and its reversals (Mora et al., 1976). This type of learning is important, for it allows organisms to respond appropriately to environmental stimuli shown by previous experience to be foods. The brain mechanisms for this type of learning are discussed below.

The responses of these neurons suggest that they participate in responses to food. Further evidence is that the responses of these neurons occur with relatively short latencies of 150-200 milliseconds, and thus precede and predict the responses of the hungry monkey to food (Rolls E.T., et al., 1979).

D. Evidence That the Responses of These Neurons Are Related to the Reward Value of Food

Given that these hypothalamic neurons respond to food when it is rewarding, that is, when the animal will work to obtain it, the possibility arises that the neurons' responses are related to the reward value of food for the hungry animal. Evidence consistent with this comes from findings that electrical stimulation of some brain regions is rewarding: animals, including man, will work to obtain it (Olds, 1977; Rolls, E.T., 1975, 1976, 1979). At some sites, including the lateral hypothalamus, electrical stimulation appears to produce reward, which is equivalent to food for the hungry animal; that is, the animal will work hard to obtain the stimulation if it is hungry, but will work much less if it has been satiated (Olds, 1977; Hoebel, 1969). There is even evidence that the reward at some sites can mimic food for a hungry animal and at other sites water for a thirsty animal, since rats chose electrical stimulation at one hypothalamic site when hungry and at a different site when thirsty (Gallistel and Beagley, 1971). It was therefore very interesting when it was discovered that some of the neurons normally activated by food when the monkey was hungry were also activated by brain stimulation reward (Rolls, E.T., 1975, 1976; Rolls, E.T., et al., 1980). Thus there was convergence of the effects of natural food reward and brain-stimulation reward at some brain sites (e.g., the orbitofrontal cortex and amygdala) onto single hypothalamic neurons. Further, it was shown that self-stimulation occurred through the recording electrode if it was near a region where hypothalamic neurons which responded to food had been recorded, and that this self-stimulation was attenuated by feeding the monkey to satiety (Rolls, E.T., et al., 1980).

The finding that these neurons were activated by brain-stimulation reward is consistent with the hypothesis that their activity is related to reward produced by food, and not to some other effect of food. Indeed, this evidence from the convergence of brain-stimulation reward and food reward onto these hypothalamic neurons, and from the self-stimulation found through the recording electrode, suggests that animals work to obtain activation of these neurons by food, and that this is what makes food rewarding. At the same time this accounts for self-stimulation of some brain sites, which would be understood as the animal's seeking to activate the neurons that it normally seeks to activate by food when it is hungry. This and other evidence (see Rolls, E.T., 1975, 1982) indicates that feeding normally occurs in order to obtain the sensory input produced by food, which is rewarding if the animal is hungry.

E. Sites in the Hypothalamus and Basal Forebrain Containing Neurons That Respond to Food

These neurons are found as a relatively small proportion of cells in a region that includes the lateral hypothalamus and substantia innominata and extends from the lateral hypothalamus posteriorly through the anterior hypothalamus and lateral preoptic area to a region ventral and anterior to the anterior commissure (see Fig. 7 of Rolls, E.T., et al., 1979). Further information about the particular populations of neurons in these regions with feeding-related activity, and about the functions of these neurons in feeding, could be provided by evidence on their output connections. It is known that some hypothalamic neurons project to brainstem autonomic regions such as the dorsal motor nucleus of the vagus (Saper et al., 1976, 1979). If some of the hypothalamic neurons with feeding-related activity projected in this way, it would be very likely that their functions would include the generation of autonomic responses to the sight of food. Some hypothalamic neurons project to the substantia nigra (Nauta and Domesick, 1978), and some neurons in the lateral hypothalamus and basal magnocellular forebrain nuclei of Meynert project directly to the cerebral cortex (Kievit and Kuypers, 1975; Divac, 1975). If some of these were feeding-related neurons, then by such routes they could influence the initiation of feeding. To determine the regions to which hypothalamic neurons with feeding-related activity project, we have stimulated various brain regions electrically in order to activate antidromically hypothalamic neurons. So far we have found in such experiments by E.T. Rolls, E. Murzi, and C. Griffiths that some of these feeding-related neurons in the lateral hypothalamus and substantia innominata project directly to the cerebral cortex, to such areas as the prefrontal cortex in the sulcus principalis and the supplementary motor cortex. This provides evidence that at least some of these neurons with feeding-related activity project this information to the cerebral cortex, where it could be used in such processes as the initiation of feeding behavior. It also indicates that at least some of these feeding-related neurons are in the basal magnocellular forebrain nuclei of Meynert, which is quite consistent with the reconstructions of the recording sites (Rolls, E.T., et al., 1979, Fig. 7). In addition, it seems quite likely that at least some of the feeding-related neurons influence the brainstem autonomic centers. This remains to be investigated further.

F. Functions of the Hypothalamus in Feeding

The functions of the hypothalamus in feeding are thus related, at least in part, to the inputs it receives from the forebrain, since it contains neurons that respond to the sight of food and are influenced by learning. (Such pattern-specific visual responses and their modification by learning require forebrain areas such as the inferior temporal visual cortex and the amygdala, as described below.) This conclusion is consistent with the anatomy of the hypothalamus and limbic structures such as the amygdala, which receives projections from the substantia innominata, which in turn receives projections from the association cortex (Nauta, 1961; Herzog and Van Hoesen, 1976) (Fig. 1). The conclusion is also consistent with the evidence that decerebrate rats retain simple controls of feeding, but do not show normal learning about foods (Grill and Norgren, 1978). These rats accept sweet solutions placed in their mouths when hungry and reject them when satiated, so that some control of responses to gustatory stimuli that depend on hunger can occur caudal to the level of the hypothalamus. However, these rats are unable to feed themselves and do not learn to avoid poisoned solutions. The importance to feeding of visual inputs and learning is that animals, especially primates, may eat many foods every day, and must be able to select foods from other visual stimuli and produce appropriate preparative responses such as salivation and the release of insulin.

III. ACTIVITY IN THE GUSTATORY PATHWAYS DURING FEEDING

We have seen that there are neurons in the hypothalamus that can respond to the taste and sight of foods but not non-foods, and that modulation of this sensory input by motivation is seen in recordings made from these hypothalamic neurons. We may now ask, are these special properties shown by

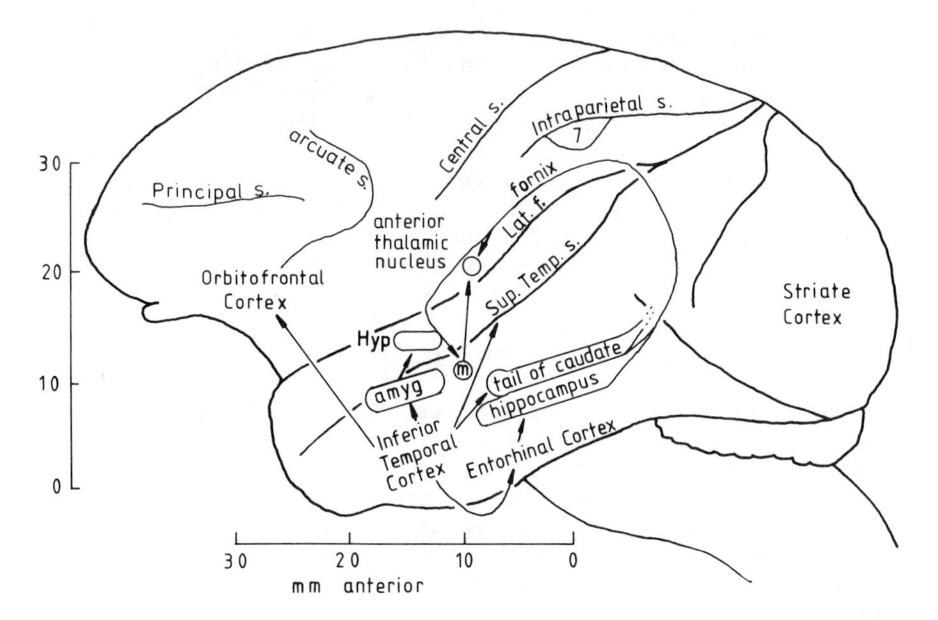

Fig. 1. Some of the pathways described in the text are shown on this lateral view of the brain of the macaque monkey. amyg = amygdala; central s = central sulcus; Hyp = hypothalamus/substantia innominata/ basal forebrain; Lat f = lateral (or Sylvian) fissure; m = mammillary body; Sup Temp s = superior temporal sulcus: 7 = posterior parietal cortex, area 7.

hypothalamic neurons because they are particularly involved in the control of motivational responses, or are the degree of specificity and type of modulation general properties that are evident throughout sensory systems? In one respect, it would be inefficient if motivational modulation were exerted far distally in sensory systems, because this would imply that sensory information was being discarded without the possibility for central processing. The impossibility of tasting food when satiated may be a subjective correspondent of such a situation. It is perhaps more efficient for most of the system to function similarly whether hungry or satiated, and to have a special system (such as the hypothalamus) follow sensory processing where motivational state influences responsiveness. Evidence on the relationship of visual processing and feeding has been summarized above. Evidence is now being obtained on the tuning of neurons in the gustatory pathways and the influence of motivation on responsiveness at different stages. The first central synapse of the gustatory system is in the rostral part of the nucleus of the solitary tract (Beckstead and Norgren,

1979; Beckstead et al., 1980). The caudal half of this nucleus receives visceral afferents, and it is possible that such visceral information, reflecting, for example, gastric distension, is used to modulate gustatory processing even at this early stage of the gustatory system.

We decided to record the activity of single neurons in the nucleus of the solitary tract to investigate the tuning of neurons, determine whether hunger influences processing at this first central opportunity in the gustatory system, and determine whether feeding to satiety influences the responsiveness of these neurons to gustatory stimuli. To ensure that our results were relevant to the normal control of feeding (and were not due, for example, to abnormally high levels of artificially administered putative satiety signals), we allowed the monkeys to feed until they were satiated. We then determined whether this normal and physiological induction of satiety influenced the responsiveness of neurons in the nucleus of the solitary tract, which were recorded throughout the feeding until satiety was reached. We found that in the nucleus of the solitary tract, the first central relay in the gustatory system, neurons are relatively broadly tuned to the prototypical taste stimuli (sweet, salt, bitter, and sour) (Scott et al., 1986a). We also found that neuronal responses to the taste of food are not influenced by whether the monkey is hungry or satiated (Yaxley et al., 1985).

To investigate whether there are neurons in the primary gustatory cortex in the primate more closely tuned to respond to foods rather than to non-foods, and whose responsiveness is modulated by hunger, we have recorded the activity of single neurons in the primary gustatory cortex of the monkey during feeding. In the primary gustatory cortex, neurons are more sharply tuned to gustatory stimuli than in the nucleus of the solitary tract; for example, some neurons respond primarily to sweet, and much less to salt, bitter, or sour stimuli (Scott et al., 1986b). However, here also, hunger did not influence the magnitude of neuronal responses to gustatory stimuli in the neurons we sampled (Rolls, E.T., et al., 1985).

A secondary cortical taste area, in the caudolateral orbitofrontal taste cortex of the primate, has recently been discovered. In this area gustatory neurons are even more sharply tuned to particular taste stimuli (Rolls, Yaxley and Sienkiewicz, 1987; Wiggins et al., 1987). In this region, it is found that the responses of taste neurons to the particular food with which a monkey is fed to satiety decrease to zero (Rolls, Sienkiewicz and Yaxley, 1987). That is, not only is motivational modulation of taste responses found in this region, but this modulation is sensory-specific.

These results were obtained during normal feeding to satiety, when a comparison was made between the hungry and the satiated condition. The results do not completely eliminate the possibility that at some considerable time into the post-satiety period, some decrease of responsiveness to foods may occur. But even if this does occur, such modulation would not then account for the change in acceptability of food, which, of course, is seen as the satiety develops, and is used to define satiety. Nor would this modulation be relevant to the decrease in the pleasantness in the taste of a food that occurs when it is eaten to satiety (Cabanac, 1971; Rolls, B.J., and Rolls et al., 1981; Rolls B.J., and Rowe et al., 1981; Rolls, B.J., et al., 1982; Rolls, E.T., and Rolls, 1977, 1982; Rolls, E.T., Rolls, and Rowe, 1983). Thus it appears that the reduced acceptance of food as satiety develops, and the reduction in its pleasantness, are not produced by a reduction in the responses of neurons in the nucleus of the solitary tract or frontal opercular or insular gustatory cortices to gustatory stimuli. (As described above, the responses of gustatory neurons in these areas do not decrease as satiety develops.) Indeed, after feeding to satiety, humans reported that the taste of the food on which they had been satiated was almost as intense as it had been when they were hungry, though much less pleasant (Rolls, E.T., Rolls and Rowe, 1983). This comparison is consistent with the possibility that activity in the frontal opercular and insular taste cortices as well as the nucleus of the solitary tract does not reflect the pleasantness of the taste of a food, but rather its sensory qualities, independently of motivational state. On the other hand, the responses of the neurons in the orbitofrontal taste area and in the lateral hypothalamus are modulated by satiety, and it is presumably in areas such as these that neuronal activity may be related to whether a food tastes pleasant, and to whether the food should be eaten.

The present results also provide evidence on the nature of the mechanisms which underlie sensory-specific satiety. Sensory-specific satiety, as noted above, is the phenomenon in which the decrease in the palatability and acceptability of a food that has been eaten to satiety are partly specific to the particular food that has been eaten (Rolls, B.J., Rolls and Rowe, 1981; Rolls, B.J., and Rowe et al., 1981; Rolls, B.J., et al., 1982; Rolls, E.T., and Rolls, 1977, 1982; Rolls, E.T., Rolls and Rowe, 1983). The results just described suggest that such sensory-specific satiety cannot be largely accounted for by adaptation at the receptor level, in the nucleus of the solitary tract, or in the frontal opercular or insular gustatory cortices, to the food which has been eaten to satiety; otherwise, modulation of neuronal responsiveness should have been apparent in the recordings made in these regions. Indeed, the findings suggest that sensory-specific satiety is not represented in the primary gustatory cortex. It is thus of particular interest that a decrease in the response of orbitofrontal cortex neurons occurs which is partly specific to the food which has just been eaten to satiety (Rolls, Sienkiewicz, and Yaxley, 1987).

These findings lead to a proposed neuronal mechanism for sensory-specific satiety. The tuning of neurons becomes more specific for gustatory stimuli through the nucleus of the solitary tract, gustatory thalamus, and frontal opercular taste cortex. Satiety, habituation, and adaptation are not features of the responses here, as has been observed (see above). The tuning of neurons becomes even more specific in the orbitofrontal cortex, but here there is some effect of satiety by internal signals such as gastric distension and glucose utilization. In addition, habituation with a time course of several minutes which lasts for 1-2 hours is a feature of the synapses which are activated. Because of the relative specificity of the tuning of orbitofrontal taste neurons, this results in a decrease in the response to that food, but different foods continue to activate other neurons. Then, the orbitofrontal cortex neurons have the required response properties, and it is only then necessary for other parts of the brain to use the activity of the orbitofrontal cortex neurons to reflect the reward value of that particular taste. Evidence that the activity of neurons here does reflect reward is described below, and includes the evidence that electrical stimulation here produces reward, which is like food in that its reward value is attenuated by satiety (Mora et al., 1979). One output of these neurons may be to the hypothalamic neurons with food-related responses, for their responses to the sight and/or taste of food show a decrease which is partly specific to a food that has just been eaten to satiety (see above).

The suggested computational significance of this architecture is as follows (see also Rolls, 1986a). If satiety were to operate at an early level of sensory analysis, then because of the broad tuning of neurons, responses to non-foods would become attenuated as well as responses to foods (and this could well be dangerous if poisonous non-foods became undetectable). This argument becomes even more compelling when we realize that satiety typically shows some specificity for the particular food eaten, with others not eaten in the meal remaining relatively pleasant (see above). Unless tuning were relatively fine, this mechanism could not operate, for reduction in neuronal firing after one food had been eaten would inevitably reduce behavioral responsiveness to other foods. Indeed, it is of interest to note that such a sensory-specific satiety mechanism can be built by arranging for tuning to particular foods to become relatively specific at one level of the nervous system (as a result of categorization processing in earlier stages), and then at this stage (but not at prior stages) to allow habituation to be a property of the synapses, as proposed above.

Thus information processing in the taste system illustrates an important principle of higher nervous system function in primates, namely, that it is only after several or many stages of sensory information processing (which produce efficient categorization of the stimulus) that there is an interface to motivational systems, to other modalities, or to systems involved in association memory (Rolls, 1986a).

IV. FUNCTIONS OF THE TEMPORAL LOBE IN FEEDING

Bilateral damage to the temporal lobes of primates leads to the Kluver-Bucy syndrome, in which lesioned monkeys may select and place in their mouths non-food as well as food items and repeatedly fail to avoid noxious stimuli (Kluver and Bucy, 1939; Jones and Mishkin, 1972). Rats with lesions in the basolateral amygdala also display altered food selection, ingesting relatively novel foods (Rolls, E.T., and Rolls, 1973; Borsini and Rolls, 1984), and failing to learn to avoid ingesting a solu- tion that has previously resulted in sickness (Rolls and Rolls, 1973). The basis for these alterations in food selection and in food-related learning are considered next.

The monkeys with temporal lobe damage have a visual discrimination deficit. They are impaired in learning to select one of two objects under which food is found, and thus fail to form a correct association between the visual stimulus and reinforcement (Jones and Mishkin, 1972; Spiegler and Mishkin, 1981). This syndrome is produced by lesions damaging the cortical areas in the anterior part of the temporal lobe and the underlying amygdala (Jones and Mishkin, 1972) or by lesions of the amygdala (Weiskrantz, 1956; Spiegler and Mishkin, 1981), or of the temporal lobe neocortex (Akert et al., 1961). Lesions to part of the temporal lobe neocortex, damaging the inferior temporal visual cortex and extending into the cortex in the ventral bank of the superior temporal sulcus, produce visual aspects of the syndrome, seen, for example, as a tendency to select non-food as well as food items (L. Weiskrantz, personal communication). Anatomically, there are connections from the inferior temporal visual cortex to the amygdala (Herzog and Van Hoesen, 1976), which in turn projects to the hypothalamus (Nauta, 1961), thus providing a route for visual information to reach the hypothalamus (see Rolls, E.T., 1981b). This evidence, together with the evidence that damage to the hypothalamus can disrupt feeding (see Rolls, E.T., 1981b; Winn et al., 1984), thus indicates that a system including the visual cortex in the temporal lobe, projections to the amygdala, and further connections to structures such as the lateral hypothalamus is involved in behavioral responses made on the basis of learned associations between visual stimuli and reinforcers such as food. Given this evidence from lesion and anatomical studies, the contribution of each of these regions to the visual analysis and learning required for these functions in food selection will be considered, using evidence from the activity of single neurons in these regions.

Recordings were made from single neurons in the inferior temporal visual cortex while rhesus monkeys performed visual discriminations in which they were shown visual stimuli associated with positive reinforcement such as food, visual stimuli associated with negative reinforcement such as aversive hypertonic saline, and neutral visual stimuli (Rolls, E.T., et al., 1977). We found that during visual discriminations, inferior temporal neurons often had sustained visual responses with latencies of 100-140 milliseconds to the discriminanda, but that these responses did not depend on whether the visual stimuli were associated with reward or punishment; that is, the neuronal responses did not alter during reversals, when the previously rewarded stimulus was made to signify aversive saline, and the previously punished stimulus was made to signify reward (Rolls, E.T., et al., 1977). The conclusion that the responses of inferior temporal neurons during visual discriminations do not code for whether a visual stimulus is associated with reward or punishment is also consistent with the findings of Ridley et al. (1977), Jarvis and Mishkin (1977), Gross et al. (1979), and Sata et al., (1980). Further, we found that inferior temporal neurons did not respond only to food-related or aversive visual stimuli and were not dependent on hunger, but that in many cases their responses depended on physical aspects of the stimuli such as shape, size, orientation, color, or texture (Rolls, E.T., et al., 1977).

These findings indicate that the responses of neurons in the inferior temporal visual cortex do not reflect the association of visual stimuli with reinforcers such as food. Given these findings and the lesion evidence described above, it is likely that the inferior temporal cortex is an input stage for this process of association. On the basis of its anatomical connections, the amygdala will be considered next.

In recordings made from 1754 amygdaloid neurons, we found that 113 neurons (6.4%), many of which were in a dorsolateral region of the amygdala known to receive directly from the inferior temporal visual cortex (Herzog and Van Hoesen, 1976), had visual responses that in most cases were sustained while the monkey looked at effective visual stimuli (Sanghera et al., 1979). The latency of the responses was 100-140 milliseconds or more. The majority (85%) of these visual neurons responded more strongly to some stimuli than to others, but physical factors accounting for the responses, such as orientation, color, and texture, usually could not be identified. It was found that 22 (19.5%) of these visual neurons responded primarily to foods and to objects associated with food, but none of these neurons responded solely to food-related stimuli, since they all responded to one or more aversive or neutral stimuli. Further, although some neurons responded in a visual discrimination to the visual stimulus indicating food reward, but not to the visual stimulus associated with aversive saline, nly minor modifications of the neuronal responses occurred when the association of the stimuli with reinforcement was reversed. Thus, even the responses of these neurons were not invariably associated with whichever stimulus was associated with reward. A comparable population of neurons whose reponses appear to be partly but not uniquely related to aversive visual stimuli was also found (Sanghera et al., 1979).

These findings suggest that the amygdala could be involved at an early stage of the processing which associates visual stimuli with reinforcement, but that neuronal responses here do not code uniquely for that association. Neurons with responses more closely related to reinforcement are found in areas to which the amygdala projects, the lateral hypothalamus and substantia innominata, as described above. Thus there is evidence that through the anatomical sequence inferior temporal visual cortex to amygdala, to lateral hypothalamus and substantia innominata, neuronal responses become more relevant to the control of feeding as a result of learning, so that finally there are neurons in the lateral hypothalamus and substantia innominata responding only to stimuli that the organism has learned are food or signify food, and that can appropriately initiate feeding when the organism is hungry.

VI. FUNCTIONS OF THE ORBITOFRONTAL CORTEX IN FEEDING

Damage to the orbitofrontal cortex alters food preferences, as shown by monkeys with damage to the orbitofrontal cortex that select and eat foods previously rejected (Butter et al., 1969). Lesions of the orbitofrontal cortex also lead to a failure to correct feeding responses when these become inappropriate. Examples of the situations in which these abnormalities in feeding responses are found include (1) extinction, where feeding responses continue to be made to the previously reinforced stimulus, (2) reversals of visual discriminations, where the monkeys make responses to the previously reinforced stimulus or object, (3) Go/Nogo tasks, where responses are made to the stimulus not associated with food reward, and (4) passive avoidance, where feeding responses are made even when they are punished (Butter, 1969; Iversen and Mishkin, 1970; Jones and Mishkin, 1972; Tanaka, 1973; see also Rosenkilde, 1979; Fuster, 1980). It may be noted that in contrast, the formation of associations between visual stimuli and reinforcement is much less affected by these lesions than by temporal lobe lesions, as tested during visual discrimination learning and reversals (Jones and Mishkin, 1972).

To investigate how the orbitofrontal cortex may be involved in feeding and in the correction of feeding responses when these become inappropriate, recordings were made of the activity of 494 orbitofrontal neurons during the performance of a Go/Nogo task, reversals of a visual discrimination task, extinction, and passive avoidance (Thorpe et al., 1983). First, neurons were found that responded either to the preparatory auditory or visual signal used before each trial (15.1%), or non-discriminatively during the period when the discriminative visual stimuli were shown (37.8%). These neurons are not considered further here. Second, 8.6% of neurons had discriminative responses during the period when the visual stimuli were shown. The majority of these neurons responded to whichever visual stimulus was associated with reward; that is, the stimulus to which they responded changed during reversal. However, six of these neurons required a combination of a particular visual stimulus in the discrimination *and* reward in order to respond. Further, none of this second group of neurons responded to all the reward-related stimuli, including different foods, which were shown, so that in general this group of neurons coded for a combination of one or several visual stimuli *and* reward. Thus, information that particular visual stimuli had previously been associated with reinforcement was represented in the responses of orbitofrontal neurons. Third, the responses of 9.7% of neurons occurred after the lick response to obtain reward was made. Some of these responded independently of whether fruit juice reward was obtained, aversive hypertonic saline was obtained in trials where the monkey licked in error, or saline was given in the first trials of a reversal. Through these neurons information that a lick had been made was represented to the orbitofrontal cortex. Other neurons in this third group responded only when fruit juice was obtained; thus, through these neurons, information that reward had been given on that trial was represented in the orbitofrontal cortex. Other neurons in this group responded when saline was obtained by an erroneous response, or when saline was obtained on the first few trials of a reversal (but not in either case when saline was simply placed in the mouth), when reward was not given in extinction, or when food was taken away instead of being given to the monkey. However, these neurons did not respond in all these situations when reinforcement was omitted or punishment was given. Thus, task-selective information that reward had been omitted or punishment given was represented in the responses of these neurons.

These three groups of neurons found in the orbitofrontal cortex could together provide for computation of whether the reinforcement previously associated with a particular stimulus was still being obtained, and generation of a signal if a match was not obtained. This signal could be partly reflected in the responses of the last subset of neurons with task-selective responses to non-reward or to unexpected punishment. It could be used to alter the monkey's behavior, leading, for example, to reversal to one particular stimulus but not to other stimuli, to extinction to one stimulus but not to others, and so on. It could also lead to the altered responses of the orbitofrontal differential neurons found as a result of learning in reversal, so that their responses indicate appropriately whether a particular stimulus is now associated with reinforcement.

Thus, the orbitofrontal cortex contains neurons that appear to be involved in altering behavioral responses when these are no longer associated with reward or become associated with punishment. In the context of feeding, it appears that without these neurons the primate is unable to suppress its behavior correctly to non-food objects, since altered food preferences are produced by orbitofrontal damage (Butter et al., 1969). It also appears that without these neurons the primate is unable to correct its behavior when breaking a learned association between a stimulus and a reward such as food becomes appropriate (Jones and Mishkin, 1972). The orbitofrontal neurons could be involved in the actual breaking of the association, or in the alteration of behavior when other neurons signal that the connection is no longer appropriate. As shown here, the orbitofrontal cortex contains neurons with responses which could provide the information necessary for the unlearning. This type of unlearning is important in enabling animals to alter the environmental stimuli toward which motivational responses such as feeding are directed, if experience shows that such responses are no longer appropriate. In this way they can ensure that their feeding and other motivational responses remain continually adapted to a changing environment.

VII. FUNCTIONS OF THE STRIATUM IN FEEDING

Damage to the nigrostriatal bundle, which depletes the striatum of dopamine, produces aphagia and adipsia associated with a sensorimotor disturbance in the rat (Ungerstedt, 1971; Marshall et al., 1974; Stricker and Zigmond, 1976; Stricker, 1984). In order to analyze how striatal function is involved in feeding, the activity of single neurons was recorded in different regions of the striatum during the initiation of feeding and during the performance of other tasks known to be affected by damage to particular regions of the striatum (see Rolls, E.T., 1979, 1984, 1986b; Rolls and Williams, 1986).

In the head of the caudate nucleus (Rolls, Thorpe, and Maddison, 1983), which receives inputs particularly from the prefrontal cortex, many neurons responded to environmental stimuli that were cues to the monkey to prepare for the possible initiation of a feeding response. Thus, 22.4% of neurons recorded from responded during a cue given by the experimenter that a food or non-food object was about to be shown, and if food, would be fed to the monkey. Comparably, in a visual discrimination task made to obtain food, 14.5% of the neurons (including some of the above) responded during a 0.5 second tone/light cue that preceded and signalled the start of each trial. It is suggested that these neurons are involved in the utilization of environmental cues for the preparation for movement, and that disruption of the function of these neurons contributes to the akinesia or failure to initiate movements, including those required for feeding, found after depletion of dopamine in the striatum (Rolls, Thorpe, and Maddison, 1983). Some other neurons (25.8%) responded if food was shown to the monkey by the experimenter immediately prior to feeding, but the responses of these neurons typically did not occur in other situations in which food-related visual stimuli were shown, such as during the visual discrimination task. Comparably, some other neurons (24.3%) responded differentially in the visual discrimination task, for example, to the visual stimulus indicating that the monkey could initiate a lick response to obtain food, yet typically did not respond when food was simply shown to it prior to feeding. The responses of these neurons occur to particular stimuli indicating that particular motor responses should be made, and are thus situation-specific, suggesting that these neurons are involved in stimulus-motor response connections. Since these responses are situation-specific, they are different from the responses of the hypothalamic neurons with visual responses to the sight of food described above (Rolls, Thorpe, and Maddison, 1983). It is thus suggested that these neurons in the head of the caudate nucleus could be involved in relatively fixed feeding responses to food made in particular, probably well-learned, situations. It is further suggested that these neurons do not provide a signal reflecting whether a visual stimulus is associated with food, thereby providing a basis for initiation of a feeding response. Rather, it is likely that the systems in the temporal lobe, hypothalamus, and orbitofrontal cortex are involved in this more flexible decoding of the food value of visual stimuli.

In the tail of the caudate nucleus, which receives inputs from the inferior temporal visual cortex, neurons were found that responded to visual stimuli such as gratings and edges, but that showed rapid and pattern-specific habituation (Caan et al., 1984). It has been suggested that these neurons are involved in orientation to patterned visual stimuli, and in pattern-specific habituation to these stimuli (Caan et al., 1984). These neurons would thus appear not to be directly involved in the control of feeding, although a disturbance in the ability to orient normally to a changed visual stimulus could have an indirect effect on the ability to react normally.

In the putamen, which receives inputs from the sensorimotor cortex, neurons were found with activity related to mouth or arm movements made by the monkey (Rolls, E.T., et al., 1984). Disturbances in the normal function of these neurons might be expected to affect the ability to initiate and execute movements, and thus indirectly affect the ability to feed normally.

Thus, in different regions of the striatum neurons are found that may be involved in orientation to environmental stimuli, in the use of such stimuli in the preparation for and initiation of movements, in the execution of movements, and in stimulus-response connections appropriate for particular responses made in particular situations to particular stimuli. Disturbances of feeding produced by damage to the striatum might be expected to occur because of disruption of these functions, and not because the striatum plays a direct role in the regulation of food intake.

VIII. SUMMARY AND CONCLUSIONS

Investigations in non-human primates have provided evidence that the lateral hypothalamus and adjoining substantia innominata are involved in the control of feeding. There is a population of neurons in these regions that responds to the sight and/or taste of food if the organism is hungry. The responses of these neurons may reflect the reward value or pleasantness of food, since it can be mimicked by stimulation in this region. It has been found that after satiation with one food these neurons no longer respond to that food, although they may still respond to other foods which have not been eaten. Following this finding, it was shown that sensory-specific satiety is an important determinant of human food intake; furthermore, variety is an important factor in determining the amount of food eaten. Analysis of activity in the gustatory pathways of the monkey during feeding has shown that in the nucleus of the solitary tract, the first central relay in the gustatory system, neurons are relatively broadly tuned to the prototypical taste stimuli (sweet, salt, bitter, and sour), and that neuronal responses to the taste of food are not influenced by whether the monkey is hungry or satiated. In the primary gustatory cortex, neurons are more sharply tuned to gustatory stimuli, but here too, hunger does not influence the magnitude of neuronal responses to gustatory stimuli. There is a secondary gustatory cortical area in the caudal orbitofrontal cortex, and in this area, neurons are more sharply tuned to taste stimuli, and it is found that satiety here does modulate the responsiveness of these taste neurons. Indeed, sensory-specific satiety is reflected in the responses of these taste neurons, and a neurophysiological mechanism for sensory-specific satiety is proposed. This analysis thus provides a clear example of where in the primate brain motivational state starts to influence sensory processing, to produce motivation-specific reward signals. Temporal lobe structures such as the inferior temporal visual cortex and amygdala, which is important for learning which visual stimuli are foods, provide a route to the hypothalamus for this information. The orbitofrontal cortex contains a population of neurons which appear to be important in correcting feeding responses as a result of learning. The striatum contains neural systems important for the initiation of different types of motor and behavioral responses, including feeding.

REFERENCES

Akert, K., R.A. Gruesen, C.N. Woolsey, and D.R. Meyer. 1961. Kluver-Bucy syndrome in monkeys with neocortical ablations of temporal lobe. *Brain* 84: 480-498.

Anand, B.K., and J.R. Brobeck. 1951. Localization of a feeding center in the hypothalamus of the rat. *Proc. Soc. Exp. Biol. Med.* 77: 323-324.

Beckstead, R.M., and R. Norgren. 1979. An autoradiographic examination of the central distribution of the trigeminal, facial, glossopharyngeal, and vagal nerves in the monkey. *J. Comp. Neurol.* 184: 455-472.

Beckstead, R.M., J.R. Morse, and R. Norgren. 1980. The nucleus of the solitary tract in the monkey: projections to the thalamus and brainstem nuclei. *J. Comp. Neurol.* 190: 259-282.

Borsini, F., and E.T. Rolls. 1984. Role of noradrenaline and serotonin in the basolateral region of the amygdala in food preferences and learned taste aversions in the rat. *Physiol. Behav.* 33. In press.

Burton, M.J., E.T. Rolls, and F. Mora. 1976. Effects of hunger on the responses of neurons in the lateral hypothalamus to the sight and taste of food. *Exp. Neurol.* 51: 668-677.

Butter, C.M. 1969. Perseveration in extinction and in discrimination reversal tasks following selective prefrontal ablations in *Macaca mulatta*. *Physiol. Behav.* 4: 163-171.

Butter, C.M., J.A. McDonald, and D.R. Snyder. 1969. Orality, preference behavior, and reinforcement value of non-food objects in monkeys with orbital frontal lesions. *Science* 164: 1306-1307.

Caan, W., D.I. Perrett, and E.T. Rolls. 1984. Responses of striatal neurons in the behaving monkey. 2. Visual processing in the caudal neostriatum. *Brain Res.* 290: 53-65.

Cabanac, M. 1971. Physiological role of pleasure. *Science* 173: 1103-1107.

Cabanac, M., and R. Duclaux. 1970. Specificity of internal signals in producing satiety for taste stimuli. *Nature* 227: 966-967.

Cabanac, M., and M. Fantino. 1977. Origin of olfaco-gustatory alliesthesia: Intestinal sensitivity to carbohydrate concentration? *Physiol. Behav.* 10: 1039-1045.

Divac, I. 1975. Magnocellular nuclei of the basal forebrain project to neocortex, brain stem, and olfactory bulb. Review of some functional correlates. *Brain Res.* 93: 385-398.

Dunnett, S.B., D.M. Lane, and P. Winn. 1985. Ibotenic acid lesions of the lateral hypothalamus: comparison with 6-hydroxydopamine-induced sensorimotor deficits. *Neuroscience*. In press.

Friedman, M.I., and E. Stricker. 1976. The physiological psychology of hunger: a physiological perspective. *Psychol. Rev.* 83: 409-431.

Fuster, J.M. 1980. *The Prefrontal Cortex*. Raven Press, New York.

Gallistel, C.R., and G. Beagley. 1971. Specificity of brain-stimulation reward in the rat. *J. Comp. Physiol. Psychol*. 76: 199-205.

Gibbs, J., S.P. Maddison, and E.T. Rolls. 1981. The satiety role of the small intestine in sham feeding rhesus monkeys. *J. Comp. Physiol. Psychol*. 95: 1003-1015.

Grill, H.J., and R. Norgren. 1978. Chronically decerebrate rats demonstrate satiation but not bait shyness. *Science* 201: 267-269.

Gross, C.G., D.B. Bender, and G.L. Gerstein. 1979. Activity of inferior temporal neurons in behaving monkeys. *Neuropsychologia* 17: 215-229.

Grossman, S.P. 1967. *A Textbook of Physiological Psychology*. John Wiley & Sons, New York.

Grossman, S.P. 1973. *Essentials of Physiological Psychology*. John Wiley & Sons, New York.

Herzog, A.G., and G.W. Van Hoesen. 1976. Temporal neocortical afferent connections to the amygdala in the rhesus monkey. *Brain Res*. 115: 57-69.

Hoebel, B.G. 1969. Feeding and self-stimulation. *Ann. N.Y. Acad Sci*. 157: 757-778.

Iversen, S.D., and M. Mishkin. 1970. Perseverative interference in monkey following selective lesions of the inferior prefrontal convexity. *Exp. Brain. Res*. 11: 376-386.

Jarvis, C.D., and M. Mishkin. 1977. Responses of cells in the inferior temporal cortex of monkeys during visual discrimination reversals. *Soc. Neurosci. Abstr*. 3: 1794.

Jones, B., and M. Mishkin. 1972. Limbic lesions and the problem of stimulus-reinforcement associations. *Exp. Neurol*. 36: 362-377.

Kievit, J., and H.G.J.M. Kuypers. 1975. Subcortical afferents to the frontal lobe in the rhesus monkey studied by means of retrograde horseradish peroxidase transport. *Brain Res*. 85: 261-266.

Kluver, H., and P.C. Bucy. 1939. Preliminary analysis of functions of the temporal lobes in monkeys. *Arch. Neurol. Psychiatry* 42: 979-1000.

Le Magnen, J. 1980. The body energy regulation: the role of three brain responses to glucopenia. *Neurosci. Biobehav. Rev*. 4, Suppl. 1: 65-72.

Marshall, J.F., J.S. Richardson, and P. Teitelbaum. 1974. Nigrostriatal bundle damage and the lateral hypothalamic syndrome. *J. Comp. Physiol. Psychol*. 87: 808-830.

Mora, F., E.T. Rolls, and M.J. Burton. 1976. Modulation during learning of the responses of neurons in the hypothalamus to the sight of food. *Exp. Neurol*. 53: 508-519.

Mora, F., D.B. Avrith, A.G. Phillips, and E.T. Rolls. 1979. Effects of satiety on self-stimulation of the orbitofrontal

cortex in the rhesus monkey. *Neurosci. Lett.* 13(2): 141-145.
Nauta, W.J.H. 1961. Fiber degeneration following lesions of the amygdaloid complex in the monkey. *J. Anat.* 95: 515-531.
Nauta, W.J.H., and V.B. Domesick. 1978. Crossroads of limbic and striatal circuitry: Hypothalamonigral connections. In *Limbic Mechanisms*, ed. K.E. Livingston and O. Hornykiewicz, 75-93. Plenum, New York.
Olds, J. 1977. *Drives and Reinforcements: Behavioral Studies of Hypothalamic Functions*. Raven Press, New York.
Ono, T., H. Nishino, K. Sasaki, M. Fukuda, and K. Muramoto. 1980. Role of the lateral hypothalamus and amygdala in feeding behavior. *Brain Res. Bull.* 5, Suppl. 4: 143-149.
Ridley, R.M., N.S. Hester, and G. Ettlinger. 1977. Stimulus- and response-dependent units from the occipital and temporal lobes of the unanesthetized monkey performing learnt visual tasks. *Exp. Brain Res.* 27, 539-552.
Rolls, B.J., and E.T. Rolls. 1973. Effects of lesions in the basolateral amygdala on fluid intake in the rat. *J. Comp. Physiol. Psychol.* 83: 240-247.
Rolls, B.J., E.T. Rolls, E.A. Rowe and K. Sweeney. 1981. Sensory specific satiety in man. *Physiol. Behav.* 27: 137-142.
Rolls, B.J., E.A. Rowe, E.T. Rolls, B. Kingston, A. Megson, and R. Gunary. 1981. Variety in a meal enhances food intake in man. *Physiol. Behav.* 26: 215-221.
Rolls, B.J., E.A. Rowe, and E.T. Rolls. 1982. How sensory properties of foods affect human feeding behavior. *Physiol. Behav.* 29: 409-417.
Rolls, B.J., P.M. Van Duijenvoorde, and E.A. Rowe. 1983. Variety in the diet enhances intake in a meal and contributes to the development of obesity in the rat. *Physiol. Behav.* 31: 21-27.
Rolls, B.J., P.M. Van Duijenvoorde, and E.T. Rolls. 1984. Pleasantness changes and food intake in a varied four course meal. *Appetite* 5: 337-348.
Rolls, E.T. 1975. *The Brain and Reward*. Pergamon, Oxford.
Rolls, E.T. 1976. The neurophysiological basis of brain-stimulation reward. In *Brain-stimulation Reward*, ed. A. Wauquier and E.T. Rolls, 65-87. North-Holland, Amsterdam.
Rolls, E.T. 1979. Effects of electrical stimulation of the brain on behavior. In *Psychology Surveys*, Vol. 2, ed. K. Connolly, 151-169. George Allen and Unwin, Hemel Hempstead, U.K.
Rolls, E.T. 1981a. Processing beyond the inferior temporal visual cortex related to feeding, memory, and striatal function. In *Brain Mechanisms of Sensation*, Ch. 16, ed.

Y. Katsuki, R. Norgren, and M. Sato, 241-269. John Wiley & Sons, New York.

Rolls, E.T. 1981b. Central nervous mechanisms related to feeding and appetite. *Br. Med. Bull.* 37: 131-134.

Rolls, E.T. 1982. Feeding and reward. In *The Neural Basi of Feeding and Reward*, ed. B.G. Hoebel and D. Novin, 323-337. Haer Institute for Electrophysiological Research, Brunswick, Maine.

Rolls, E.T. 1983. Feeding. In *Advances in Vertebrate Neuroethology* ed. J.-P. Ewert, R.R. Capranica, and D.J. Ingle, 1067-1086. Plenum Press, New York.

Rolls, E.T. 1984. Activity of neurons in the basal ganglia of the behaving monkey. In *The Basal Ganglia: Structure and Function*, ed. J. McKenzie and L. Wilcox. Plenum Press, New York.

Rolls, E.T. 1986a. Information representation, processing and storage in the brain: analysis at the single neuron level. In *Neural and Molecular Mechanisms of Learning*, ed. J.-P. Changeux and M. Konishi. Springer-Verlag, Berlin.

Rolls, E.T. 1986b. Investigations of the functions of different regions of the basal ganglia. In *Parkinson's Disease*, ed. G. Stern. Chapman and Hall, London.

Rolls, E.T., and B.J. Rolls. 1973. Altered food preferences after lesions in the basolateral region of the amygdala in the rat. *J. Comp. Physiol. Psychol.* 83: 248-259.

Rolls, E.T., and B.J. Rolls. 1977. Activity of neurones in sensory, hypothalamic, and motor areas during feeding in the monkey. In *Food Intake and Chemical Senses*, ed. Y. Katsuki, M. Sato, S.F. Takagi, and Y. Oomura, 525-549. University of Tokyo Press, Tokyo.

Rolls, E.T., and B.J. Rolls. 1982. Brain mechanisms involved in feeding. In *Psychobiology of Human Food Selection*, Ch. 3, ed. L.M. Barker, 33-62. AVI Publishing Co., Westport, Connecticut.

Rolls, E.T., and A.W.L. de Waal. 1985. Long-term sensory-specific satiety: evidence from an Ethiopian refugee camp. *Physiol. Behav.* 34: 1017-1020.

Rolls, E.T., and G.V. Williams. 1986. Sensory and movement-related neuronal activity in different regions of the striatum of the primate. In *Sensory Considerations for Basal Ganglia Functions*, ed. J.S. Schneider and T.I. Lidsky. Haber.

Rolls, E.T., M.J. Burton, and F. Mora. 1976. Hypothalamic neuronal responses associated with the sight of food. *Brain Res.* 111: 53-66.

Rolls, E.T., S.J. Judge, and M.K. Sanghera. 1977. Activity of neurons in the inferotemporal cortex of the alert monkey. *Brain Res.* 130: 229-238.

Rolls, E.T., M.K. Sanghera, and A. Roper-Hall. 1979. The latency of activation of neurons in the lateral hypothalamus and substantia innominata during feeding in the monkey. *Brain Res*. 164: 121-135.

Rolls, E.T., M.J. Burton, and F. Mora. 1980. Neurophysiological analysis of brain-stimulation reward in the monkey. *Brain Res*. 194: 339-357.

Rolls, E.T., B.J. Rolls, and E.A. Rowe. 1983. Sensory-specific and motivation-specific satiety for the sight and taste of food and water in man. *Physiol. Behav*. 30: 185-192.

Rolls, E.T., T.R. Scott, S. Yaxley, and Z.J. Sienkiewicz. 1986. The responsiveness of neurons in the frontal opercular gustatory cortex of the macaque monkey is independent of hunger. Submitted.

Rolls, E.T., S.J. Thorpe, and S.P. Maddison. 1983. Responses of striatal neurons in the behaving monkey. 1. Head of the caudate nucleus. *Behav. Brain Res*. 7: 179-210.

Rolls, E.T., S.J. Thorpe, M. Boytim, I. Szabo, and D.I. Perrett. 1984. Responses of striatal neurons in the behaving monkey. 3. Effects of iontophoretically applied dopamine on normal responsiveness. *Neuroscience* 12: 1202-1212.

Rolls, E.T., S. Yaxley, Z.J. Sienkiewicz, and T.R. Scott. 1985. Gustatory responses of single neurons in the orbitofrontal cortex of the macaque monkey. *Chem. Senses* 10: 443.

Rolls, E.T., E. Murzi, S. Yaxley, S.J. Thorpe, and S.J. Simpson. 1986. Sensory-specific satiety: food-specific reduction in responsiveness of ventral forebrain neurons after feeding in the monkey. *Brain Res*. 368: 79-86.

Rolls, E.T., T.R. Scott, S. Yaxley, and Z.J. Sienkiewicz. 1986. The responsiveness of neurones in the frontal opercular gustatory cortex of the macaque monkey is independent of hunger. Submitted.

Rolls, E.T., Z.J. Sienkiewicz, and S. Yaxley. 1987. Hunger modulates the responses to gustatory stimuli of single neurons in the orbitofrontal cortex. In preparation.

Rolls, E.T. S. Yaxley, And Z.J. Sienkiewicz. 1987. Gustatory responses of single neurons in the orbitofrontal cortex of the macaque monkey. In preparation.

Rosenkilde, C.E. 1979. Functional heterogeneity of the prefrontal cortex in the monkey: a review. *Behav. Neural Biol*. 25: 301-345.

Russchen, F.T., D.G. Amaral, and J.L. Price. 1985. The afferent connections of the substantia innominata in the monkey, Macaca fascicularis. *J. Comp. Neurol*. 242: 1-27.

Sanghera, M.K., E.T. Rolls, and A. Roper-Hall. 1979. Visual responses of neurons in the dorsolateral amygdala of the alert monkey. *Exp. Neurol*. 63: 610-626.

Saper, C.B., A.D. Loewy, L.W. Swanson, and W.M. Cowan. 1976. Direct hypothalamo-autonomic connections. *Brain Res.* 117: 305-312.

Saper, C.B., L.W. Swanson, and W.M. Cowan. 1979. An autoradiographic study of the efferent connections of the lateral hypothalamic area in the rat. *J. Comp. Neurol.* 183: 689-706.

Sato, T., T. Kawamura, and E. Iwai. 1980. Responsiveness of inferotemporal single units to visual pattern stimuli in monkeys performing discrimination. *Exp. Brain Res.* 38, 313-319.

Scott, T.R., S. Yaxley, Z.J. Sienkiewicz, and E.T. Rolls. 1986a. Gustatory responses in the nucleus tractus solitarius of the alert cynomolgus monkey. *J. Neurophysiol.* 55: 182-200.

Scott, T.R., S. Yaxley, Z.J. Sienkiewicz, and E.T. Rolls. 1986b. Gustatory responses in the anterior operculum of the alert cynomolgus monkey. *J. Neurophysiol.* 56. In press.

Spiegler, B.J., and M. Mishkin. 1981. Evidence for the sequential participation of inferior temporal cortex and amygdala in the acquisition of stimulus-reward associations. *Behav. Brain Res.* 3: 303-317.

Stellar, E. 1954. The physiology of motivation. *Psychol. Rev.* 61: 5-22.

Stricker, E.M. 1984. Brain monoamines and the control of food intake. *Int. J. Obes.* 8, Suppl. 1. In press.

Stricker, E.M., and M.J. Zigmond. 1976. Recovery of function after damage to central catecholamine-containing neurons: a neurochemical model for the lateral hypothalamic syndrome. *Prog. Psychobiol. Physiol. Psychol.* 6: 121-188.

Tanaka, D. 1973. Effects of selective prefrontal decortication on escape behavior in the monkey. *Brain Res.* 53: 161-173.

Thorpe, S.J., E.T. Rolls, and S. Maddison. 1983. Neuronal activity in the orbitofrontal cortex of the behaving monkey. *Exp. Brain Res.* 49: 93-115.

Ungerstedt, U. 1971. Adipsia and aphagia after 6-hydroxydopamine induced degeneration of the nigrostriatal dopamine system. *Acta Physiol. Scand.* 81, Suppl. 367: 95-122.

Weiskrantz, L. 1956. Behavioral changes associated with ablation of the amygdaloid complex in monkeys. *J. Comp. Physiol. Psychol.* 49: 381-391.

Wiggins, L.L., G.C. Baylis, E.T. Rolls, and S. Yaxley. 1986. Afferent connections of the orbitofrontal cortex taste area of the primate. In preparation.

Winn, P., A. Tarbuck, and S.B. Dunnett. 1984. Ibotenic acid

lesions of the lateral hypothalamus: comparison with electrolytic lesion syndrome. *Neuroscience*. In press.

Woods, S.C., L.D. McKay, L.J. Stein, D.B. West, and D. Porte. 1980. Neuroendocrine regulation of food intake and body weight. *Brain Res. Bull.* 5, Suppl. 4: 1-5.

Yaxley, S., E.T. Rolls, Z.J. Sienkiewicz, and T.R. Scott. 1985. Satiety does not affect gustatory activity in the nucleus of the solitary tract of the alert monkey. *Brain Res.* 347: 85-93.

Chapter 7

NEUROCHEMICAL CONTROLS OF APPETITE

Sarah F. Leibowitz and B. Glenn Stanley

I. INTRODUCTION

Multiple behavioral and physiological mechanisms have evolved to ensure that animals obtain and maintain sufficient energy stores and specific nutrients to survive and function efficiently across diverse environmental conditions. These energy stores and nutrients are obtained through ingestion of specific foods. The ability to modify feeding behavior according to environmental conditions suggests that multiple brain mechanisms exist for the control of feeding behavior. These control mechanisms undoubtedly incorporate considerable redundancy, involving hierarchical fail-safe mechanisms to permit maintenance of energy balance in diverse nutritional environments. While controls of the onset, maintenance, and termination of eating and the choice of foods eaten are complex, involving multiple central and peripheral mechanisms, evidence from neuropharmacology and neurochemistry over the past two decades has contributed to the understanding of how the brain, through its various neurotransmitters, tells us what, when, and how much to eat.

It is the goal of this laboratory to determine (1) which neurotransmitters control the onset and termination of feeding and of macronutrient selection, (2) the brain areas and neural pathways that mediate the effects of these transmitters, (3) the conditions under which these neural pathways and their transmitters are stimulated, and (4) the interaction of the peripheral nervous system and endocrine systems with these control mechanisms of feeding behavior. To this end, we have measured the impact on feeding behavior of microinjections of neurotransmitters, their agonists, antagonists, and presynaptic releasers, directly into brain tissue. Studies of this nature have revealed that microinjection of specific neurotransmitters into discrete brain areas may elicit complex behaviors, such as feeding, whose sequence and topography appear normal, and that also appear to be under normal stimulus control. Thus, with the appropriate controls for behavioral, anatomical, and pharmacological specificity, and in conjunction with other techniques, we can determine which neural pathway stimulates feeding behavior, the neurotransmitter that activates it, and the receptor subtype through which the transmitter acts. More recently, we have also employed a complementary approach, measuring the levels of neurotransmitters and receptors to determine the behavioral and physiological conditions causing these neurotransmitters to be released. In these experiments, instead of manipulating brain neurochemistry and measuring the effect on behavior, we manipulate the behavior and measure the effect on brain neurochemistry.

These studies have focused primarily on the hypothalamus. While animals are capable of eating independent of hypothalamic neural control (Ellison et al., 1970; Grill and Norgren, 1978), the hypothalamus, in response to sensory and metabolic information about the nutritional status of the animal, does play a major role in the initiation and termination of feeding and in appetite for specific nutrients. The hypothalamus contains many of the classical neurotransmitters and is particularly rich in the more recently discovered neuropeptides. Thus it is not surprising that several neurotransmitters and peptides, injected directly into specific nuclei of the hypothalamus, are found to be particularly effective in stimulating feeding in satiated animals or in terminating feeding in hungry animals.

There are three critical issues in central microinjection studies that must be addressed to evaluate the physiological role of particular neurotransmitters. The first is *anatomical specificity*; that is, what specific brain areas mediate the effect of the centrally injected substances on feeding? The second is *behavioral specificity*. Is the effect due to action directly on neural substrates of feeding behavior or is it secondary to other behavioral or physiological effects or to general activation or debilitation? This issue is of particular concern when a suppression of feeding behavior has been demonstrated. However, it cannot be ignored when a stimulation of feeding is demonstrated, since stress induced by nonspecific manipulations, such as tail pinch, may stimulate feeding. The third issue is *pharmacological specificity*. Is the effect on feeding behavior mediated through specific receptors rather than through nonspecific effects induced by changes in parameters such as pH and osmotic pressure? More specifically, is the neurotransmitter that affects feeding the actual endogenous ligand for the receptors being activated and mediating the response? Without information on these three basic issues, any conclusions concerning the physiological significance of central microinjection studies must be viewed with great caution.

II. CLASSICAL NEUROTRANSMITTERS

Numerous neurotransmitters are now believed to play specific roles in the control of food intake and appetite for specific macronutrients. These neurotransmitters include the monoamines, namely, norepinephrine, epinephrine, dopamine, and serotonin; the amino acid gamma-aminobutyric acid; and, more recently, a variety of brain-gut peptides and neuropeptides.

These specific roles are most evident in the neurotransmitters norepinephrine, acting in the medial hypothalamus to stimulate feeding behavior, and dopamine and epinephrine, acting further laterally to suppress feeding. Thus, we will focus initially on these neurotransmitters and related mechanisms.

A. Norepinephrine

Now-classic experiments conducted from the 1930s to the 1950s have focused on the lateral hypothalamus (LH) and the ventromedial hypothalamus (VMH) in the control of feeding behavior. The LH was proposed to be a "feeding center" while the VMH was believed to be a "satiety center" (Stellar, 1984). These ideas were primarily derived from experiments demonstrating that lesions of the LH produced aphagia, while stimulation of this structure elicited feeding, and, conversely, that lesions of the VMH increased food consumption, while stimulation of this area decreased eating (Anand and Brobeck, 1951; Beltt and Keesey, 1975; Hetherington and Ranson, 1940; Hoebel and Leibowitz, 1981). However, the specific involvement of these hypothalamic "centers" in feeding and satiety was questioned, and the alternative hypothesis that these lesion effects were due to destruction or stimulation of fibers of passage, rather than to effects on cell bodies (Ungerstedt, 1971), was advanced. In 1960, S.P. Grossman discovered that norepinephrine (NE) administered in the hypothalamus elicited a strong feeding response in satiated animals. This finding confirmed earlier ideas about the involvement of hypothalamic neurons in the control of feeding behavior and also expanded our thinking about the role of specific neurotransmitters in this behavioral response. The phenomenon of NE-stimulated eating has stood the test of time, generating a broad hypothesis that attributes to brain NE an important and specific role in the short-term maintenance of body energy stores.

Specifically, this hypothesis states that (1) a function of medial hypothalamic NE is to signal a condition of negative energy balance and then to initiate a powerful feeding response specifically oriented toward ingestion of foods, especially carbohydrates, that can be rapidly metabolized to correct that imbalance; (2) a critical neural component of this specific feeding response is noradrenergic neurons that ascend from specific hindbrain cell groups to innervate the paraventricular nucleus of the hypothalamus (PVN); (3) under conditions of low energy stores, these neurons release NE, which through $\alpha 2$-adrenergic receptors inhibits PVN efferents that descend medially to terminate in specific hindbrain cell groups; (4) this feeding response is partially dependent upon vagally-mediated neurogenic release of pancreatic insulin; (5) adrenal corticosterone plays a permissive role in this feeding response by controlling the number of $\alpha 2$-adrenergic receptors in the PVN; and (6) this NE feeding system is believed to be particularly active during food deprivation and the initial phase of the feeding period, when short-term energy stores are low. Of course, NE does not work alone but in concert with other neurochemical, neuroanatomical, and peripheral systems to integrate a complex array of metabolic, humoral, and neural information.

1. Site of Action and Neural Substrates

Since the discovery that NE potentiates feeding in the rat, a variety of studies have attempted to find the neural substrates of this effect. Evidence points to the PVN, and possibly the paraventricular nucleus, as the brain area that most likely mediates this stimulation of feeding. Extensive cannula-mapping studies have revealed that hypothalamic, but not extra-hypothalamic, NE stimulates feeding behavior; that the amount eaten increases, and the latency to eat decreases, with increased proximity of the injection site to the PVN; and that injections of near physiological doses of NE into this nucleus effectively stimulate feeding behavior (Leibowitz, 1978a,b) and inhibit neural firing (Moss et al., 1972). Moreover, electrolytic lesions of the PVN strongly attenuate feeding elicited by ventricular injection of NE, and severing PVN efferents with brainstem knife cuts attenuates feeding produced by PVN injection of NE (Leibowitz, et al., 1983; Weiss and Leibowitz, 1985).

An extensive series of neuropharmacological studies focused on the type of receptor mediating this response suggests that it is α2-noradrenergic in nature. Feeding is elicited by PVN or, in some cases, peripheral injections of α2-noradrenergic agonists, but not by ß-adrenergic, dopaminergic, serotonergic, or cholinergic agonists. Conversely, food intake stimulated by PVN injections of exogenous NE, release of endogenous NE, or α2-noradrenergic agonists is attenuated by α2-noradrenergic antagonists, but not by other antagonists (Goldman et al., 1985; Leibowitz, 1980; McCabe et al., 1984). Moreover, the PVN contains particularly high levels of α2-adrenergic receptors (Leibowitz et al., 1982). Findings to suggest that NE is the endogenous ligand for these receptors are (1) the PVN contains high levels of NE-containing presynaptic terminals (Sawchenko and Swanson, 1982) and (2) feeding behavior is elicited by PVN injection of drugs that release endogenous NE. The feeding behavior elicited by presynaptic releasers of NE, but not that induced by α-adrenergic agonists which act postsynaptically, is blocked by inhibitors of endogenous NE synthesis, and the feeding elicited by both types of drug is blocked by α-adrenergic receptor blockers (Leibowitz et al., 1978a,b). Further evidence supporting a role for endogenous NE in feeding behavior is that feeding is elicited by PVN injection of low, near physiological, doses (4 ng; 25 pmol) of NE (Leibowitz, 1978a).

Other studies have focused on identifying the afferent and efferent neural systems that mediate this response. Anatomical and behavioral studies suggest that the NE feeding response is mediated by catecholaminergic (CA) PVN afferents that may arise from the locus coeruleus (and possibly dorsal medullary cell groups) and project to the PVN via the dorsal central tegmental tract. It has been shown that electrolytic lesions or catecholamine neurotoxin injection in the locus coeruleus and along the dorsal central tegmental pathway to the PVN reduce the level of endogenous NE in the PVN and, more important, abolish the stimulation of feeding produced by drugs that act presynaptically to release endogenous NE (Halperin et al., 1979; Leibowitz and Brown, 1980a). In contrast to the ineffective presynaptic-acting drugs, postsynaptic NE agonists actually become more effective after denervation of the PVN, thus elegantly demonstrating the behavioral specificity of this effect.

Anatomical and behavioral studies of the behaviorally relevant PVN efferents suggest that the feeding stimulation is produced by activation of neurons projecting through a descending medial paraventricular system to innervate pontine/medullary nuclei, most likely the dorsal motor nucleus of the vagus and nucleus of the solitary tract. This conclusion is based on the findings that feeding induced by PVN injection of NE and by peripheral α2-agonists is strongly attenuated by knife cuts that sever at various levels a pathway projecting dorsocaudally from the PVN, through the thalamic paraventricular region, and then along the periaqueductal gray at the level of the midbrain and pons (McCabe et al., 1984; Weiss and Leibowitz, 1985). The involvement of the vagal nuclei is suggested by the finding that feeding elicited by PVN injection of NE is blocked by vagotomy or by peripheral cholinergic antagonists (Sawchenko et al., 1981).

2. Behavioral Analysis and Diet Selection

The next question is, What are the behavioral effects of stimulating this pathway? Norepinephrine (between 0.025 and 50.0 nmol) injected into the PVN of satiated rats produces a dose-dependent stimulation of feeding behavior: the average latency to eat is 2-3 minutes, with feeding frequently occurring within 20-30 seconds in some animals. When eating pelleted food, the subjects eat virtually without interruption for about 15-20 minutes and consume an average of 3-4 g of food (Leibowitz, 1975a; 1980). This effect may be elicited at any time during the light/dark phase, but it appears to be particularly effective just prior to the beginning of the nocturnal feeding period, when the animals' short-term energy stores are at their lowest (Bhakthavatsalam and Leibowitz, in press). It is clear that a meal may be initiated by stimulation of this NE feeding control system; however, meal size may also be increased, since PVN injections of NE into food-deprived rats actually increase the length of time the animals eat, as well as the amount of food eaten (Ritter and Epstein, 1975). This effect on meal size may be more characteristic of this system, since satiated animals given PVN injections of NE, either acutely or chronically, are more likely to eat bigger meals than to eat more meals (Leibowitz et al., 1984a; Shor-Posner et al., 1985b), and since the threshold dose for NE's effect on meal size is lower than that for its effect on meal frequency (Ritter and Epstein, 1975). For this and other reasons, it has been suggested that NE may act to stimulate feeding by inhibiting "satiety" neurons in the PVN (Leibowitz, 1980; 1985).

Recent studies have also indicated that this PVN noradrenergic system may function to stimulate appetite specifically for carbohydrate. When given acute or chronic PVN injections of NE or drugs that release NE, subjects with simultaneous access to three diets of pure protein, fat, and carbohydrate markedly increase consumption of the carbohydrate diet with little or no consumption of the other diets (Leibowitz et al., 1985a,b). A crucial question is whether the increased consumption of carbohydrate is due to sensory factors, such as taste, texture, or odor, or to the postingestive (i.e., metabolic) consequences of the diet. Evidence to date clearly favors metabolic over sensory factors as critical in the diet selection, since animals given PVN injection of NE consume a variety of sweet and non-sweet carbohydrate diets and drink water containing sucrose but not saccharin. This is not to say that sensory factors are unimportant, and in fact we suggest that one mechanism of this effect may be an NE-induced change in the hedonic value of the sensory properties of the carbohydrate diet that have, through experience, been associated with this diet's postingestive metabolic consequences (Leibowitz, 1982).

Considerations of the function of carbohydrates provided one of the first clues to a possible function of this PVN NE feeding control system. The major function of pure carbohydrates is to provide a source of energy for metabolism. Although both protein and fat may be used for energy, and in fact fat is more calorically dense than carbohydrate, they are not immediately available for metabolism because of factors such as slow gastric transit times. In contrast to ingestion of pure fat, which has no effect on blood glucose, ingestion of carbohydrate may double blood glucose during the meal; the increase is apparent within 2 minutes of ingestion, rapidly repleting low energy stores (Steffens, 1969). Therefore, the NE injections may elicit carbohydrate ingestion because this diet provides energy in the form which can most rapidly be utilized. According to this formulation, a function of PVN NE is to respond to low energy availability by initiating feeding behavior oriented toward ingestion of that food which can be metabolized most rapidly.

The idea that the PVN NE feeding control system may have a function specifically in relation to the organism's short-term energy stores is supported by the dependence of this phenomenon on plasma levels of glucocorticoids (Leibowitz et al., 1984b; Roland et al., 1985), and on vagally-mediated release of pancreatic insulin (Sawchenko et al., 1981), both of which are known to play critical roles in the disposition and utilization of short-term energy stores. It has been shown that plasma adrenal glucocorticoids play a permissive role in NE-induced feeding, with a strong positive correlation between the levels of circulating corticosterone and the magnitude of feeding produced by PVN injections of NE (Leibowitz et al., 1984b; Roland et al., 1985). This correlation may be due to the impact of plasma corticosterone on the number of α2-adrenergic receptors available in the PVN, with high levels of circulating corticosterone causing increases in the number of receptors and low levels of plasma corticosterone decreasing the number of α2-adrenergic receptors in the PVN (Jhanwar-Uniyal et al., 1984a; Leibowitz et al., 1984c).

Further evidence for a relationship to energy availability is that α-adrenergic antagonists attenuate feeding elicited by glucoprivation, which has been shown to increase NE turnover and release unless the animals are allowed to feed (Bellin and Ritter, 1981; Berthoud and Mogenson, 1977; Muller et al., 1972; McCaleb and Myers, 1982; Smythe et al., 1984).

3. When Is the System Active?

Studies employing central microinjections may demonstrate that activation of a neural system can elicit eating; however, they cannot show under what conditions, if any, the system is normally active. One way to show this is to measure changes in the activity of the neurotransmitters or their receptors within discrete brain areas as a function of conditions normally producing hunger or satiety. Studies of this nature support a role for PVN NE in the normal daily control of eating.

Depletion of body energy stores by 6 to 48 hours of food deprivation causes increased turnover of NE and consequent down-regulation of α2-adrenergic receptors, specifically in the PVN. These effects are reversed subsequent to the ingestion of food (Jhanwar-Uniyal et al., 1980, 1982). Similarly, it has been reported that medial hypothalamic release of NE is increased during food deprivation, remains high during eating, and is inhibited by gastric intubation of carbohydrate (Martin and Myers, 1975; McCaleb et al., 1979; Myers and McCaleb, 1980; Van Der Gutgen and Slanger, 1977). Thus, the release of medial hypothalamic NE appears to be enhanced by conditions that normally induce eating and is suppressed by factors associated with satiety. These findings, in conjunction with studies demonstrating that medial hypothalamic injections of NE stimulate feeding, strongly argue for a role of the PVN NE system in the control of normal food intake.

Further support for a role of the PVN in control of normal eating comes from the findings that electrolytic lesions of this nucleus produce hyperphagia and obesity in rats (Leibowitz et al., 1981) and similarly, that rats given chronic repeated injection of NE overeat and gain weight on a mixed macronutrient diet (Lichtenstein et al., 1984). Interestingly, when these subjects are given three diets consisting of a pure carbohydrate, fat, and protein, the hyperphagia is expressed only by increased consumption of the carbohydrate diet, while intake of the protein and fat diets may be unchanged or actually decreased (Leibowitz et al., 1981, 1985a; Lichtenstein et al., 1984; Sclafani and Aravich, 1983). Furthermore, subjects with neurotoxin-induced NE depletion in the PVN exhibit a deficit in carbohydrate ingestion, under ad libitum feeding conditions and in response to deprivation (Azar et al., 1984; Leibowitz et al., 1980).

More recent findings argue for a specific role for this control system in feeding at the beginning of the active cycle when short-term reserves of energy are low and circulating levels of corticosterone are high (Krieger and Hauser, 1978; LeMagnen, 1981). Recent studies have shown that during this period, feeding is characterized by increased meal size and, more important, a strong preference for carbohydrate (Tempel et al., 1985). The possibility that this consumption of carbohydrate is normally elicited by the release of endogenous PVN NE at the beginning of the active cycle is suggested by the finding that PVN $\alpha 2$-adrenergic receptors are strongly (65%) down-regulated by as little as 3 hours of food deprivation, specifically during the beginning of the nocturnal feeding period (Jhanwar-Uniyal et al., 1985) and that during this period PVN NE is more effective that at any other time in stimulating feeding behavior (Bhakthavatsalam and Leibowitz, in press).

B. Dopamine and Epinephrine

Another phenomenon that has generated considerable research is the suppression of feeding caused by perifornical hypothalamic (PFH) injection of dopamine (DA) and epinephrine (EPI). When injected into the PFH, these two neurotransmitters have little effect on the food intake of satiated animals, in contrast to PVN injection of NE. However, in food-deprived animals, DA, EPI, and to a lesser extent, NE produce a powerful suppression of eating. While evidence for the specific function of these neurotransmitters in feeding is not as strong as for PVN NE, it has allowed an integrated hypothesis to be developed (Leibowitz, 1985).

According to this hypothesis, axons from DA cell bodies in the midbrain tegmentum and EPI neurons from cell groups in the ventrolateral medulla ascend in the ventral central tegmental tract to terminate on feeding-stimulatory neurons in the PFH. We believe that the ingestion of food, and perhaps ingestion of food high in protein, stimulates the release of DA and EPI, which then act on DA and ß-adrenergic receptors, respectively, to inhibit PFH "feeding neurons" and thereby suppress feeding behavior. Conversely, under conditions of food deprivation and perhaps specifically in response to low levels of available protein, the release of DA and EPI is decreased, thereby disinhibiting PFH "feeding neurons" and consequently initiating feeding behavior.

1. Site(s) of Action and Neural Substrates

Since the discovery that DA and EPI suppress feeding (Leibowitz, 1970; Leibowitz and Rossakis, 1978), a variety of studies have attempted to explain the neural substrates of this effect, and evidence points to the PFH as the most likely brain site. An extensive cannula-mapping study has demonstrated that the PFH is the hypothalamic site most sensitive to the anorectic effects of locally injected DA and EPI (Leibowitz and Rossakis, 1979a). Injections of either of these transmitters specifically into the PFH can produce a 70-90% suppression of feeding by 18-hour food-deprived animals. As the distance between the injection site and the PFH increases, the effect diminishes, and extra-hypothalamic injections are completely ineffective. Doses of DA and EPI as low as 30 ng produce a significant inhibition of eating when injected specifically into the PFH (Leibowitz and Rossakis, 1978a; 1979a). This brain site is also responsive to locally injected amphetamine, which inhibits feeding through the release of endogenous catecholamines (Leibowitz, 1975b). The PFH also appears to be crucial in the anorectic effects of peripheral amphetamine, since DA and ß-adrenergic antagonists injected directly into the PFH, as well as electrolytic lesions of this area, attenuate the anorectic effect of peripherally injected amphetamine (Blundell and Leshem, 1974; Leibowitz, 1975b,c; McCabe et al., 1985).

An extensive series of neuropharmacological studies suggests that DA and ß-adrenergic receptors in the PFH mediate the anorexia produced by DA and EPI. Specifically, anorexia induced by PFH injection of either DA or EPI is attenuated by DA and ß-adrenergic receptor antagonists, respectively, but not by a variety of other antagonists. Conversely, specific DA and ß-adrenergic agonists generally suppress feeding (Leibowitz and Rossakis, 1978b; 1979b), and further, the PFH is rich in dopaminergic and to a lesser extent ß-adrenergic receptors (Leibowitz et al., 1982a). The EPI neurons may terminate on presynaptic DA terminals, since DA antagonists block the anorectic effect of EPI, while ß-adrenergic antagonists do not block DA's effect (Leibowitz and Rossakis, 1978b; 1979b). Dopamine and EPI appear to be the endogenous ligands for these receptors, since drugs (such as amphetamine) that release endogenous DA and EPI mimic their effects, and these effects are blocked by inhibitors of DA and EPI synthesis (Leibowitz, 1975c; Leibowitz and Rossakis, 1979c).

The origins of the EPI and DA neurons mediating feeding have recently been described. Anatomical studies have shown that catecholamine (CA)-containing PFH efferents arise from, among others, DA-containing cell groups in the midbrain tegmentum dorsal to the substantial nigra, and from EPI- or NE-containing cells in the lateral reticular nucleus or nucleus of the solitary tract projecting to the PFH through the ventral tegmental tract (Leibowitz and Brown, 1980b). Electrolytic lesions or CA neurotoxins given at various points along this pathway produce depletion of CA terminals and, more important, reduce the anorectic effect of PFH as well as peripheral injection of DA- and EPI-releasing drugs. In contrast to the ineffective presynaptic releasers, DA and EPI acting postsynaptically actually become more effective, suggesting that DA and ß-adrenergic receptors are up-regulated by denervation of the PFH (Ahlskog and Hoebel, 1973; Leibowitz and Brown, 1980b; McCabe and Leibowitz, 1984b).

2. Behavioral Analysis and Macronutrient Selection

The next question is, What are the specific behavioral effects that result from activation of this pathway? Dopamine or EPI injected into the PFH of food-deprived rats produce a dose-dependent suppression of feeding behavior. The magnitude of the suppression is 70-90% for optimally placed injections, and the effect may be seen as soon as 1 or 2 minutes postinjection. Meal pattern analysis has revealed that PFH or peripheral administration of amphetamine, presumably through activation of hypothalamic CA neurons, decrease food intake by increasing the latency to eat and the intermeal interval (Blundell, 1984; Blundell et al., 1976; Grinker et al., 1982). The behavioral specificity of this effect is demonstrated most clearly by the finding that DA antagonists injected into the PFH of satiated rats actually cause an increase in food intake (Hoebel and Leibowitz, 1981). This suggests that the system is tonically active in satiated rats, acting to prevent meal initiation.

Recent studies also indicate that the PFH catecholamine feeding system may function to control appetite for specific macronutrients, particularly protein. Dopamine, EPI, or amphetamine injected into the PFH cause a specific suppression of protein consumption in food-deprived rats given access to three diets of pure macronutrients. The opposite effect, a specific increase in protein intake, is seen in satiated rats given PFH injections of DA antagonists (Leibowitz et al., 1982b, and unpublished results). Recent findings show that rats switch from meals composed mainly of carbohydrate early in the active period to meals of protein later in this period, and that they also alternate between predominantly protein and predominantly carbohydrate meals (Tempel et al., 1985). The specific effects of PFH CA on protein consumption suggest that the alternation between carbohydrate and protein meals may be mediated in part by PFH CA (Leibowitz and Shor-Posner, 1985), consistent with the increased synthesis of CA induced by consumption of protein (Fernstrom and Faller, 1978). Thus, DA and EPI in the PFH may play a role not only in determining when a meal is initiated, and how much is eaten, but also in determining the specific types of food eaten and the sequence in which they are consumed.

3. When Is the System Active?

Recent studies measuring the changes in the levels of DA and ß-adrenergic receptors in the PFH as a function of the subject's nutritional status have helped to clarify the conditions that activate this system. These studies suggest that DA and EPI are released within the PFH to prevent meal initiation (possibly of protein meals) or to cause meal termination under conditions in which stores of nutrients are sufficient to meet short-term need. In contrast, release is inhibited, allowing animals to eat, under conditions (such as food deprivation) in which the short-term stores of nutrients are low. Specifically, we have shown that in the PFH DA and ß-adrenergic receptors are up-regulated, indicating a decrease in the release of CA (Jhanwar-Uniyal et al., 1980). Conversely, gastric intubation of nutrients into hungry animals increases PFH release of CA (Myers and McCaleb, 1980), and the release of DA from the hypothalamus is increased during feeding (Heffner et al., 1980). These findings support the idea that PFH CA activity may be positively correlated with the availability of ingested nutrients.

Further evidence for a suppressive role of endogenous PFH CA in normal feeding is that destruction of PFH afferent neurons by brainstem knife cuts, electrolytic lesions, or CA neurotoxin injection causes hyperphagia and obesity. Acute blockade of endogenous DA and EPI activity by local injections of DA antagonists also increases eating behavior and, conversely, inhibiting PFH neural output by lesioning this area causes hypophagia and weight loss (Ahlskog and Hoebel 1973; Hoebel and Leibowitz, 1981; Leibowitz et al., 1980; McCabe et al., 1985). Like lesions, chronic PFH injections of DA and EPI cause hypophagia and weight loss, indicating that these neurotransmitters suppress feeding by inhibiting PFH neural output, which is important in normal feeding and maintenance of normal body weight (Leibowitz and Roossin, unpublished results). Consistent with this idea are findings that electrical stimulation of the PFH elicits an immediate and powerful feeding response in satiated rats (Hoebel and Leibowitz, 1981).

A series of studies employing electrophysiological recordings of changes in LH neuronal response to food-related stimuli in awake monkeys has greatly expanded our understanding of hypothalamic function in relation to the control of feeding behavior (Burton et al., 1976; Mora et al., 1976; Rolls et al., 1976, 1979; Rolls, 1982). These studies have shown that cells in the LH respond to food-related stimuli presented in several different sensory modalities. For example, food presented visually changes the firing rate of some LH neurons within milliseconds. These changes are seen only in hungry animals, and are related to the degree of preference for the specific food presented. Since the firing of these LH neurons precedes and predicts the animal's reaction to the food, it has been suggested that they play a role in the initiation of the feeding responses. Electrophysiological studies in rats have also shown that some LH cells are inhibited by both local and hepatoportal glucose injection. It is particularly interesting that these glucose-sensitive cells are inhibited by local NE and DA administration, as well as by electrical stimulation of the ventral central tegmental pathway. Surprisingly, α-adrenergic rather than ß-adrenergic antagonists block the electrophysiological effects of hepatic, but not local, glucose administration (Miyahara and Oomura, 1982; Shimizu et al., 1983; Sikdar et al., 1985). While further study to clarify the specific involvement of particular CAs and their receptor subtypes is clearly needed, these studies represent an important effort to integrate information obtained through electrophysiological and neuropharmacological approaches. Taken together with the evidence for PFH CA mediation of nutrient availability on feeding, these findings suggest the interesting possibility that CA neurons terminating in the PFH/LH may mediate the satiating effects of ingested foods on the response of LH feeding neurons.

C. Gamma-Aminobutyric Acid

In contrast to most neurotransmitters, which affect feeding in only one direction, the amino acid gamma-aminobutyric acid (GABA) has been shown to either stimulate or inhibit feeding, depending on the site of injection and deprivation state of the animal. In particular, these effects appear to center on the PVN and the PFH, respectively, where GABA interneurons may mediate the stimulatory effects of PVN NE on feeding, and the inhibitory effects of DA and EPI on feeding behavior.

1. Sites of Action and Neural Substrates

The levels of GABA and its synthesizing enzymes are dense within the hypothalamus, and evidence suggests that hypothalamic GABAergic neurons in the hypothalamus are generally short interneurons, inhibitory in nature (Tappaz and Brownstein, 1977). Injection of GABA into the medial hypothalamus elicits feeding behavior in satiated rats (Grandison and Guidotti, 1977), whereas injection further laterally suppresses feeding in deprived rats (Kelly, 1978). Analysis of the magnitude of these effects as a function of cannula placement indicates that the focus of these stimulatory and inhibitory effects is on the PVN and PFH, respectively. Eating elicited by PVN injections of GABA appears to be mediated by activation of GABA receptors, since injection of low doses of GABA or its agonists into the PVN stimulates feeding, and GABA antagonists attenuate this effect. These antagonists also attenuate feeding elicited by PVN injection of NE, indicating the possible dependence of this neurotransmitter's function on GABAergic neurons. Furthermore, PVN injections of GABA antagonists actually reduce deprivation-induced feeding, implicating GABA receptors in normal eating behavior (Kelly, 1978; Kelly and Grossman, 1979; Kelly et al., 1977; 1979). It may be noted that benzodiazepines, which may act in part via GABA receptors (Sanger, 1985), stimulate feeding behavior after either peripheral or PVN injections (Anderson-Baker et al., 1979; Cooper, 1983; Cooper and Estall, 1985; Kelly, 1978).

In the lateral areas of the hypothalamus (PFH/LH), changes opposite to those observed medially are obtained with GABAergic drugs. When injected into the PFH or further ventrally just lateral to the VMH, GABA, its agonists, or drugs that enhance endogenous GABA synthesis suppress feeding. GABA receptor antagonists reduce the suppression produced by GABA and actually increase feeding behavior when injected into satiated animals, suggesting that in the LH, endogenous GABA inhibits normal feeding (Kelly and Grossman 1979; Kelly et al., 1977, 1979;

Panksepp and Meeker, 1980). Evidence that the inhibitory effect of PFH CA on feeding is mediated, in part, through GABA interneurons is provided by the finding that the anorectic effect of amphetamine is dependent upon synthesis of endogenous GABA in the PFH (Kelly, 1978).

Extra-hypothalamic GABA may also be involved in feeding. Injection of GABA agonists into the dorsal raphe nucleus produces an intense feeding response, which is attenuated by local injection of GABA antagonists (Przewlocka et al., 1979). Feeding elicited by midbrain GABA may be mediated by α-adrenergic receptors, since this feeding response is attenuated by an α-adrenergic antagonist, but not by a serotonin neurotoxin (Borsini et al., 1983).

2. Possible Functional Significance

The possibility that medial and lateral hypothalamic GABA interneurons provide a functional intermediary link between glucose metabolism and feeding is suggested by a variety of biochemical studies. The levels of GABA in the hypothalamus appear to be coupled to glucose availability and utilization. Opposite changes in medial and lateral hypothalamic GABA levels occur in response to hyper- and hypoglycemic conditions, with hypoglycemia increasing GABA synthesis in the medial hypothalamus and decreasing it in the LH (Kimura and Kuriyama, 1975). Furthermore, medial hypothalamic injections of GABA antagonists attenuate eating induced by peripheral hypoglycemia (Kamatchi et al., 1984). Conversely, in hyperglycemic conditions, as well as during food ingestion, endogenous GABA activity is enhanced ventrolateral to the ventromedial nucleus as opposed to in the nucleus itself (Meeker and Myers, 1980). Injection of glucose and specific metabolic substrates of GABA synthesis into the PFH is effective in inhibiting feeding (Panksepp and Meeker, 1980). This evidence implicates LH GABA as a potential "satiety" signal, consistent with the circadian analysis of hypothalamic GABA levels and the inhibitory impact of enhanced brain synthesis on food ingestion (Cattabeni et al., 1978; Coscina, 1983; Olgiatti et al., 1980). The finding that PVN GABA affects feeding in the opposite direction, causing a potentiation of food consumption, indicates that GABA may function similarly to the CAs in the PVN and PFH. The coupling of GABA synthesis to energy metabolism may provide an intermediary step in the monitoring of energy and nutrient utilization by hypothalamic catecholamine projections and their subsequent control of eating behavior.

D. Serotonin

Numerous studies have suggested that activation of serotonergic receptors reduces eating in food-deprived animals. Central and peripheral injections of serotonin (5HT), or its precursors, agonists, and presynaptic releasers, have consistently produced a suppression of feeding (Blundell, 1984; Leibowitz, 1985); conversely, depletion of central 5HT has been shown to increase feeding (Breish et al., 1976; Saller and Stricker, 1976). However, there are as yet unresolved questions concerning the specificity and site(s) of action of these effects (Blundell, 1984; Coscina et al., 1978; Hoebel et al., 1978). There is recent evidence to suggest that medial hypothalamic 5HT activity may act to suppress feeding by acting in opposition to medial hypothalamic NE. More specifically, the general hypothesis proposed here is that an increase in 5HT synthesis resulting from a meal high in carbohydrate acts to reduce subsequent ingestion of food, in particular carbohydrate, by counteracting the stimulatory effect of PVN NE on ingestion of this macronutrient. Serotonin in other brain areas may control consumption of other macronutrients.

1. Sites of Action and Neural Substrates

The weight of evidence suggests a central site of action for 5HT, possibly in addition to a peripheral site. Intraventricular injections of 5HT, or drugs which mimic its effects, suppress feeding behavior (Blundell, 1984; Kruk, 1973; Leibowitz, 1985). Central injections of 5HT neurotoxins have been shown to reduce the anorectic effects of drugs that release 5HT presynaptically, but not of those acting directly on 5HT receptors (Clineschmidt et al., 1974). These findings, however, have not been confirmed in all studies (Blundell, 1984; Carlton and Rowland, 1984).

Further insight into the action of 5HT comes from behavioral and pharmacological studies of its inhibitory actions within the hypothalamus. The medial hypothalamus appears to be important, since medial hypothalamic injection of a drug that releases 5HT produces hyperphagia and weight gain (Blundell and Leshem, 1973; Waldbillig et al., 1981). One site that appears to be particularly sensitive to 5HT is the PVN (Leibowitz and Papadakos, 1978), which is innervated by 5HT-containing neurons originating from cell groups in the midbrain raphe (Sawchenko et al., 1983). Although cannula mapping studies have not been performed, injection of 5HT into the PVN produces a suppression of feeding behavior. This effect appears to be mediated by 5HT receptors, since serotonergic antagonists, as opposed to others, reduce 5HT's effectiveness. Furthermore, drugs which increase the synthesis of endogenous 5HT or which act presynaptically to release 5HT are effective in low doses within the PVN, suggesting that release of endogenous medial hypothalamic 5HT may have functional significance in feeding behavior (Leibowitz, 1985).

2. Behavioral Analysis and Diet Selection

Analysis of meal patterns reveals that injection of 5HT into the PVN of food-deprived rats causes a significant decrease in the duration of the meal, the rate of eating, and the amount eaten, without any effect upon the latency to eat or the number of meals eaten (Leibowitz and Shor-Posner, 1986). This pattern of results is consistent with the effect on meal patterns of peripheral or central administration of drugs that enhance or mimic 5HT synaptic activity. Specifically, peripheral administration of fenfluramine, as well as PVN injection of norfenfluramine, decreases meal size and rate of eating with no effect upon the number of meals eaten or the latency to eat (Blundell, 1984; Blundell and Leshem, 1973; Leibowitz and Shor-Posner, 1986). These findings suggest that centrally and peripherally administered 5HT agonists may to some extent share a common site of action.

This proposal is consistent with results obtained from studies of 5HT's impact on consumption of specific macronutrients. These studies reveal that with PVN injection of 5HT or drugs that enhance its synaptic availability, there is a specific reduction in ingestion of carbohydrate, and either no effect on protein intake, or an enhancement of it (Leibowitz and Shor-Posner, 1986). Like PVN injection of 5HT, peripheral administration of drugs that release 5HT significantly reduces consumption of carbohydrates (or spare protein) in a macronutrient self-selection paradigm (Blundell and McArthur, 1979; Li and Anderson, 1984; Shor-Posner and Leibowitz, in preparation; Wurtman and Wurtman, 1977; 1979). These results, suggesting a specific suppressive role for 5HT in carbohydrate consumption, are supported by the increase in carbohydrate consumption induced by manipulations that block 5HT function (Shor-Posner and Leibowitz, in preparation). These findings have been extended to humans, and, as in animals, enhanced 5HT activity is associated with a decrease in the consumption of carbohydrate-rich snacks (Wurtman and Wurtman, 1984). Other studies indicate that under certain conditions fat consumption may also be affected (Orthen-Gambill and Kanarek, 1982; Shor-Posner and Leibowitz, in preparation).

3. How do PVN NE and 5HT interact?

It may be noted that PVN administration of NE and 5HT have opposite effects on meal patterns and macronutrient selection (see section II,A,2). Whereas PVN 5HT decreases meal size, meal duration, rate of eating, and carbohydrate intake, PVN NE increases each of these parameters. This supports the proposal that these neurotransmitter systems interact (Samanin and Garattini, 1975), perhaps within the PVN, in an antagonistic fashion to control food intake, specifically, carbohydrate intake or the carbohydrate/protein ratio (Hoebel and Leibowitz, 1981; Leibowitz and Shor-Posner, 1986). This possibility is supported by studies that directly examined the impact of PVN injections of 5HT on feeding elicited by NE injected into the PVN (Leibowitz and Papdakos, 1978; Leibowitz and Shor-Posner, 1986). These studies reveal that 5HT is particularly effective in suppressing feeding elicited by PVN injection of NE. Enhanced release of endogenous 5HT, through administration of its precursor 5-hydroxytryptophan, is also effective in suppressing NE's effect. In terms of the specific relationship of these neurotransmitters in the PVN, it is suggested that 5HT may suppress feeding by directly acting to increase the firing of PVN "satiety" neurons, and that α2-adrenergic receptors on 5HT presynaptic terminals reduce 5HT release and consequently disinhibit eating (Leibowitz and Papadakos, 1978; Leibowitz, 1985; Shor-Posner, 1986). This proposal is supported by the findings that, within the hypothalamus, α2-adrenergic receptors act to inhibit 5HT release; 5HT innervation of the PVN is most dense where noradrenergic terminals are believed to act in the control of feeding; and fenfluramine acts to enhance electrical activity in the medial hypothalamus (Chesselet, 1984; Foxwell et al., 1969; Galzin et al., 1984; Leibowitz, 1978b; Sawchenko et al., 1983).

This proposed antagonistic relationship between medial hypothalamic 5HT and NE, perhaps in conjunction with CA of the LH, provides a potential neural substrate for controlling the initiation and termination of eating behavior, particularly for balancing the ingestion of specific macronutrients (Leibowitz and Shor-Posner, 1986). During the early part of the nocturnal active period, food intake by rats is characterized by the preferential consumption of carbohydrate (Tempel et al., 1985). It has been proposed (see section II,A) that in response to low energy stores at the beginning of the active cycle, the consumption of carbohydrate is initiated in part by increased release of NE in the PVN. We now suggest that ingestion of carbohydrates, perhaps through the increased ratio of tryptophan to other amino acids (Wurtman et al., 1981), may increase 5HT activity in the PVN and thereby antagonize PVN NE. This may account for the decrease in preference for carbohydrates, and the switch to consumption of proteins, which may be mediated by opiate peptides and/or by PFH DA and EPI (see sections II,B and III,A).

III. FEEDING-STIMULATORY PEPTIDES

Our understanding of brain function has been dramatically enhanced by the discovery of endogenous brain peptides and by subsequent evidence that these substances may function as neurotransmitters, neuromodulators, and neurohormones. A role in stimulating feeding behavior has so far been demonstrated for only three peptides or peptide families, whereas numerous peptides have been shown to inhibit feeding (see section IV). Those that increase feeding are the opiate peptides, and the more recently discovered pancreatic polypeptides and growth hormone-releasing factor.

A. Opiate Peptides

Evidence is accumulating to support the proposition that endogenous opiate peptides play an important role in the expression of eating behavior. Central and peripheral opioid agonists generally stimulate eating in satiated animals, whereas opioid antagonists reduce feeding in food-deprived animals. These effects on feeding behavior do not appear to be expressed through a single mechanism, but rather through several different, possibly independent, mechanisms. Opioid-elicited feeding behavior involves multiple hypothalamic and extra-hypothalamic brain sites, multiple opiate receptor subtypes, as well as CA- and adrenal-dependent and independent mechanisms. Evidence for opiate involvement in control of feeding behavior has recently been extensively reviewed (Morley et al., 1983); therefore, our focus here will be on recent findings and results obtained in this laboratory.

1. Neural Substrates

Cannula mapping studies have revealed that opioids injected into several different brain areas produce large increases in food intake (Grandison and Guidotti, 1977; Lanthier et al., 1985; Stanley et al., 1984; Woods and Leibowitz, 1985). A brain site which appears to be particularly sensitive in the rat is the PVN. In this hypothalamic area, food intake is increased by an average of 7-9 g over baseline, 3 hours after morphine (25 nmol) injection (Lanthier et al., 1985). Conversely, the opioid antagonist naloxone, as well as antibodies to endogenous opiates, are found to suppress deprivation-induced feeding when injected into this brain site (Gosnell et al., 1984; Schulz et al., 1984; Woods and Leibowitz, 1985). These results, suggesting that the PVN may partially mediate opioid-elicited feeding, are supported by the findings that electrolytic lesions of the PVN significantly attenuate, but do not abolish, feeding elicited by peripheral morphine (Shor-Posner et al., 1985a) and that morphine, like NE, inhibits the firing of PVN neurons (Pittman et al., 1980). Furthermore, injections of a met-enkephalin analogue as well as ß-endorphin into the PVN have been shown to enhance feeding behavior (Leibowitz and Hor, 1982; McLean and Hoebel, 1983; Stanley et al., 1984). The PVN has been reported to contain dense concentrations of opiate terminals and several different types of opiate receptor sites (DiFiglia and Aronin, 1984; for review see Leibowitz, 1986).

In a number of studies, the eating-stimulatory effect of medial hypothalamic opiate injection is reversed by α-adrenergic receptor blockade, as well as by opiate antagonists (Leibowitz and Hor, 1982; Tepperman et al., 1981). This raises the possibility that some opioid peptides in the PVN may interact with NE to produce eating. An association between the opiates and NE is supported by the findings that peripheral morphine increases the turnover of NE in the PVN (Jhanwar-Uniyal et. al., 1984b) and that, as with NE, the feeding stimulatory effect of PVN and peripheral morphine injection is attenuated by adrenalectomy and restored by corticosterone (Bhakthavatsalam and Leibowitz, 1984). Further, the magnitude of feeding stimulation produced in individual subjects by PVN injection of NE and opioid peptides is positively correlated (Leibowitz and Hor, 1982), and both exhibit a similar diurnal pattern of responsiveness, with peak effects observed early in the dark phase when circulating corticosterone, is at its highest (Bhakthavatsalam and Leibowitz, 1984; and in press; Krieger and Hauser, 1978).

It is clear that opioid receptor systems may also function independently of the PVN α-adrenergic feeding system. In contrast to α2-noradrenergic receptors, the different opioid receptors (μ, δ, κ) appear to be active at multiple brain sites to stimulate feeding (Lanthier et al., 1985; Scott et al., 1984; Stanley et al., 1984; Woods and Leibowitz, 1985). The feeding stimulatory effects of κ and δ agonists, in contrast to μ opioid and α2-adrenergic agonists, do not appear to be dependent on circulating corticosterone, since they are not blocked by adrenalectomy (Levine and Morley, 1983; McLean and Hoebel, 1982; Bhakthavatsalam and Leibowitz, 1984; Bhakthavatsalam and Leibowitz, in press). Finally, the functional NE input to the PVN does not seem to be necessary for morphine stimulation of feeding, as a CA neurotoxin injected into the PVN did not affect the morphine elicited feeding response (Shor-Posner et al., 1985a).

2. Diet Selection

Of particular interest are the different effects of morphine and NE on diet selection. As described above (section II,A,2), PVN injection of NE selectively enhances carbohydrate intake. Similar studies indicate that, in contrast to NE, morphine injected either peripherally or into the PVN preferentially stimulates protein and fat consumption, while sometimes causing a reduction of carbohydrate consumption (Leibowitz, unpublished studies; Marks-Kaufman, 1982; Shor-Posner et al., 1985a). These data, suggesting a common site of action for peripheral and PVN morphine in macronutrient selection, are supported by studies of PVN lesion effects on morphine-elicited intake of these macronutrients. It has been shown that PVN lesions significantly reduce but do not abolish the enhancement of protein and fat consumption induced by peripheral morphine injection in freely feeding rats (Shor-Posner et al., 1985a).

3. When Is This System Active?

These and other findings suggest that medial hypothalamic opiate peptides may be involved in the specific pattern of macronutrient consumption normally eliciting protein ingestion after a meal high in carbohydrates. Specifically, it has been proposed (Shor-Posner et al., 1985a) that increased medial hypothalamic opiate peptide release mediates the switch in macronutrient consumption from initial carbohydrate consumption to protein consumption, which occurs either at the end of the first meal or at the beginning of the second meal of the active cycle. In freely feeding rats with access to three diets of pure macronutrients, there is a dramatic increase in feeding behavior at the beginning of the nocturnal feeding period. This is characterized by a large meal consisting almost entirely of carbohydrates with little or no protein. This first carbohydrate meal is followed by a second meal with a much higher proportion of protein (Tempel et al., 1985). As previously described, the first (carbohydrate) meal is proposed to be initiated by PVN release of NE, which is later counteracted by increased release of 5HT (see section II,D,3). The protein intake occurring near the end of the first meal, or more commonly during the second meal of the dark phase, is proposed to be initiated, in part, by increased release of PVN opiates, perhaps in conjunction with decreased PFH release of CA (see section II,B,2). It is also possible that later in the night period, when protein is preferred over carbohydrate, the opiates may also become activated. Opiate peptides in other brain areas may play complementary or antagonistic roles in macronutrient selection.

B. Pancreatic Polypeptides

Recent evidence suggests that the pancreatic polypeptides, a family of structurally-related 36-amino acid peptides, play an important role in the control of feeding behavior. This family of peptides, consisting of neuropeptide Y (NPY), peptide YY (PYY) and pancreatic polypeptide, have all been shown to elicit feeding behavior in satiated rats (Clark et al., 1984; Levine and Morley, 1984; Morley et al., 1985; Stanley and Leibowitz, 1984a; Stanley and Leibowitz, 1985b). Clark (1984) first demonstrated that injections of NPY, and to a lesser extent, pancreatic polypeptide, into the third ventricle elicit a feeding response in satiated rats. Levine and Morley (1984) observed similar effects in response to lateral ventricular injection of NPY.

1. Site(s) of Action

We have recently shown that administration of NPY directly into the PVN elicits feeding behavior in satiated rats, that this response is larger, and that it occurs at lower doses, than ventricular injections of NPY (Stanley and Leibowitz, 1984a, 1985b). These findings, suggesting a hypothalamic site of action for NPY, are supported by studies demonstrating that NPY-like immunoreactivity and receptor binding sites are most abundant in the PVN (Allen et al., 1983; Unden et al., 1983). They are also consistent with the results of a cannula-mapping study showing that NPY injections into several hypothalamic areas (the LH, VMH, and PVN) produce large increases in food intake, while injections into sites anterior, posterior, lateral, or dorsal to the hypothalamus are ineffective (Stanley et al., 1985a). While the exact site within the hypothalamus where NPY is most effective remains to be determined, these results suggest that the hypothalamus contains the critical receptors mediating the powerful eating response elicited by NPY.

2. Behavioral Analysis and Macronutrient Selection

We have recently shown that NPY is the most powerful chemical stimulant of feeding behavior and is several times more powerful than any other putative neurotransmitter tested to date (Stanley and Leibowitz, 1985b). Injection of NPY into the PVN of satiated rats causes a dose-dependent stimulation of feeding behavior: a low dose of 24 pmol produces a significant increase and a 235 pmol dose causes a peak effect. The subjects begin to eat after a relatively short latency, within an average of about 10 minutes after the injection, and some animals consistently begin to eat within 1 or 2 minutes. In the lower end of the dose range, the feeding consists of a single meal, completed within 1 hour. Higher doses, however, not only produce greater intakes initially (15 g in 1 hour), but also cause a continued eating response, such that food intake 4 hours postinjection is equivalent to normal total daily food intake. There is also a significant increase in total daily food intake after a single PVN injection of NPY. These results, suggesting that hypothalamic stimulation by NPY can override mechanisms of satiety, may implicate NPY not only in the control of food intake, but also in the regulation of body weight.

A recent study employing chronic PVN injections of NPY supports this possibility (Stanley et al., 1985b). This study demonstrates that animals injected with NPY 3 times per day can double their daily food consumption and gain weight at a rate of more than 10 g per day (over a 10-day period), which is comparable to that of rats with electrolytic lesions in the VMH (Hetherington and Ranson, 1940). These results demonstrate that chronic injection of NPY into the PVN can produce hyperphagia and obesity, presumably by overriding long- and short-term signals of satiety and control mechanisms for body weight, and they suggest the possibility that elevated release of endogenous NPY may contribute to some types of natural or experimentally-induced obesity. In this regard, it may be noted that the PVN has one of the most abundant supplies of NPY-containing presynaptic terminals in the brain (Everitt et al., 1984; Olschowka, 1984); to date, NPY is the most abundant peptide in the brain of rats and humans, with a similar distribution in the brains of both species (Adrian et al., 1983; Allen et al., 1983). Whether some cases of bulimia or obesity in humans may be mediated by NPY remains to be determined.

To study the behavioral specificity of NPY injection in the PVN, we administered into the PVN a dose of NPY that elicits somewhat greater feeding than occurs in a normal meal and then observed its impact on the animals' behavioral pattern (Stanley and Leibowitz, 1984a). In response to the peptide stimulation, the rats are found to eat a single large meal that lasts approximately 5 minutes. At the termination of the meal, they exhibit the sequence of behaviors typical of rats, which includes grooming and sleeping (Antin et al., 1975). Except for a small increase in water intake, no alterations in any other behaviors (e.g., levels of activity, grooming, resting, sleeping) are observed in response to NPY. These findings demonstrate that NPY's effect is specific to ingestive behavior, producing a strong orientation toward the ingestion of food without other behavioral effects. This specificity supports a role for NPY in the control of feeding behavior.

It has also been shown that PYY, a peptide structurally related to NPY, produces feeding when injected into the PVN (Stanley and Leibowitz, 1985a; Stanley et al., 1985d). To the extent that PYY and NPY have been compared, they appear to exhibit almost identical effects, except that PYY is approximately three times more potent than NPY. Thus PYY, although not more powerful than NPY in terms of maximal response, does appear to be more potent, eliciting feeding at doses at least as low as 8 pmol. Since the effects of these two structurally-related peptides are almost identical, it has been suggested that NPY and PYY stimulate feeding by acting on the same receptor subtype.

Recently, in a macronutrient self-selection study, we examined the stimulatory effects of NPY and PYY on consumption of pure macronutrients (Stanley and Leibowitz, 1985a; Stanley et al., 1985d). Both NPY and PYY are found to produce a dramatic and selective increase in consumption of carbohydrate, with no change in ingestion of either protein or fat. This pattern occurs whether the carbohydrate diet is presented with just one or both of the other two macronutrients. In the absence of the carbohydrate diet, NPY and PYY are found to potentiate protein and fat intake; however, these responses are significantly smaller than those potentiating carbohydrate ingestion.

What factors account for the carbohydrate preference, as well as the shift to protein and fat in the absence of this diet? One possibility is that carbohydrate is strongly preferred because it most rapidly provides usable energy (Steffens, 1969). In the absence of this preferred diet, however, the animals may turn to the protein or fat diets as alternative, albeit less rapid, sources of usable energy. This interpretation, which focuses on metabolic features of these diets, leads to the suggestion that endogenous NPY may translate signals of negative energy balance to elicit compensatory feeding behavior. A similar function has been proposed for the PVN noradrenergic feeding system (see section II,A,2), and given the similar effects of NE and NPY, it is possible that NPY may perform this function in concert with this amine.

A functional interaction between NPY and NE in relation to feeding is suggested by recent studies demonstrating a close functional relationship between these two neurotransmitters in other systems, and by the remarkable similarities in their effects on feeding behavior. Neuropeptide Y and NE coexist in central neurons and in nerves of the sympathetic nervous system (Everitt et al., 1984). They are released from sympathetic nerves apparently simultaneously, interacting to produce the same complex pattern of responses produced by nerve stimulation (Lundberg et al., 1984). Neuropeptide Y has also been shown to specifically up-regulate α2-noradrenergic receptors (Agnati et al., 1983). Both NE and NPY elicit a strong feeding response and a small drinking response when injected into the medial hypothalamus (see section II,A). Both produce virtually identical effects on macronutrient consumption, and the feeding response produced by either is attenuated by adrenalectomy (Stanley and Leibowitz, 1984b).

While these results suggest that NPY and NE may interact in some manner to control feeding, it is clear that the effects are not mediated through activation of identical neural substrates. This is demonstrated by the findings that feeding elicited by NE, but not NPY, is attenuated by PVN injection of an α-adrenergic receptor blocker. Feeding elicited by NPY is also unaffected by midbrain knife cuts that strongly attenuate feeding elicited by NE (Stanley and Leibowitz 1984b; 1985b). Thus, while it is possible that NPY and NE may interact to control feeding, the precise nature of such interaction remains to be elucidated.

C. Growth Hormone-Releasing Factor

A recent study has demonstrated that within 30 minutes of injection into the lateral ventricles, growth hormone-releasing factor (GRF) potentiates food intake by rats on a 23-hour food-deprivation schedule (Vaccarino et al., 1985). This peptide is effective at doses at least as low as 0.2 pmol. In contrast to the intact peptide, fragments of GRF have no significant impact on food intake, suggesting that the effect is pharmacologically specific. The stimulation of feeding is observed in the absence of changes in locomotor activity, indicating that the effect is behaviorally specific. Further, the increase in feeding appears to be independent of GRF's effect on growth hormone, as well as its peripheral effects, since peripheral injection of this peptide, which stimulates growth hormone release, has no influence on food ingestion. Taken together, these results point to GRF as a third potential candidate for endogenous peptide-induced stimulation of eating behavior.

IV. FEEDING-INHIBITORY PEPTIDES

The list of centrally-injected peptides that inhibit food intake is growing rapidly and now includes at least fifteen peptides, namely cholecystokinin, bombesin, neurotensin, adrenocorticotropin, anorexigenic peptide, corticotropin-releasing factor, vasopressin, insulin, calcitonin, somatostatin, thyrotropin-releasing hormone, substance P, satietin, and glucagon (Leibowitz, 1985; Morley et al., 1985). Since in most studies these peptides were injected intraventricularly and tested at high doses, it is at present difficult to evaluate the site(s) of action and mechanism by which they produce their effects. Further, there is little evidence regarding behavioral specificity for the majority of these peptides. The peptides for which these issues have been addressed most extensively include cholecystokinin (CCK) and neurotensin (NT).

A. Cholecystokinin

Although studies originally focused on a role for circulating CCK in satiety (Gibbs et al., 1973), recently a role for central CCK has been suggested. Experiments conducted with CCK in ruminants are particularly convincing, showing a suppressive effect on feeding with ventricular CCK injection in low, perhaps near physiological, doses (Della-Fera and Baile, 1979). Continuous infusions are considerably more effective than bolus injections, and the doses required to reduce meal size in food-deprived sheep increase with the length of the deprivation, revealing an adaptive interaction between brain CCK and energy depletion (Della-Fera and Baile, 1980a,b). These investigators have also demonstrated that only biologically active forms of CCK reduce eating in sheep (Della-Fera and Baile, 1981). Conversely, ventricular infusion of CCK antibodies have been shown to increase food intake (Della-Fera et al., 1981). These findings, suggesting that endogenous CCK may act to inhibit feeding, also imply that CCK may act in a primary and behaviorally-specific manner to inhibit eating. Since CCK is actively taken up from cerebrospinal fluid (Della-Fera et al., 1982), these investigators suggest that during eating, this peptide may be released into the ventricular system and transported to the brain site(s) of action. Studies of ventricular CCK injection in the rat have been less successful in revealing a reliable effect on feeding than studies of sheep, where consistent effects were observed.

Although extensive cannula-mapping studies have not been conducted, evidence to date points to the hypothalamus, and possibly other brain structures (Ritter and Ladenheim, 1984), as the site mediating central CCK's effect on feeding behavior. It has been shown that CCK injected into either the PVN or VMH causes a significant decrease in eating and, further, that as little as 5 ng of CCK injected directly into the PVN inhibits eating (Faris et al., 1983; Willis et al., 1984). Conversely, PVN injection of proglumide, a CCK antagonist, potentiates feeding (Dorfman et al., 1984). It is of interest that this hypothalamic nucleus may also be involved in relaying information, via ascending vagal afferents, from peripheral CCK receptors to central satiety systems (Crawley and Knas, 1984). These findings support the PVN as a brain site for CCK-induced satiety and suggest that endogenous PVN CCK receptors may have a function in satiety.

Further evidence for hypothalamic CCK receptors and endogenous CCK in satiety is provided by investigations examining the impact of eating and fasting on endogenous CCK levels and receptors. For example, it has been shown that fasting decreases hypothalamic levels of CCK, while causing an increase in CCK receptor number. Feeding, in contrast, increases CCK levels exclusively in the hypothalamus (McLaughlin et al., 1985; Saito et al., 1981; Scallet et al., 1984).

Studies have suggested that hypothalamic CCK may exert its inhibitory effect on feeding by altering the effect of CA in both the medial and lateral hypothalamus. Specifically, the stimulation of feeding produced by PVN injection of NE is strongly inhibited by CCK (75 ng) injected into the PVN but not other hypothalamic areas (Myers and McCaleb, 1981). Further, it has been shown that CCK infused into the PVN may suppress the release of NE in the PVN. Of particular interest is the observation that CCK suppresses PVN release of NE in food-deprived rats but actually increases NE release in satiated animals (Myers, 1985). This suggests that PVN terminals which release NE, presumably in response to low energy stores (see section II,A,2), are affected by activation of CCK receptors. This pattern of results contrasts with that observed in the PFH, where CCK increases the release of NE in deprived but not satiated animals. A consequence of this CA release in the PFH would be expected to be a reduction of food intake (see section II,B) similar to that seen with central CCK injection. Since CCK has also been suggested to act as an opioid antagonist

(Faris, 1983), an additional mode of action may be via its influence on opioid peptide systems that potentiate eating (see section III,A).

B. Neurotensin

Accumulating evidence suggests that neurotensin (NT), a tridecapeptide found in the gut and brain, may induce satiety when injected centrally in the rat (Luttinger et al., 1982; Stanley et al., 1982). It has been shown that intraventricular injection of NT causes a dose-related reduction in deprivation-induced and nocturnal feeding (Levine et al., 1983; Luttinger et al., 1982). We have demonstrated that injection of NT directly into the PVN is effective in suppressing feeding in deprived rats, and that this effect may be mediated by an action on the PVN noradrenergic system for feeding (Stanley et al., 1983; 1985c).

A convergence of evidence suggests that this effect on NT is behaviorally specific. Specifically, PVN injection of NT suppresses consumption of solid and liquid food, but it does not affect drinking induced by water deprivation (Stanley et al., 1983). Also, doses of NT that cause a large suppression of feeding do not produce a conditioned taste aversion, suggesting that NT does not induce gastrointestinal malaise (Luttinger et al., 1982). We have shown that PVN injections of NT, which produce large decreases in feeding behavior, do not affect other behaviors, namely drinking, grooming, resting, rearing, sleeping, or levels of locomotor activity (Stanley et al., 1983). Furthermore, NT appears to act within the general vicinity of the PVN injection site, presumably on the NE feeding-stimulatory system; this is reflected by the finding that unilaterally administered NT is significantly more effective in suppressing PVN NE-induced feeding when administered ipsilateral, as opposed to contralateral, to the NE injection (Stanley et al., 1985c). These results, arguing for behavioral specificity, are supported and extended by the finding that ventricular injection of NT suppresses feeding elicited by central injection of NE or dynorphin but not feeding elicited by central muscimol or peripheral insulin injection (Levine et al., 1983).

There is evidence to suggest that exogenous NT produces its satiating effect through activation of endogenous NT receptors. This is indicated by a study employing NT fragments to evaluate the amino acid structure required for this peptide's suppressive action on feeding (Stanley et al., 1985c). It has been found that the structural requirements for NT to suppress feeding closely parallel the structure required to activate NT receptors in other biological systems. Specifically, the C-terminal hexapeptide is critical; the absence of a single C-terminal amino acid abolishes NT's suppression of feeding. This evidence argues for pharmacological specificity, despite the relatively high dose-range (0.25 to 1.25 nmol) required to observe a reliable response with NT (Stanley et al., 1983; 1985c). Since this peptide has an extremely short half-life (less than 45 seconds; Checler et al., 1983), it is logical to conclude that these high doses are needed to overcome the counteracting effects of the brain's degrading enzymes. Evidence also points to a central site of action for NT. Intravenous injections of NT at doses comparable to those effective centrally have no effect on feeding. Only at higher doses does peripheral NT suppress feeding, and, in contrast to the behaviorally specific effects of central NT, this effect of peripheral NT is characterized by a general non-specific suppression of behavior (Stanley et al., 1983). The conclusion that NT is acting via central receptors is strengthened by the finding (see above) that NT is anatomically site-specific in its suppressive effect on feeding elicited by PVN NE injection. Although these results suggest that NT may act in the medial hypothalamus, preliminary cannula-mapping studies indicate that NT, in addition to its effects in the PVN, also acts in other hypothalamic as well as extra-hypothalamic brain sites to suppress feeding behavior (Stanley and Leibowitz, unpublished results).

The finding that NT has an inhibitory effect on feeding induced by exogenous NE is consistent with evidence indicating that NT may act in part via inhibition of *endogenous* NE. Specifically, it has recently been shown that VMH infusion of NT, like that of CCK, inhibits the release of NE from the VMH of food-deprived, but not satiated, animals (Myers, 1985). While this may be only one mode of action for NT, together these biochemical and pharmacological results present a strong case for a satiety function of medial hypothalamic NT receptors. These peptides may induce satiety, possibly in concert with CCK, through antagonism of the well-documented stimulatory influence of NE on food intake and meal size (section II,A).

V. CONCLUSION

It is clear from this review that many neurotransmitters are highly active in altering food ingestion when injected directly into the brain. This provides an important step in our quest to understand the neurochemical and neuroanatomical substrates of feeding behavior. As emphasized in this review, however, it is imperative that we concern ourselves with such critical questions as the behavioral, anatomical, and pharmacological specificity of these effects. At the same time, we must advance beyond these injection studies, focusing our attention on *endogenous* neurotransmitters to determine how they act to control normal feeding processes. In view of the complexity of the control mechanisms of feeding behavior, this will be a formidable task. However, progress has never been more rapid, and research over the next few years should establish many critical features of the neurochemical control mechanisms of feeding behavior.

REFERENCES

Adrian, T.E., J.M. Allen, S.R. Bloom, M.A Ghatei, M.N. Rossor, T.J. Crow, K. Tatemoto, and J.M. Polak. 1983. Neuropeptide Y distribution in human brain. *Nature* 306: 584-586.

Agnati, L.F., K. Fuxe, F. Benfenati, N. Battestini, K. Harfstrand, K. Tatemoto, T. Hokfelt, and V. Mutt. 1983. Neuropeptide Y in vitro selectively increases the number of a2-adrenergic binding sites in membranes of the medulla oblongata of the rat. *Acta. Physiol. Scand.* 118: 293-295.

Ahlskog, J.E., and B.G. Hoebel. 1973. Overeating and obesity from damage to a noradrenergic system in the brain. *Science* 182: 166-169.

Allen, Y.S., T.E. Adrian, J.M. Allen, K. Tatemoto, T.J. Crow, S.R. Bloom, and J.M. Polak. 1983. Neuropeptide Y distribution in the rat brain. *Science* 221: 877-879.

Anand, B.K., and J.R. Brobeck. 1951. Hypothalamic control of food intake in rats and cats. *Yale J. Biol. Med.* 24: 123-140.

Anderson-Baker, W.C., C.L. McLaughlin, and C.A. Baile. 1979. Oral and hypothalamic injections of barbiturates, benzodiazepines and cannabinoids and food intake in rats. *Biochem. Behav.* 11: 487-491.

Antin, J., J. Gibbs, J. Holt, R.C. Young, and G.P. Smith. 1975. Cholecystokinin elicits the complete behavioral sequence of satiety in rats. *J. Comp. Physiol. Psychol.* 87: 784-790.

Azar, A.P., G. Shor-Posner, R. Filart, and S.F. Leibowitz. 1984. Impact of medial hypothalamic 6-hydroxydopamine injections on daily food intake, diet selection, and body weight in freely-feeding and food restricted rats. *Proc. Eastern Psychol. Assoc.* 55: 106.

Bellin, S.I., and S. Ritter. 1981. Insulin-induced elevation of hypothalamic norepinephrine turnover persists after glucorestoration unless feeding occurs. *Brain Res*. 217: 327-337.

Beltt, B.M., and R.E. Keesey. 1975. Hypothalamic map of stimulation current thresholds for inhibition of feeding in rats. *Am. J. Physiol*. 229: 1124-1133.

Berthoud, H.R., and G.J. Mogensen. 1977. Ingestive behavior after intracerebral and intracerebroventricular infusions of glucose and 2-deoxy-D-glucose. *Am. J. Physiol*. 233: R127-R133.

Bhakthavatsalam, P., and S.F. Leibowitz. 1984. Studies on the influence of adrenal glucocorticoids on the alpha-noradrenergic and opiate feeding effects of different drugs in rats. *Proc. Eastern Psychol. Assoc*. 55: 106.

Bhakthavatsalam, P., and S.F. Leibowitz. 1985. a2-Noradrenergic feeding rhythm in paraventricular nucleus: Relation to corticosterone. *Am. J. Physiol*. In press.

Bhakthavatsalam, P., and S.F. Leibowitz. Morphine-elicited feeding: diurnal rhythm, circulating corticosterone and macronutrient selection. Submitted.

Blundell, J.E. 1984. Serotonin and appetite. *Neuropharmacology* 23: 1537-1552.

Blundell, J.E., and M.B. Leshem. 1973. Dissociation of the anorexic effects of amphetamine and fenfluramine following intrahypothalamic injection. *Br. J. Pharmacol*. 47: 81-88.

Blundell, J.E., and R.A. McArthur. 1979. Investigation of food consumption using a dietary self-selection procedure: effects of pharmacological manipulation and feeding schedule. *Br. J. Pharmacol*. 67: 436P-438P.

Blundell, J.E., C.J. Latham, and M.B. Leshem. 1976. Differences between the anorexic actions of amphetamine and fenfluramine--possible effects on hunger and satiety. *J. Pharm. Pharmacol*. 28: 471-477.

Blundell, J.E., and M.B. Leshem. 1974. Central action of anorexic agents: effects of amphetamine and fenfluramine in rats with lateral hypothalamic lesions. *Eur. J. Pharmacol*. 28: 81-88.

Borsini, F., C. Bendotti, B. Przewlocka, and R. Saminin. 1983. Monoamine involvement in the overeating caused by muscimol injection in the rat nucleus raphe dorsalis and the effects of d-fenfluramine and d-amphetamine. *Eur. J. Pharmacol*. 94: 109-115.

Breisch, S.T., F.P. Zemlan, and B.G. Hoebel. 1976. Hyperphagia and obesity following serotonin depletion with intraventricular parachlorophenylalanine. *Science* 192: 283-285.

Burton, M.J., E.T. Rolls, and F. Mora. 1976. Effects of hunger on the responses of neurons in the lateral hypothalamus to the sight and taste of food. *Exp. Neurol.* 51: 668-667.

Carlton, J., and N. Rowland. 1984. Anorexia and brain serotonin: development of tolerance to the effects of fenfluramine and quipazine in rats with serotonin-depleting lesions. *Pharmacol. Biochem. Behav.* 20: 739-745.

Cattabeni, F., A. Maggi, M. Monduzzi, L. DeAngelis, and G. Racagni. 1978. GABA: Circadian fluctuations in rat hypothalamus. *J. Neurochem.* 31: 565-567.

Checler, F., J.P. Vincent, and P. Kitabgi. 1983. Neurotensin analogs [D-Tyr^{11}] and [D-Phe^{11}] neurotensin resist degradation by brain peptidases in vitro and in vivo. *J. Pharmacol. Exp. Ther.* 227: 743-748.

Chesselet, M.F. 1984. Presynaptic regulation of neuro-transmitter release in the brain: facts and hypothesis. *Neurosci.* 12: 347-375.

Clark, J.T., P.S. Kalra, W.R. Crowley, and S.P. Kalra. 1984. Neuropeptide Y and human pancreatic polypeptide stimulate feeding behavior in rats. *Endocrinology* 115: 427-429.

Clineschmidt, B.V., J.C. McGriffin, and A.B. Werner. 1974. Role of monoamines in the anorexigenic actions of fenfluramine, amphetamine, and p-chloromethamphetamine. *Eur. J. Pharmacol.* 27: 313-323.

Cooper, S.J. 1983. GABA and endorphin mechanisms in relation to the effects of benzodiazepines on feeding and drinking. *Prog. Neuropsychopharmacol. Biol. Psychiat.* 7: 495-503.

Cooper, S.J., and L.B. Estall. 1985. Behavioral pharmacology of food, water and salt intake in relation to drug actions at benzodiazipine receptors. *Neurosci. Biobehav. Rev.* 9: 5-19.

Coscina, D.V. 1983. GABA and feeding: reversal by central GABA-transaminase inhibition. *Prog. Neuropsychopharmacol. Biol. Psychiat.* 7: 463-467.

Coscina, D.V., J.V. Daniel, P. Li, and J.J. Warsh. 1978. Potential nonserotonergic basis of hyperphagia elicited by intraventricular p-chlorophenylalanine. *Pharmacol. Biochem. Behav.* 9: 791-797.

Crawley, J.N., and J.Z. Knas. 1984. Tracing the sensory pathway from gut to brain regions mediating the actions of cholecystokinin on feeding and exploration. *Soc. Neurosci. Abstr.* 10: 533.

Della-Fera, M.A., and C.A. Baile. 1979. Cholecystokinin octapeptide: continuous picomole injections into the cerebral ventricles of sheep suppress feeding. *Science* 206: 471-473.

Della-Fera, M.A., and C.A. Baile. 1980a. CCK-octapeptide injected in CSF decreases meal size and daily food intake in sheep. *Peptides* 1: 51-54.

Della-Fera, M.A., and C.A. Baile. 1980b. Cerebral ventricular injection of CCK-octapeptide and food intake: the importance of continuous injection. *Physiol. Behav.* 24, 1133-1138.

Della-Fera, M.A., and C.A. Baile. 1981. Peptides with CCK-like activity administered intracranially elicit satiety in sheep. *Physiol. Behav.* 26: 979-983.

Della-Fera, M.A., C.A. Baile, B.S. Schneider, and J.A. Grinker. 1981. Cholecystokinin antibody injected in cerebral ventricles stimulates feeding in sheep. *Science* 212: 687-689.

Della-Fera, M.A., C.A. Baile, and M.C. Beinfeld. 1982. Cerebral ventricular transport and uptake: importance of CCK-mediated satiety. *Peptides* 3: 963-96

DiFiglia, M., and N. Aronin. 1984. Immunoreactive Leu-enkephalin in the monkey hypothalamus including observations on its ultra-structural localization in the paraventricular nucleus. *J. Comp. Neurol.* 225: 313-326.

Dorfman, D., P. Scott, and B.G. Hoebel. 1984. Feeding induced by the cholecystokinin antagonist, proglumide, injected in the paraventricular region of the hypothalamus. *Soc. Neurosci. Abstr.* 10: 65.

Ellison, G.D., C.A. Sorenson, and B.L. Jacobs. 1970. Two feeding syndromes following surgical isolation of the hypothalamus in the rat. *J. Comp. Physiol. Psychol.* 70: 173-188.

Everitt, B.J., T. Hokfelt, L. Terenius, K. Tatemoto, V. Mutt, and M. Goldstein. 1984. Differential coexistence of neuropeptide Y (NPY)-like immunoreactivity with catecholamine in the central nervous system of the rat. *Neurosci.* 11: 443-462.

Faris, P.L., B.R. Komisaruk, L.R. Watkins, and D.J. Mayer. 1983. Evidence for the neuropeptide cholecystokinin as an antagonist of opiate analgesia. *Science* 219: 310-312.

Faris, P.L., A.C. Scallet, J.W. Olney, M.A. Della-Fera, and C.A. Baile. 1983. Behavioral and immunohistochemical analysis of the function of cholecystokinin in the hypothalamic paraventricular nucleus. *Soc. Neurosci. Abstr.* 9: 184.

Fernstrom, J.D., and D.V. Faller. 1978. Neutral amino acids in the brain: changes in response to food ingestion. *J. Neurochem.* 30: 1531-1538.

Foxwell, M.H., W.H. Funderburk, and J.W. Ward. 1969. Studies on the site of action of a new anorectic agent, fenfluramine. *J. Pharmacol. Exp. Ther.* 165: 60-70.

Galzin, A.M., C. Moret, and S.C. Langer. 1984. Evidence that exogenous but not endogenous norepinephrine activates the presynaptic alpha-2 adrenoceptors on serotonergic nerve endings in the rat hypothalamus. *J. Pharmacol. Exp. Ther.* 228: 725-732.

Gibbs, J., R.C. Young, and G.P. Smith. 1973. Cholecystokinin elicits satiety in rats with open gastric fistulas. *Nature* 245: 323-325.

Goldman, C.K., L. Marino, and S.F. Leibowitz. 1985. Post-synaptic a2-noradrenergic receptors in the paraventricular nucleus mediate feeding induced by norepinephrine and clonidine. *Eur. J. Pharmacol.* In press.

Gosnell, B.G., J.E. Morley, and A.S. Levine. 1984. Localization of naloxone-sensitive brain areas in relation to food intake. *Soc. Neurosci. Abstr.* 10: 306.

Grandison, L., and A. Guidotti. 1977. Stimulation of food intake by muscimol and beta endorphin. *Neuropharmacology* 16: 533-536.

Grill, H.J., and R. Norgren. 1978. Chronically decerebrate rats demonstrate satiation but not bait shyness. *Science* 201: 267-269.

Grinker, J., C. Marinescu, and S.F. Leibowitz. 1982. Effects of central injections of neurotransmitters and drugs on freely-feeding rats. *Soc. Neurosci. Abstr.* 8: 604.

Grossman, S.P. 1960. Eating or drinking by direct adrenergic or cholinergic stimulation of the hypothalamus. *Science* 132, 301-302.

Halperin, R., L.L. Brown, and S.F. Leibowitz. 1979. Analysis of noradrenergic projections mediating feeding response elicited through paraventricular hypothalamic drug stimulation. *Soc. Neurosci. Abstr.* 5: 218.

Heffner, T.G., J.A. Hartman, and L.S. Seiden. 1980. Feeding increases dopamine metabolism in rat brain. *Science* 208: 1168-1170.

Hetherington, A.W., and S.W. Ranson. 1940. Hypothalamic lesions and adiposity in the rat. *Anat. Rec.* 78: 149-172.

Hoebel, B.G., and S.F. Leibowitz. 1981. Brain monoamines in the modulation of self-stimulation, feeding, and body weight. In *Brain, Behavior, and Bodily Disease*, ed. H. Weiner, M.A. Hofer, and A.J. Stunkard, 103-142. Raven Press, New York.

Hoebel, B.G., F.P. Zemlan, M.E. Trulson, R.G. MacKenzie, R.P. DuCret, and C. Norelli. 1978. Differential effects of p-chlorophenylalanine and 5,7-dihydroxytryptamine on feeding in rats. In *Serotonin Neurotoxins*, ed. J.H. Jacoby and L.D. Lytle. Ann. N.Y. Acad. Sci. 305: 590-594.

Jhanwar-Uniyal, M., B. Dvorkin, M.H. Makman, and S.F. Leibowitz. 1980. Distribution of catecholamine receptor binding sites in discrete hypothalamic regions and the influence of food deprivation on binding. *Soc. Neurosci. Abstr.* 6: 2.

Jhanwar-Uniyal, M., F. Fleisher, B.E. Levin, and S.F. Leibowitz. 1982. Impact of food deprivation on hypothalamic a-adrenergic receptor activity and norepinephrine (NE) turnover in the rat brain. *Soc. Neurosci. Abstr.* 8: 711.

Jhanwar-Uniyal, M., C.R. Roland, and S.F. Leibowitz. 1984a. Influence of adrenalectomy an a1- and a2- noradrenergic receptors in discrete hypothalamic and extrahypothalamic areas. *Soc. Neurosci. Abstr.* 10: 302.

Jhanwar-Uniyal, M., J.S. Woods, B.E. Levin, and S.F. Leibowitz. 1984b. Cannula mapping and biochemical studies of the hypothalamic opiate and noradrenergic systems in relation to eating behavior. *Proc. Eastern Psychol. Assoc.* 55: 26.

Jhanwar-Uniyal, M., A.D. Factor, and M. Bailo. 1985. Impact of adrenalectomy on a1- and a2-noradrenergic receptors in rat brain and on daily food intake. *Proc. Eastern Psychol. Assoc.* 56: 36.

Kamatchi, G.L., P. Bhakthavatsalam, and D. Chandra. 1984. . Inhibition of insulin hyperphagia by gamma-aminobutyric acid antagonists in rats. *Life Sci.* 34: 2297-2301.

Kelly, J. 1978. *GABA: a possible neurochemical substrate for hypothalamic feeding mechanisms.* Ph.D. Thesis, University of Chicago.

Kelly, J., and S.P. Grossman. 1979. GABA and hypothalamic feeding systems: a comparison of GABA, glycine, and acetylcholine agonists and their antagonists. *Pharmacol. Biochem. Behav.* 11: 649-652.

Kelly, J., G.F. Alheid, A. Newberg, and S.P. Grossman. 1977. GABA stimulation and blockage in the hypothalamus and midbrain: effects on feeding and locomotor activity. *Pharmacol. Biochem. Behav.* 7, 537-541.

Kelly, J., J. Rothstein, and S.P. Grossman. 1979. GABA and hypothalamus feeding system I. Topographic analysis of the effects of microinjections of muscimol. *Physiol. Behav.* 23: 1123-1134.

Kimura, H., and K. Kuriyama. 1975. Distribution of gamma-aminobutyric acid (GABA) in the rat hypothalamus: functional correlates with activities of appetite-controlling mechanisms. *J. Neurochem.* 24: 903-907.

Krieger, D., and H. Hauser. 1978. Comparison of synchronization of circadian corticosteroid rhythms by photoperiod and food. *Proc. Natl. Acad. Sci.* 75: 1577-1581.

Kruk, Z.L. 1973. Dopamine and 5-hydroxytryptamine inhibit feeding in rats. *Nature (London)* 246: 52-53.

Lanthier, D., B.G. Stanley, and S.F. Leibowitz. 1985. Feeding elicited by central morphine injection: sites of action in the brain. *Proc. Eastern Psychol. Assoc.* 56: 30.

Leibowitz, S.F. 1970. Hypothalamic ß-adrenergic "satiety" system antagonizes an a-adrenergic "hunger" system in the rat. *Nature* 226: 963-964.

Leibowitz, S.F. 1975a. Pattern of drinking and feeding produced by hypothalamic norepinephrine injection in the satiated rat. *Physiol. Behav.* 14: 731-742.

Leibowitz, S.F. 1975b. Amphetamine: possible site and mode of action for producing anorexia in rat. *Brain Res.* 84: 160-167.

Leibowitz, S.F. 1975c. Catecholaminergic mechanisms of the lateral hypothalamus: their role in the mediation of amphetamine anorexia. *Brain Res.* 98: 529-545.

Leibowitz, S.F. 1978a. Adrenergic stimulation of the paraventricular nucleus and its effects on ingestive behaviors as a function of drug dose and time of injection in the light-dark cycle. *Brain Res. Bull.* 3: 357-363.

Leibowitz, S.F. 1978b. Paraventricular nucleus: a primary site mediating adrenergic stimulation of feeding and drinking. *Pharmacol. Biochem. Behav.* 8: 163-175.

Leibowitz, S.F. 1980. Neurochemical systems of the hypothalamus: control of feeding and drinking behavior and water-electrolyte excretion. In *Handbook of the Hypothalamus*, Vol. 3A, ed. P.J. Morgane and J. Panksepp, 299-437. Marcel Dekker, Inc., New York.

Leibowitz, S.F. 1982. Hypothalamic catecholamine systems in relation to control of eating behavior and mechanisms of reward. In *The Neural Basis of Feeding and Reward*, ed. B.G. Hoebel and D. Novin, 241-257. Haer Institute for Electrophysiological Research, Brunswick, Maine.

Leibowitz, S.F. 1985. Brain neurotransmitters and appetite regulation. *Psychopharmacol. Bull.* 21: 412-418.

Leibowitz, S.F. 1985. Brain monoamines and peptides: role in the control of eating behavior. *Fed. Proc.* In press.

Leibowitz, S.F. 1986. Opiate, a2-noradrenergic and adrenocorticotropin systems of hypothalamic paraventricular nucleus. In *Perspectives on Behavioral Medicine* 5, ed. A. Baum. Academic Press, New York. In press.

Leibowitz, S.F., and L.L. Brown. 1980a. Histochemical and pharmacological analysis of noradrenergic projections to the paraventricular hypothalamus in relation to feeding stimulation. *Brain Res.* 210: 289-314.

Leibowitz, S.F., and L.L. Brown. 1980b. Histochemical and pharmacological analysis of catecholamine projections to the perifornical hypothalamus in relation to feeding inhibition. *Brain Res.* 201: 315-345.

Leibowitz, S.F., and L. Hor. 1982. Endorphinergic and a-adrenergic systems in the paraventricular nucleus: effects on eating behavior. *Peptides* 3: 421-428.

Leibowitz, S.F., and P.J. Papadakos. 1978. Serotonin-norepinephrine interaction in the paraventricular nucleus: antagonistic effects on feeding behavior in the rat. *Soc. Neurosci. Abstr.* 4: 177.

Leibowitz, S.F., and C. Rossakis. 1978a. Analysis of feeding suppression produced by perifornical hypothalamic injection of catecholamines, amphetamines and mazindol. *Eur. J. Pharmacol.* 53: 69-81.

Leibowitz, S.F., and C. Rossakis. 1978b. Pharmacological characterization of perifornical hypothalamic ß-adrenergic receptors mediating feeding inhibition in the rat. *Neuropharmacol.* 17: 691-702.

Leibowitz, S.F., and C. Rossakis. 1979a. Mapping study of brain dopamine- and epinephrine-sensitive sites which cause feeding suppression in the rat. *Brain Res.* 172: 101-113.

Leibowitz, S.F., and C. Rossakis. 1979b. Pharmacological characterization of perifornical hypothalamic dopamine receptors mediating feeding inhibition in the rat. *Brain Res.* 172: 115-130.

Leibowitz, S.F., and C. Rossakis. 1979c. L-DOPA feeding suppression: effect on catecholamine neurons of the perifornical lateral hypothalamus. *Psychopharmacol.* 61: 273-280.

Leibowitz, S.F., and G. Shor-Posner. 1986. Hypothalamic monoamine systems for control of food intake: analysis of meal patterns and macronutrient selection. In *Psychopharmacology of Eating Disorders: Theoretical and Clinical Advances.* In press.

Leibowitz, S.F., A. Arcomano, and N.J. Hammer. 1978a. Potentiation of eating associated with tricyclic antidepressant drug activation of a-adrenergic neurons in the paraventricular hypothalamus. *Prog. Neuropsychopharmacol.* 2: 349-358.

Leibowitz, S.F., A. Arcomano, and N.J. Hammer. 1978b. Tranylcypromine: stimulation of eating through a-adrenergic neuronal system in the paraventricular nucleus. *Life Sci.* 23: 749-758.

Leibowitz, S.F., N.J. Hammer, and L.L. Brown. 1980. Analysis of behavioral deficits produced by lesions in the dorsal and ventral midbrain tegmentum. *Physiol. Behav.* 25: 829-843.

Leibowitz, S.F., N.J. Hammer, and K. Chang. 1981. Hypothalamic paraventricular nucleus lesions produce overeating and obesity in the rat. *Physiol. Behav.* 27: 1031-1040.

Leibowitz, S.F., M. Jhanwar-Uniyal, B. Dvorkin, and M.H. Makman. 1982a. Distribution of a-adrenergic, ß-adrenergic and dopaminergic receptors in discrete hypothalamic areas of rat. *Brain. Res.* 233: 97-114.

Leibowitz, S.F., O. Brown, and J.R. Tretter. 1982b. Peripheral and hypothalamic injections of a-adrenergic and dopaminergic receptor drugs have specific effects on nutrient selection in rats. *Proc. Eastern Psychol. Assoc.* 53: 136.

Leibowitz, S.F., N.J. Hammer, and K. Chang. 1983. Feeding behavior induced by central norepinephrine injection is attenuated by discrete lesions in the hypothalamic paraventricular nucleus. *Pharmacol. Biochem. Behav.* 19: 945-950.

Leibowitz, S.F., P. Roossin, and M. Rosenn. 1984a. Chronic norepinephrine injection into the hypothalamic paraventricular nucleus produces hyperphagia and increased body weight in the rat. *Pharmacol. Biochem. Behav.* 21: 801-808.

Leibowitz, S.F., C.R. Roland, L. Hor, and V. Suillari. 1984b. Noradrenergic feeding elicited via the paraventricular nucleus is dependent upon circulating corticosterone. *Physiol. Behav.* 32: 857-864.

Leibowitz, S.F., M. Jhanwar-Uniyal, and C.R. Roland. 1984c. Circadian rhythms of circulating corticosterone and a2-noradrenergic receptors in discrete hypothalamic and extra-hypothalamic areas of the brain. *Soc. Neurosci. Abstr.* 10: 294.

Leibowitz, S.F., O. Brown, J.R. Tretter, and A. Kirschgesser. 1985a. Norepinephrine, clonidine, and tricyclic antidepressants selectively stimulate carbohydrate ingestion through noradrenergic system of the paraventricular nucleus. *Pharmacol. Biochem. Behav.* 23: In press.

Leibowitz, S.F., G. Weiss, F. Yee, and J.R. Tretter. 1985b. Noradrenergic innervation of the paraventricular nucleus: specific role in control of carbohydrate ingestion. *Brain Res. Bull.* 14: 561-567.

LeMagnen, J. 1981. The metabolic basis of dual periodicity of feeding in rat. *Behav. Brain Sci.* 4: 561-607.

Levine, A.S., and J.E. Morley. 1983. Adrenal modulation of opiate-induced feeding. *Pharmacol. Biochem. Behav.* 19: 403-406.

Levine, A.S., and J.E. Morley. 1984. Neuropeptide Y: a potent inducer of consummatory behavior in rats. *Peptides* 5: 1025-1029.

Levine, A.S., J. Kneip, M. Grace, and J.E. Morley. 1983. Effect of centrally administered neurotensin on multiple feeding paradigms. *Pharmacol. Biochem. Behav.* 18: 19-23.

Li, E.T.S., and G.H. Anderson. 1984. 5-Hydroxytryptamine: a modulator of food composition but not quantity? *Life Sci*. 34: 2453-2460.

Lichtenstein, S.S., C. Marinescu, and S.F. Leibowitz. 1984. Chronic infusion of norepinephrine and clonidine into hypothalamic paraventricular nucleus. *Brain Res. Bull*. 13: 591-595.

Lundberg, J.M., L. Terenius, T. Hokfelt, and K. Tatemoto. 1984. Catecholamines, neuropeptide Y (NPY), and the pancreatic polypeptide family: coexistence and interaction in the sympathetic response. In *Catecholamines: Neuropharmacology and Central Nervous System--Theoretical Aspects*, 179-189. Alan R. Liss, Inc., New York.

Luttinger, D., R.A. King, D. Sheppard, J. Struff, C.B. Nemeroff, and A.J. Prange, Jr. 1982. The effect of neurotensin on food consumption in the rat. *Eur. J. Pharmacol*. 81: 499-503.

Marks-Kaufman, R. 1982. Increased fat consumption induced by morphine administration in rats. *Pharmacol. Biochem. Behav*. 16: 949-955.

Martin, G.E., and R.D. Myers. 1975. Evoked release of ^{14}C-norepinephrine from rat hypothalamus during feeding. *Am. J. Physiol*. 229: 1547-1555.

McCabe, J.T., and S.F. Leibowitz. 1984. Determination of the course of brainstem catecholamine fibers mediating amphetamine anorexia. *Brain Res*. 311: 211-224.

McCabe, J.T., M. DeBellis, and S.F. Leibowitz. 1984. Clonidine-induced feeding: analysis of central sites of action and fiber projections mediating the response. *Brain Res*. 309: 85-104.

McCabe, J.T., D. Bitran, and S.F. Leibowitz. 1985. Amphetamine-induced anorexia: analysis with hypothalamic lesions and knife cuts. *Pharmacol. Biochem. Behav*. In press.

McCaleb, M.L., and R.D. Myers. 1982. 2 Deoxy-D-glucose and insulin modify release of norepinephrine from rat hypothalamus. *Am. J. Physiol*. 242: R596-R601.

McCaleb, M.L., R.D. Myers, G. Singer, and G. Willis. 1979. Hypothalamic norepinephrine in the rat during feeding and push-pull perfusion with glucose, 2-DG, or insulin. *Am. J. Physiol*. 236: R312-R321.

McLaughlin, C.L., C.A. Baile, M.A. Della-Fera, and T.G. Kasser. 1985. Meal-stimulated increased concentrations of CCK in the hypothalamus of Zucker obese and lean rats. *Physiol. Behav*. 35: 215-220.

McLean, S., and B.G. Hoebel. 1982. Opiate and norepinephrine-induced feeding from the paraventricular nucleus of the hypothalamus are dissociable. *Life Sci*. 31: 2379-2382.

McLean, S., and B.G. Hoebel. 1983. Feeding induced by opiates into the paraventricular hypothalamus. *Peptides* 4: 287-292.

Meeker, R.G., and R.D. Myers. 1980. GABA and glutamate: possible metabolic intermediaries involved in the hypothalamic regulation of food intake. GABA neurotransmission. *Brain Res. Bull.*, Suppl. 2: 253-259.

Miyahara, S., and Y. Oomura. 1982. Inhibitory action of the ventral noradrenergic bundle on the lateral hypothalamic neurons through alpha-noradrenergic mechanisms in the rat. *Brain Res*. 234: 459-463.

Mora, F., E.T. Rolls, and M.J. Burton. 1976. Modulation during learning of the responses of neurons in the lateral hypothalamus to the sight of food. *Exp. Neurol*. 53: 508-519.

Morley, J.E., A.S. Levine, G.K. Yim, and M.T. Lowy. 1983. Opioid modulation of appetite. *Neurosci. Biobehav. Rev.* 7: 281-305.

Morley, J.E., A.S. Levine, B.A. Gosnell, and D.D. Krahn. 1985. Peptides as central regulators of feeding. *Brain Res. Bull.* 14: 511-519.

Moss, R.L., I. Urban, and B.A. Cross. 1972. Microelectrophoresis of cholinergic and aminergic drugs on paraventricular neurons. *Am. J. Physiol*. 223: 310-318

Muller, E.E., D. Cocchi, and P. Mantegazza. 1972. Brain adrenergic systems in the feeding response induced by 2-deoxy-D-glucose. *Am. J. Physiol*. 223: 945-950.

Myers, R.D. 1985. Peptide-catecholamine interactions: feeding and satiety. *Psychopharmacol. Bull.* 21: 406-411.

Myers, R.D., and M.L. McCaleb. 1980. Feedings: satiety signal from intestine triggers brain's noradrenergic mechanism. *Science* 209: 1035-1037.

Myers, R.D., and M.L. McCaleb. 1981. Peripheral and intrahypothalamic cholecystokinin acts on the noradrenergic "feeding circuit" in the rat's diencephalon. 1981. *Neurosci*. 6: 645-655.

Olgiatti, V.R., C. Netti, F. Guidobono, and A. Pecile. 1980. The central GABAergic system and control of food intake under different experimental conditions. *Psychopharmacol*. 68: 163-167.

Olschowka, J.A. 1984. Neuropeptide Y innervation of the rat paraventricular and supraoptic nuclei. *Soc. Neurosci. Abstr*. 10: 437.

Orthen-Gambill, N., and R.B. Kanarek. 1982. Differential effects of amphetamine and fenfluramine on dietary self-selection in rats. *Pharmacol. Biochem. Behav.* 16: 303-309.

Panksepp, J., and R.B. Meeker. 1980. The role of GABA in the ventromedial hypothalamic regulation of food intake. *GABA Neurotransmission*. Suppl. 2, *Brain Res. Bull.* 5: 453-460.
Pittman, Q.J., J.D. Hatton, and F.E. Bloom. 1980. Morphine and opioid peptides reduce paraventricular neuronal activity: studies on the rat hypothalamic slice preparation. *Proc. Natl. Acad. Sci. U.S.A.* 77: 5527-5531.
Przewlocka, B., L. Stala, and Sheel-Kruger. 1979. Evidence that GABA in the nucleus dorsalis raphe induces stimulation of locomotor activity and eating behavior. *Life Sci.* 25: 937-946.
Ritter, R.C., and A.N. Epstein. 1975. Control of meal size by central noradrenergic action. *Proc. Natl. Acad. Sci. U.S.A.* 72: 3740-3743.
Ritter, R.C., and E.E. Ladenhein. 1984. Fourth ventricular infusion of cholecystokinin suppresses feeding in rats. *Soc. Neurosci. Abstr.* 10: 652.
Roland, C.R., P. Bhakthavatsalam, P., and Leibowitz, S.F. 1985. Interaction between corticosterone and a2-noradrenergic system of the paraventricular nucleus in relation to feeding behavior. *Neuroendocrinol.* In press.
Rolls, E.T. 1982. Feeding and reward. In *The Neural Basis of Feeding and Reward*, ed. D. Novin and B.G. Hoebel, 323-337. Haer Institute for Electrophysiological Research, Brunswick, Maine.
Rolls, E.T., M.J. Burton, and F. Mora. 1976. Neuronal responses associated with the sight of food. *Brain Res.* 111: 53-66.
Rolls, E.T., M.K. Sanghera, and A. Roper-Hall. 1979. The latency of activation of neurons in the lateral hypothalamus and substantia innominata during feeding in the monkey. *Brain Res.* 164: 121-135.
Saito, A., J.A. Williams, and I.D. Goldfine. 1981. Alterations in brain cholecystokinin receptors after fasting. *Nature* 289: 599-600.
Saller, C.F., and E.M. Stricker. 1976. Hyperphagia and increased growth in rats after intraventricular injection of 5,7-dihydroxytryptamine. *Science* 192: 385-387.
Samanin, R., and S. Garattini. 1975. Serotoninergic system in the brain and its possible functional connections with other aminergic system. *Life Sci.* 17: 1201-1210.
Sanger, D.G. 1985. GABA and the behavioral effects of anxiolytic drugs. *Life Sci.* 36: 1503-1513.
Sawchenko, P.E., and L.W. Swanson. 1982. The organization of noradrenergic pathways from the brainstem to the paraventricular and supraoptic nuclei in the rat. *Brain Res. Rev.* 4: 275-325.

Sawchenko, P.E., R.M. Gold, and S.F. Leibowitz. 1981. Evidence for vagal involvement in the eating elicited by adrenergic stimulation of the paraventricular nucleus. *Brain Res*. 225: 249-269.

Sawchenko, P.E., L.W. Swanson, H.W.M. Steinbusch, and A.A.J. Verhofstad. 1983. The distribution and cells of origin of serotonergic input to the paraventricular and supraoptic nuclei of the rat. *Brain Res*. 227: 355-360.

Scallet, A.C., M.A. Della-Fera, and C.A. Baile. 1984. Changes in cholecystokinin (CCK) content of specific hypothalamic areas of sheep with feeding and fasting. *Soc. Neurosci. Abstr*. 10: 652.

Schulz, R., A. Wilhelm, and G. Dirlich. 1984. Intracerebral injection of different antibodies against endogenous opioids suggests a-neoendorphin participation in control of feeding behavior. *Naunyn-Schmiedeberg's Arch. Pharmacol*. 326: 222-226.

Sclafani, A., and P.F. Aravich. 1983. Macronutrient self-selection in three forms of hypothalamic obesity. *Am. J. Physiol*. 224: R686-R694.

Scott, P., K. Jawaharlal, and B.G. Hoebel. 1984. Feeding induced with kappa and mu opiate agonists injected in the region of the paraventricular nucleus (PVN) in rats. *Proc. Eastern Psychol. Assoc*. 55: 106.

Shimizu, N., Y. Oomura, D. Novin, V. Grijalva, and P.H. Cooper. 1983. Functional correlations between lateral hypothalamic glucose-sensitive neurons and hepatic portal glucose-sensitive units in rat. *Brain Res*. 265: 49-54.

Shor-Posner, G., A. Azar, R. Filart, D. Tempel, and S.F Leibowitz. 1985a. Morphine-stimulated feeding: analysis of macronutrient selection and paraventricular nucleus lesions. *Pharmacol. Biochem. Behav*. In press.

Shor-Posner, G., J.A. Grinker, C. Marinescu, and S.F. Leibowitz. 1985b. Role of hypothalamic norepinephrine in control of meal patterns. *Physiol. Behav*. 35: 209-214.

Sikdar, S.K., Y. Oomura, and A. Irokuchi. 1985. Effects of mazindol on rat lateral hypothalamic neurons. *Brain Res. Bull*. 15: 33-38.

Smythe, G.A., H.S. Grunstein, J.E. Bradshaw, M.V. Nicholson, and P.J. Compton. 1984. Relationship between brain noradrenergic activity and blood glucose. *Nature* 308: 65-67.

Stanley, B.G., and S.F. Leibowitz. 1984a. Neuropeptide Y: stimulation of feeding and drinking by injection into the paraventricular nucleus. *Life Sci*. 35: 2635-2642.

Stanley, B.G., and S.F. Leibowitz. 1984b. Neuropeptide Y injected into the hypothalamus elicits feeding behavior in the rat. In *The Neural and Metabolic Bases of Feeding*,

7. Satellite Symposium of the 1984 Neuroscience Society Meeting, Napa, California.
Stanley, B.G., and S.F. Leibowitz. 1985a. Regulation of feeding behavior by neuropeptide Y and peptide YY. In *Neural and Endocrine Peptides and Receptors '85*, ed. T.W. Moody, 67. Fifth International Washington Spring Symposium, Washington, D.C.
Stanley, B.G., and S.F. Leibowitz. 1985b. Neuropeptide Y injected into the paraventricular hypothalamus: a powerful stimulant of feeding behavior. *Proc. Natl. Acad. Sci. U.S.A.* 82: 3940-3943.
Stanley, B.G., N.E. Eppel, and B.G. Hoebel. 1982. Neurotensin injected into the paraventricular hypothalamus suppresses feeding in rats. In *Neurotensin, a Brain and Gastrointestinal Peptide*, ed. C.B. Nemeroff and A.J. Prange, Jr. *Ann. N. Y. Acad. Sci.* 400: 425-427.
Stanley, B.G., B.G. Hoebel, and S.F. Leibowitz. 1983. Neurotensin: effects of hypothalamic and intravenous injections on eating and drinking in rats. *Peptides* 4: 493-500.
Stanley, B.G., D. Lanthier, and S.F. Leibowitz. 1984. Feeding elicited by the opiate peptide D-Ala-2-Met-enkephalinamide: site of action in the brain. *Soc. Neurosci. Abstr.* 10: 1103.
Stanley, B.G., A.S. Chin, and S.F. Leibowitz. 1985a. Feeding and drinking elicited by central injection of neuropeptide Y: evidence for a hypothalamic site(s) of action. *Brain Res. Bull.* 14: 521-524.
Stanley, B.G., S.E. Kyrkouli, S. Lampert, and S.F. Leibowitz. 1985b. Hyperphagia and obesity induced by neuropeptide Y injected chronically into the paraventricular hypothalamus of the rat. *Soc. Neurosci. Abstr.* 11: 36.
Stanley, B.G., S.F. Leibowitz, N. Eppel, S. St-Pierre, and B.G. Hoebel. 1985c. Suppression of norepinephrine-elicited feeding by neurotensin: evidence for behavioral, anatomical and pharmacological specificity. *Brain Res.* In press.
Stanley, B.G., D.R. Daniel, A.S. Chin, and S.F. Leibowitz. 1985d. Paraventricular nucleus injections of peptide YY and neuropeptide Y preferentially enhance carbohydrate ingestion. *Peptides*. In press.
Steffens, A.B. 1969. Rapid absorption of glucose in the intestinal tract of the rat after ingestion of a meal. *Physiol. Behav.* 4: 829-832
Stellar, E. 1954. The physiology of motivation. *Psychol. Rev.* 61: 5-22.
Tappaz, M.L., and M.J. Brownstein. 1977. Origin of glutamate decarboxylase (GAD)-containing cells in discrete hypothalamic nuclei. *Brain Res.* 132: 95-106.

Tempel, D.L., P. Bhakthavatsalam, G. Shor-Posner, G. Dwyer, and S.F. Leibowitz. 1985. Nutrient self-selection at different periods of the light-dark cycle in free feeding and food deprived rats. *Proc. Eastern Psychol. Assoc.* 56: 10.

Tepperman, F.S., M. Hirst, and C.W. Gowdey. 1981. A probable role for norepinephrine in feeding after hypothalamic injection of morphine. *Pharmacol. Biochem. Behav.* 15: 555-558.

Unden, A., K. Tatemoto, and T. Bartfai. 1983. Receptors for neuropeptide Y in rat brain. *Soc. Neurosci. Abstr.* 9: 170.

Ungerstedt, U. 1971. Adipsia and aphagia after 6-hydroxy-dopamine induced degeneration of the nigrostriatal dopamine system. *Acta Physiol. Scand.*, Suppl. 367: 95-122.

Vaccarino, F.J., F.E. Bloom, J. Rivier, W. Vale, and G.F. Koob. 1985. Stimulation of food intake in rats by centrally administered hypothalamic growth hormone-releasing factor. *Nature* 314: 167-168.

van der Gutgen, J.V., and J.L. Slangen. 1977. Release of endogenous catecholamines from rat hypothalamus in vivo related to feeding and other behaviors. *Pharmacol. Biochem. Behav.* 7: 211-219.

Waldbillig, R.J., T.J. Bartness, and B.G. Stanley. 1981. Increased food intake, body weight, and adiposity following regional neurochemical depletion of serotonin. *J. Comp. Physiol. Psychol.* 95: 391-405.

Weiss, G.F., and S.F. Leibowitz. 1985. Efferent projections from the paraventricular nucleus mediating a2-noradrenergic feeding. *Brain Res.* In press.

Willis. G.L., J. Hansky, and G.C. Smith. 1984. Ventricular, paraventricular and circumventricular structures involved in peptide-induced satiety. *Regul. Peptides* 9: 87-99.

Woods, J.S., and S.F. Leibowitz. 1985. Hypothalamic sites sensitive to morphine and naloxone: effects on feeding behavior. *Pharmacol. Biochem. Behav.* In press.

Wurtman, J.J., and R.J. Wurtman. 1977. Fenfluramine and fluoxetine spare protein consumption while suppressing calorie intake by rats. *Science* 198: 1178-1180.

Wurtman, J.J., and R.J. Wurtman. 1979. Drugs that enhance central nervous system 5-HT transmission diminish elective carbohydrate consumption by rats. *Life Sci.* 24: 895-904.

Wurtman, R.J., F. Heftl, and E. Melamed. 1981. Precursor control of neurotransmitter synthesis. *Pharmacol. Rev.* 32: 315-335.

Wurtman, J.J., and R.J. Wurtman. 1984. D-Fenfluramine selectively decreases carbohydrate but not protein intake in obese subjects. *Int. J. Obesity* 8, Suppl. 1: 79-84.

Chapter 8

NEUROPHYSIOLOGY OF CONTROL: METABOLIC RECEPTORS

Yutaka Oomura

I. INTRODUCTION

Somewhere in the process of acquiring the resources to provide energy and maintain structural integrity, an organism must recognize its needs. Furthermore, this recognition must occur before the store of essential materials is altogether depleted. Early in the sequence of functions that lead to feeding behavior are the sensors, beyond the traditional five senses, that monitor the levels of various strategic chemicals in the body. These chemosensors respond to the absolute and relative (balance) levels of chemicals in the metabolic chain, and are discussed here as metabolite receptors.

II. LOCATIONS OF METABOLITE RECEPTORS

A. Forebrain

Metabolite receptors were first identified in the lateral hypothalamus (LHA) when it was found that local electrophoretic application of glucose depressed the activity of about 30% of the neurons in that area. These are now referred to as glucose-sensitive (GS) neurons, although, as discussed later, they also respond to other metabolites (Oomura, 1976; Shimazu et al., 1984). They also contribute to control of other functions that are related to feeding, as well as the feeding process itself (Aou et al., 1984), and their responses to some materials have been shown to depend on the hunger-satiety state of the animal (Nishino et al., 1982; Lénard et al., 1985). Since activity in the LHA is positively related to feeding behavior and its lesion depresses feeding activity, it is sometimes referred to as *the* feeding center. It should more properly be called *a* feeding center, since other regions have also been identified that have somewhat similar relations to the feeding process. These will be discussed later.

In freely moving rats, extracellular activity of single LHA neurons and feeding were measured before and after third intracerebral ventricle injection of 2-deoxy-D-glucose (2DG), an analogue of glucose. Half of the neurons that were initially excited by 2DG remained excited for a long time and then reversed to depressed activity. Most of these neurons also increased activity during meals. The few neurons that were inhibited by 2DG tended to increase activity transiently, just before meals. Feeding was initially induced by 2DG injection and then suppressed so that the total intake in the first 24 hours after injection was significantly below the pre-test base line (Katafuchi et al., 1985a). These and related effects provide further strong evidence that LHA neurons are closely, perhaps causally, related to feeding behavior.

A counterpart to the LHA is the ventromedial nucleus of the hypothalamus (VMH), where about 30% of the neurons are known as glucoreceptor (GR) neurons since their activity is increased by electrophoretic application of glucose. Analogous to the GS neurons of the LHA, GR neurons of the VMH also respond to metabolites other than glucose, and control feeding. Because of the positive relation between VMH activity and degree of satiation, and since lesion of the VMH leads to hyperphagia and obesity, the VMH has sometimes been referred to as a satiety center.

B. Medulla

Because of evidence suggesting glucose-responsive GS and GR neurons in the nucleus tractus solitarius (NTS) (Kadekaro et al., 1980), chemosensitivity of neurons was tested in rat brain slices of that region. In the caudal region of the NTS, 28% of the neurons are GS and 8% are GR neurons; in the rostral region, 11% are GS and 5% are GR (Mizuno and Oomura, 1984). Thus the NTS, located between the hypothalamus and the periphery, complements the functions of the hypothalamus. The proximity of the NTS to the fourth ventricle is comparable to the LHA proximity to the third ventricle. Hence, it might be that the NTS consolidates taste, hepatic, pancreatic, gastric, and other peripheral information with intraventricular and systemic chemical levels and sends the resultant information to the LHA. The LHA then reacts similarly with the frontal and association areas of the cortex.

Adachi and Kobashi (1985) have reported chemoresponsive neurons in the area postrema (AP). Most of these were GS-type neurons, and they responded to electrophoretic application and systemic change of glucose. The authors suggested that these particular neurons might contribute to maintenance of homeostasis by regulatory reflex control of glucose level. These GS neurons also responded to electrical stimulation of the hepatic vagal nerve (Adachi, personal communication).

C. Periphery

The effects of glucose, 2DG, and insulin on the firing rate of vagal pancreatic afferents in rabbits, guinea pigs, and rats have been reported by Niijima (1980). The activity of pancreatic afferents decreases after administration of glucose and increased after injection of 2DG. Their activity also increases in response to free fatty acids. Insulin increased and glucagon decreased the activity of pancreatic afferents. Thus, these afferent elements have some characteristics that are similar to those of GS neurons in the LHA (see below).

GS afferent units in the hepatic branch of the vagus nerve decrease their discharge rates when glucose is injected into the hepatic portal vein. Several investigators have suggested that these hepatic GS units may also influence the regulation of food intake (Novin et al., 1973; Russek, 1970). Intraportal injection of glucose into the liver reduces food consumption in proportion to glucose concentration. The importance of vagal afferents in reducing food intake has been further documented: Intraperitoneally administered CCK elicited satiety in normal rats, but was ineffective in vagotomized animals (Smith et al., 1981). Injected into the hepatic portal vein, CCK inhibited hepatic vagal afferents in a way similar to glucose (Niijima, 1981).

To determine relations between central and peripheral GS elements, the effects of portal vein glucose injections on LHA neurons were investigated after the LHA neurons were categorized by electrophoretic application of glucose. Evidence clearly demonstrates functional correlation between GS neurons in the LHA and vagal hepatic GS units. Most GS neurons in the LHA receive excitatory inputs from the vagal afferents of the hepatic GS units, but no comparable inhibitory inputs have been observed (Shimizu et al., 1983). It is important to note that these inputs to the LHA are relayed by GS neurons in the NTS and most go through the parabrachial nucleus (PBN) (Adachi et al., 1984), but some may go directly from the NTS to the LHA. These are discussed further in Section IV,B.

There are glucose responsive neuronal elements in pancreatic vagal afferents, duodenal vagal afferents (Mei, 1978), jejuno-ileal splanchnic afferents (Perrin et al., 1981), and in the dorsal motor nucleus of the vagus (DMV) (Mizuno et al., 1985). Therefore, information about peripheral blood glucose and other metabolites that influence the activity of autonomic efferent nerves might reflect sensing functions in these lower structures as well as in the hypothalamus.

III. MORPHOLOGY OF GLUCOSE RESPONSIVE UNITS

A. Forebrain

In brain slices that included the VMH and LHA, two morphologically different neurons have been observed. One of these is bipolar with few dendrites. The other is multipolar with extensive dendritic arborization and many spines. It has been determined by intracellular recording during application of glucose followed by intracellular injection of HRP that the multipolar neurons are GR and GS neurons, while the bipolar neurons are not (Minami et al., 1985; Ono et al., 1982; Oomura, 1983) (Fig. 1 and Fig. 2).

Stimulation with depolarizing electrical pulses induced action potentials in both GR and non-GR types, but with different wave shapes. Hyperpolarization following the action potential induced in the VMH bipolar neurons was slight or absent, but was definite after the action potential in the multipolar neurons. The after-hyperpolarization reached about -90mV, the K^+ equilibrium potential, E_K. The membrane potential of the multipolar neuron returned slowly to the resting level after application of a hyperpolarizing pulse. This is a characteristic K^+ conductance increase (I_A current). Both the I_A current and the after-hyperpolarization tended to reduce the firing rate. The non-GR neurons fired at a higher rate, and exhibited interneuron properties.

In the presence of the Na^+ channel blocker tetrodotoxin (TTX), a depolarizing pulse produced a graded response only, and no spike potential in the VMH bipolar neuron. An all-or-none slow action potential with a higher threshold was evoked in the multipolar neuron. In the presence of TTX plus tetraethylammonium, a K^+ channel blocker that facilitates Ca^{++} spikes, slow action potentials were induced in both neuron types. This indicates that the Ca^{++} spike is easily produced in the dendrites, i.e., in the multipolar neurons. These results, plus the report by Llinas and Sugimori (1980) that Ca^{++}-induced K^+ permeability is much greater in dendrites than in other neuron membrane areas, indicates that the after-hyperpolarization was due to Ca^{++}-induced K^+ conductance increase.

As seen in Figure 1, C-a and b, glucose application to a GR neuron depolarized the membrane and decreased the membrane conductivity. This increased both the frequency and amplitude of the spikes. The depolarization and decrease in conductivity were due to a decrease of K^+ permeability, since the reversal potential for the glucose depolarization was -90mV, close to E_K. On the other hand, as shown in Figure 2-B, GS neurons were hyperpolarized by glucose without membrane conductance

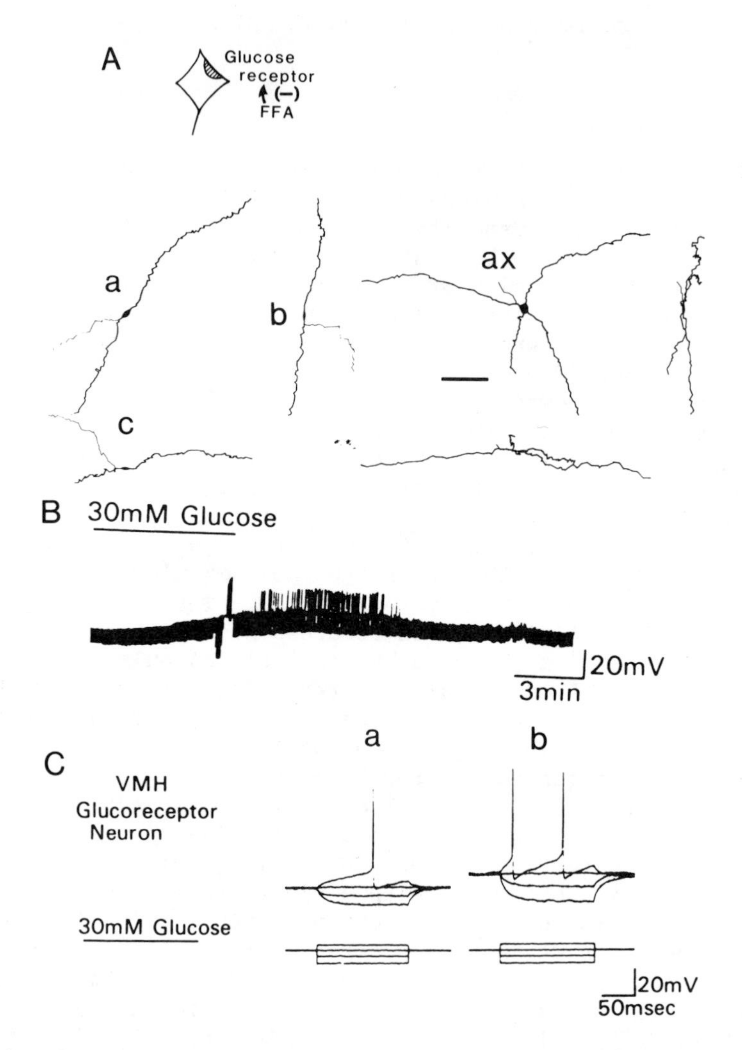

Fig. 1. Characteristics of typical glucoreceptor neuron. (A) Computer reconstruction of two rat VMH neurons stained by horseradish peroxidase. Left, typical fusiform non-GR neuron. Right, typical multipolar GR neuron. ax, axon. Cell processes generally lie in or close to coronal plane, a. b, lateral view. c, ventral view. (B) Glucose application depolarized membrane and decreased membrane conductance. Upper and lower deflections at the beginning of membrane depolarization caused by outward and inward current applications for (C). Original resting potential, -60 mV. (C) a, membrane responses to inward (down) and outward (up) current pulses before glucose application. b, after glucose application (Minami et al., 1985).

change. This was due to enhanced activity of the ATP-dependent Na-K pump caused by energy available from the additional glucose. These effects were not evident in non-GS neurons.

B. Medulla

Relations between morphology and glucose responses of NTS neurons were analogous to those of LHA and VMH neurons (Mizuno and Oomura, 1984). Most non-glucose responding neurons had their long axes parallel or slightly oblique to the dorsal surface of the medulla. The GS neurons existed mostly in the lateral part of the NTS with the non-GS neurons more medial. Some of the dendrites arising from GS and GR neurons were observed to extend dorsally to the fourth ventricle. Almost all NTS neurons sent dendrites laterally toward the tractus solitarius with axons running ventral to the dorsal motor nucleus of the vagus.

IV. NEURONAL NETWORKS INVOLVED IN FEEDING

A. Forebrain

Important information relevant to the regulation of feeding goes to the hypothalamus from internal and external sources. Visual (Ono et al., 1981; Oomura, 1980), olfactory (Takagi, 1984), and gustatory (Kita and Oomura, 1981; Norgren, 1978) signals, plus internal visceral information such as that from hepatic and intestinal glucose, Na^+, and osmotic sensors (Berthoud et al., 1980; Mei, 1978; Shimizu et al., 1983) and stomach mechanoreceptor afferents (Oomura, 1973) converge in the limbic system (Ono et al., 1984a), especially the VMH and the LHA, to integrate sensory and effector information. The LHA, VMH, and other parts of the limbic system then mediate feeding processes (Fig. 3).

Monkey LHA responses to olfactory, taste, and visual discrimination of food have been reported (Nishino et al., 1982; Ono et al., 1981). Ono et al. (1984a) divided the feeding task into three phases: discrimination, procurement, and ingestion.

Inset in (A) suggests mechanism of action. An inhibitor, such as free fatty acid, blocks glucose binding by the GR site while insulin, a well-known facilitator of glucose uptake, enhances glucose effect (Mizuno and Oomura, 1984).

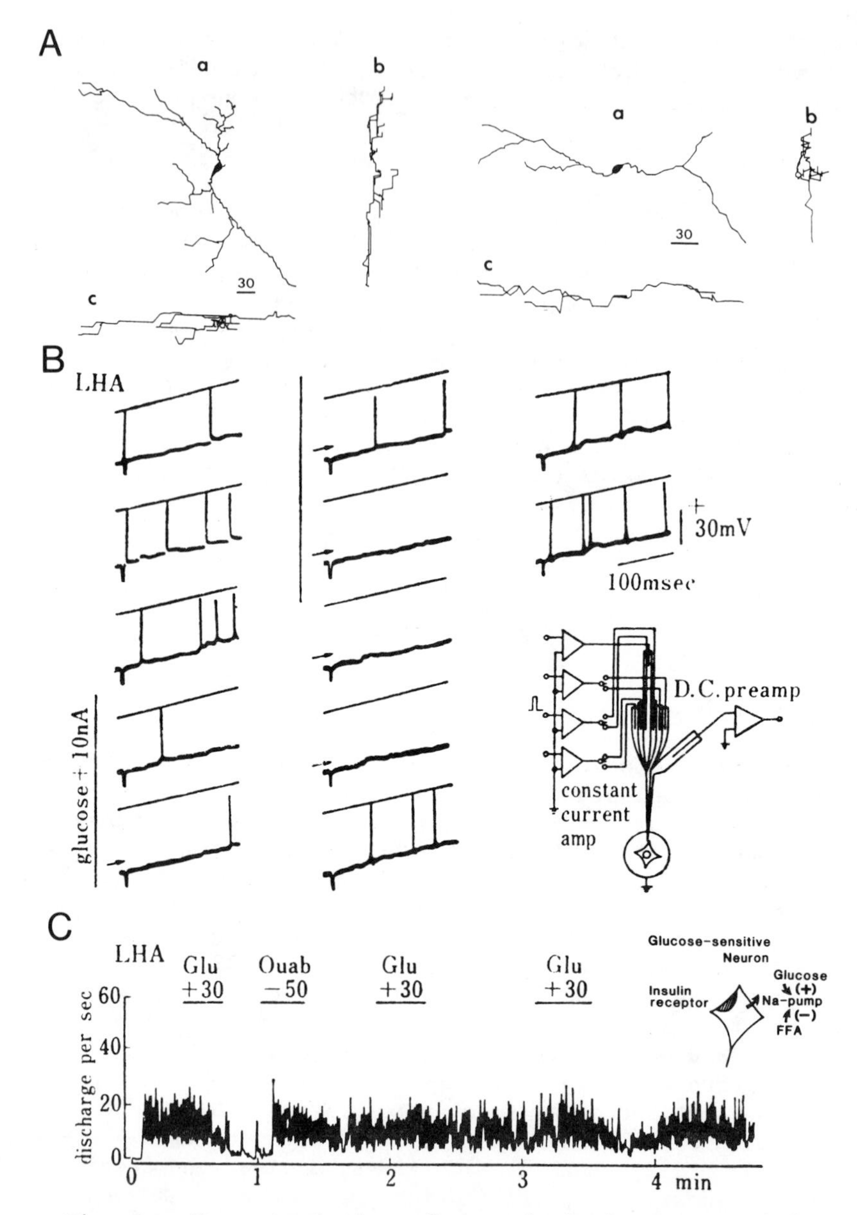

Fig. 2. Characteristics of a typical glucose-sensitive neuron. (A) Computer reconstruction of two rat neurons. GS neuron at left and non-GS at right complementary to their respective counterparts in VMH (Fig. 1). Other descriptions as for Fig. 1 (Mizuno and Oomura, 1984). (B) Intracellular recording from GS neurons in the LHA. Upper trace, electrode potential before membrane penetration. Thick vertical line

visual responses of the LHA neurons appeared to be related to discrimination of food from non-food. LHA neurons changed during bar press for food in high fixed-ratio tasks. LHA activity changes were consistent with the monitoring of internal conditions such as satiation, and the integration of these with sensory inputs to initiate feeding. Figure 4 shows the relationship between glucose level, behavior, and activity of an LHA GS neuron. Systemically administered glucose decreased the spontaneous firing rate of the neuron, extended the bar press latency, and inhibited the decrease in discharge rate during the bar press and reward periods. In statistically rare cases, i.v. injection of glucose appears to increase the activity of a GS neuron under observation. This may be due to the triggering of excess insulin secretion by the injected glucose. The resulting hyperinsulinemia might well mask the effects of the glucose and temporarily enhance LHA activity. Systemic administration of naloxone blocked the reward response, but did not affect other responses (not shown, see Fig. 5D). The fact that the ingestion reward response disappeared or was inverted upon the ingestion of aversive salty food, and that it became weaker and finally disappeared when the monkey became satiated, might be explained by gustatory afferents to the hypothalamus and the involvement of endogenous opioids during the reward period (Nishino et al., 1982; Lénard et al., 1985).

GS neurons in the monkey LHA responded more than non-GS neurons during feeding behavior. The responses were most often decreased in activity during bar press and reward, and rarely at the onset of cue stimuli, so GS neurons in the LHA might be

---

(left and middle columns), duration of extracellular glucose application. Arrows, level of membrane potential before glucose application. Read from left upper record down left column, then middle and right columns. Hyperpolarization is seen after start of glucose application and lasts for about 3 sec after termination. Resistance of membrane did not change since amplitude of 1-msec hyperpolarizing pulses remained constant throughout. Time between successive tracings, 1.3 sec. Original resting potential, −45 mV. (C) Rate meter plot of effects of ouabain on GS LHA neuron. First application of glucose inhibited activity; ouabain blocked effects of next two glucose applications. Inset, Schema for mechanism of suppression. Electrogenic Na^+-K^+ pump accelerated by glucose and slowed by FFA, which inhibits neuron uptake of glucose. Membrane has insulin receptors (Oomura et al., 1974). Parts B and C reprinted by permission from *Nature* 247: 284-286.

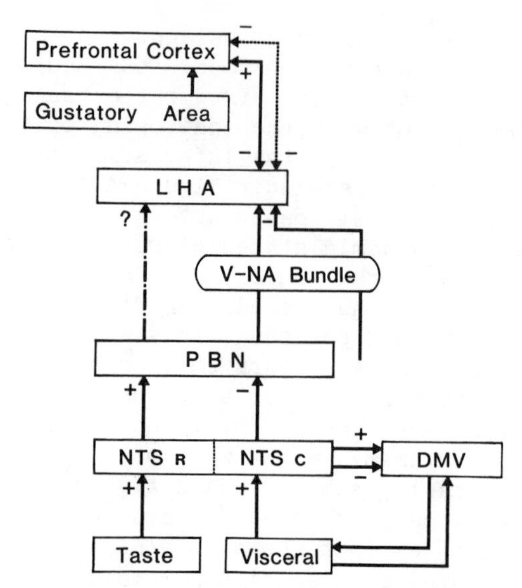

Fig. 3. Hierarchy of chemical information processing sites. Visceral includes hepatic, pancreatic, and intestinal sites. Solid lines, monosynaptic; dotted line, polysynaptic. Broken line not fully characterized. Connections: +, excitatory; -, inhibitory; ?, unknown. DMV, dorsal motor nucleus of vagus. NTS_R and NTS_C, rostral and caudal nucleus tractus solitarii. PBN, parabrachial nucleus. V-NA Bundle, ventral noradrenergic bundle. LHA, lateral hypothalamus. Arrows point othodromic. Arrowheads indicate afferent or efferent transmission (Oomura, 1985).

Fig. 4. Effect of blood glucose on response of GS neuron in LHA of chronic monkey. (A, B, D), same neuron. (A) Identification of GS neuron; activity suppressed by electrophoretic application of glucose. (B) Upper dot raster and histogram, sum of 7 trials of fixed ratio (FR30) bar press task. Lower histogram, bar presses. Arrowhead, reward. Lower rate during bar press and reward periods (Aou et al., 1984). (C) Increase of blood glucose level after meal (dotted line) and after 4g or 10g glucose injection i.v. (D) Responses for cookie in FR schedule at times indicated by a, b, c in (C). Responses similar in a and c when blood glucose was 90 mg%; depressed in b when blood glucose was 175 mg% (Nishino et al., 1982).

LHA

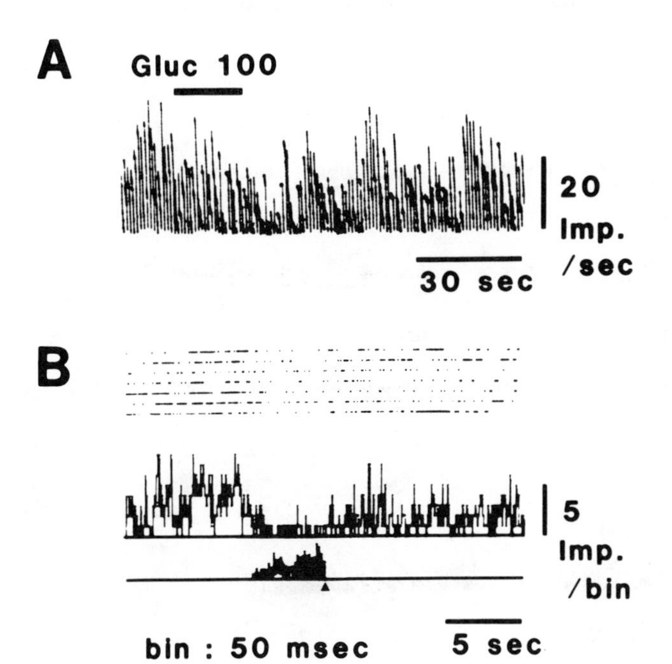
A
Gluc 100
20
Imp.
/sec
30 sec
B
5
Imp.
/bin
bin : 50 msec
5 sec

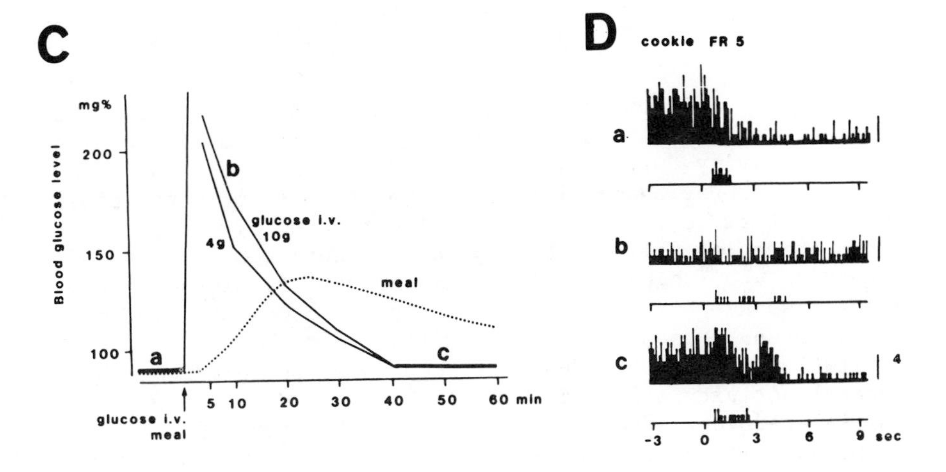
C
mg%
Blood glucose level
200
150
100
a
b
c
glucose i.v.
10g
4g
meal
5
10
20
30
40
50
60 min
glucose i.v.
meal
D
cookie FR 5
a
b
c
4
-3
0
3
6
9
sec

Figure 4.

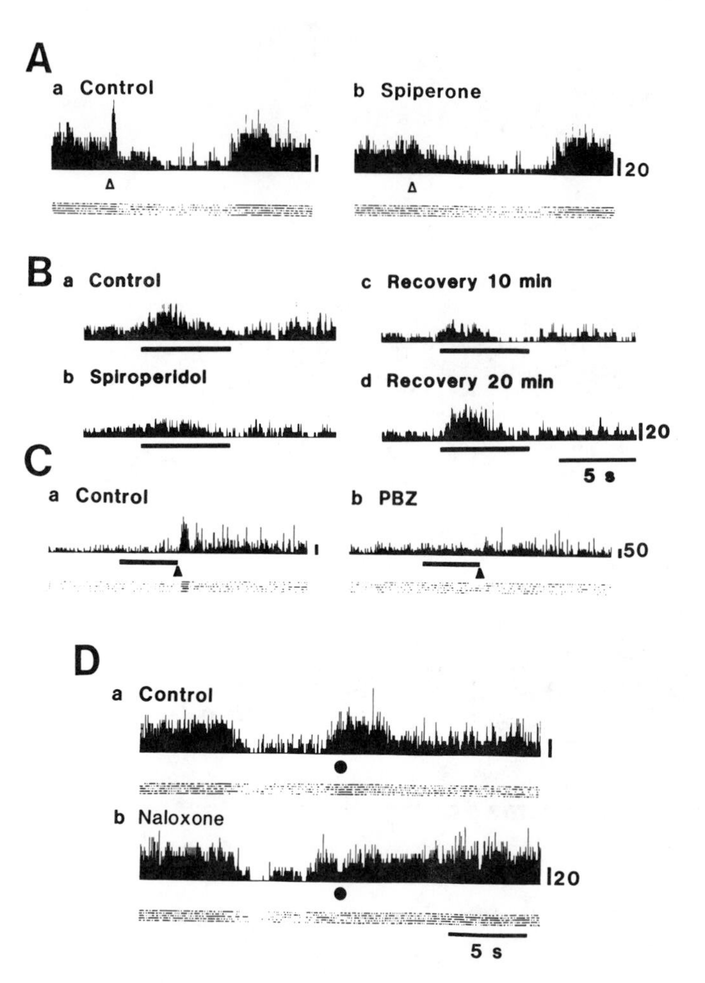

Fig. 5. Effects of some endogenous agents on responses of lateral hypothalamic (LHA) neurons. Histograms of LHA single neuron activity during fixed ratio bar press trials for food (cookie). Upper, each histogram, sum of three or more trials. Lower, dot raster. (A) a, excitation induced by cue light (open triangle) followed by suppression during bar press and ingestion. b, spiperone (dopamine antagonist) blocked cue light response but did not affect bar press or ingestion phase of response. Response recovered later (not shown). (B) a, excitation during bar press (a) suppressed by spiperone, b; general activity affected only slightly, or not at all. c, partial recovery after 10 min; d, complete after 20 min. Under bars, duration of bar pressing. (C) Strong response to short cue tone (a), (arrowhead at end of bar press period) blocked by a blockers, phenoxybenzamine (PBZ) (b). Recovered later

involved in drive and/or reward, and in hunger-motivated food acquisition (Aou et al., 1984). More GS than non-GS neurons responded to electrophoretic application of noradrenaline (NA) and morphine; their firing rates decreased. Catecholamines and opiates apparently modulate LHA control of feeding.

Dopamine (DA) sensitive neurons responded more often at the cue-light-on and the initiation of bar pressing than DA-insensitive neurons. The DA antagonist spiperone blocked cue-light-on responses (Fig. 5A) and responses associated with the initiation of bar presses (Fig. 5B). The ß-adrenoceptive blocker sotalol had no clear effect, but it did attenuate the decreased firing during bar pressing in some cases. NA-sensitive neurons responded more often in feeding tasks than NA insensitive ones, especially at the cue tone that indicated reward. NA enhanced, and the α-adrenoceptive blocker phenoxybenzamine reduced, cue-tone-related responses (Fig. 5C). Naloxone reduced ingestion-related responses, especially inhibitory ones, but did not affect other responses (Fig. 5D). Thus, DA might be important to task initiation and opiates to reward perception. These authors originally suggested that the α-adrenoceptive system might be involved in drive reduction (motor termination). It now appears, however, that the NA system may instead be associated with anticipation of reward. GS neurons in the LHA are thus involved in drive performance and reward perception (Nishino et al., 1985; Lénard et al., 1985). Results indicate that DA sensitive neurons integrate more external cue signals and GS neurons more internal cue signals. GS neurons may thus act as integrating elements that combine the various aspects of internal information and then project to effector control neurons to initiate or terminate overt feeding-related behavior. Mutual interrelations between the DA and GS neurons require further investigation.

The orbitofrontal cortex (OBF) of the monkey and the dorsolateral part (DL) of the prefrontal area contribute to control of emotional and motivated behavior. The prefrontal cortex is involved in emotion, reward, arousal, learning, cognition, and memory. The reward perception system contains the OBF (Aou et al., 1983; Nakano et al., 1984; Yamamoto et al., 1984), and the rostral DL (Ono et al., 1984b; Inoue et al., 1985) which has the mutual connections with the LHA, and influence of this system on feeding has been reported. The caudal part of the DL

(not shown). (D) Reponse during ingestion-reward (a) blocked by naloxone (b). Filled circles, time food entered mouth. Suppression during bar press not affected by naloxone. Recovered later (not shown) (Nishino et al., 1985).

responded to stimuli in the external environment, and seemed to be more cue-related; Inoue et al. (1985) observed that the rostral DL seemed to respond more in reward-related situations, but responded differently to aversive food. This may reflect both differences in the connections between these and other regions, and functional heterogenity of the DL neurons.

Nakano et al. (1984) found that i.v. glucose only slightly affected spontaneous activity in the OBF until after the onset of satiation; it then prolonged the duration of suppression. OBF responses during bar pressing and reward tended to be suppressed by i.v. glucose. Since there are no GS neurons nor glucose-responding neurons in the OBF, but there are direct connections between the OBF and the LHA (Fig. 3) (Oomura et al., 1980), OBF responses to i.v. glucose were through the LHA. Yamamoto et al. (1984) found that OBF neurons changed activity upon presentation of food or aversive stimulation during bar pressing.

B. Medulla

The NTS is a relay nucleus of vagal afferents (Fig. 3). Ascending NTS fibers project to the PBN, where excitatory inputs transmit taste signals; both excitatory and inhibitory inputs have been suggested for the transmission of signals from general visceral afferents. The inhibitory pathway from the NTS through the PBN to the LHA could be responsible for decreasing the activity of GS neurons in the LHA when decrease in NTS activity causes disinhibition at the level of the PBN. The PBN has been shown to have a high concentration of NA and is one of the origins of the ventral NA pathway.

Connections from the PBN to the LHA have been traced by horseradish peroxidase (Kita and Oomura, 1982; Norgren, 1978) and by electrophysiology (Adachi and Kobashi, 1985). Monosynaptic connections extend from the caudal NTS to the paraventricular nucleus (PVN), the dorsomedial nucleus of the hypothalamus (DMH) and the arcuate hypothalamic nucleus (see Oomura, 1983; Oomura and Yoshimatsu, 1984). There is evidence that splanchnic afferents also converge on the forebrain areas that receive afferents from the NTS (Jeanningros and Mei, 1980). Thus, reduction in food intake could be caused by specific modification of the activity of LHA feeding-related neurons, either directly or through these systems.

Niijima found that decerebrate rats retained the ability to control blood glucose level if the AP was intact, but not after lesion of the AP (personal communication, 1985). Flynn and Grill (1983) concluded from tests of decerebrate rats that the neural system caudal to the forebrain is sufficient to control food intake. Ritter and Edwards (1984) found that lesion of the AP in decerebrate rats affected only the ratio of highly palatable to normal lab chow consumed; total calorie intake was unaffected. They suggested that the hindbrain makes important contributions to initiation and termination of feeding behavior and may modulate orosensory perception (Ritter et al., 1985).

A hierarchical arrangement of sensory and processing points in the control of feeding is shown in Figure 3. Based on the information now known, it seems probable that the NTS and more peripheral points constitute a reflex network that could maintain feeding behavior, while the more central regions probably integrate information to produce the more sophisticated control necessary to maintain homeostasis.

C. Periphery

Stimulation of the ventral part of the LHA, the major locus of GS neurons in the hypothalamus, increases single unit activity in the pancreatic vagus (VPN). Stimulation of the dorsal LHA inhibits VPN activity. Stimulation of the VMH also inhibits vagal activity. Inhibition of pancreatic splanchnic nerve (SPN) activity during LHA stimulation depends on both stimulus magnitude and frequency. The inhibition is sometimes preceded by a transient activity increase. Either inhibition or facilitation of SPN activity can be observed in response to VMH stimulation (Oomura and Kita, 1981).

SPN activity decreased in VMH-lesioned rats. VMH lesion induced either a gradual increase or a sudden increase followed by restoration to normal VPN activity. In DMH-lesioned animals, activity increased persistently in the VPN and decreased in the SPN. In animals with PVN lesions, VPN activity increased immediately and maintained the higher level, while SPN activity gradually decreased. LHA lesion produced a sudden decrease in the VPN, and about equally, either decreased or increased SPN activity. Lesions outside of these areas produced almost no response (Yoshimatsu et al., 1984). Similar complex SPN responses were induced by LHA stimulation.

D. Effectors

The effects of LHA stimulation on plasma insulin level are also complex; it might elicit increase or decrease, or have no effect. It has been reported that insulin secretion elicited by LHA stimulation was further facilitated by infusion of an α-adrenoceptive blocking agent (Berthoud et al., 1980). Since the LHA has direct connections with both the DMV and the intermediolateral cell column in the spinal cord (IML) it can affect insulin secretion by either facilitation through the VPN, or inhibition through the SPN, or both (Fig. 6A). Thus, these excitatory and inhibitory components can complicate insulin secretion and autonomic nervous activity in response to manipulation of the LHA.

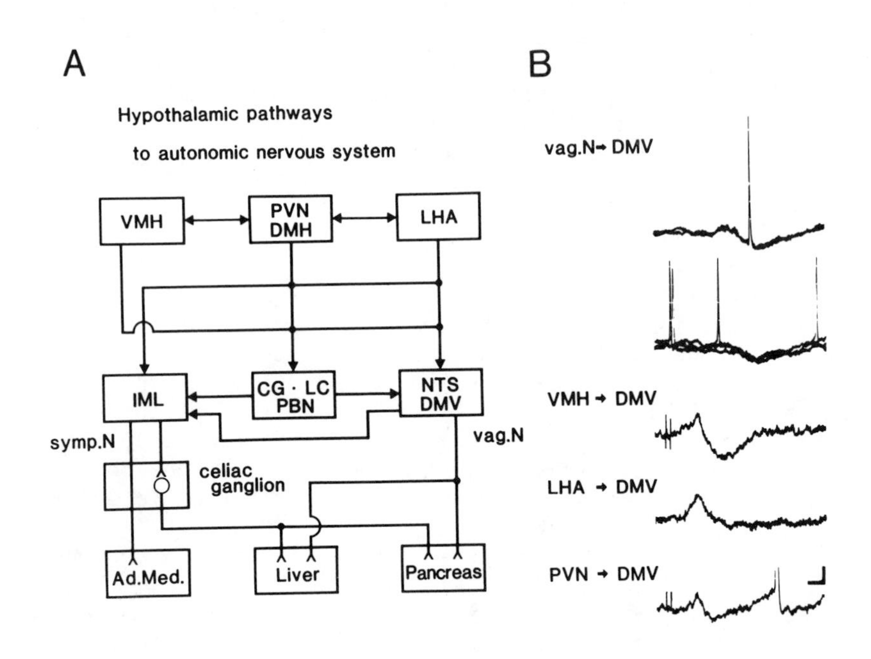

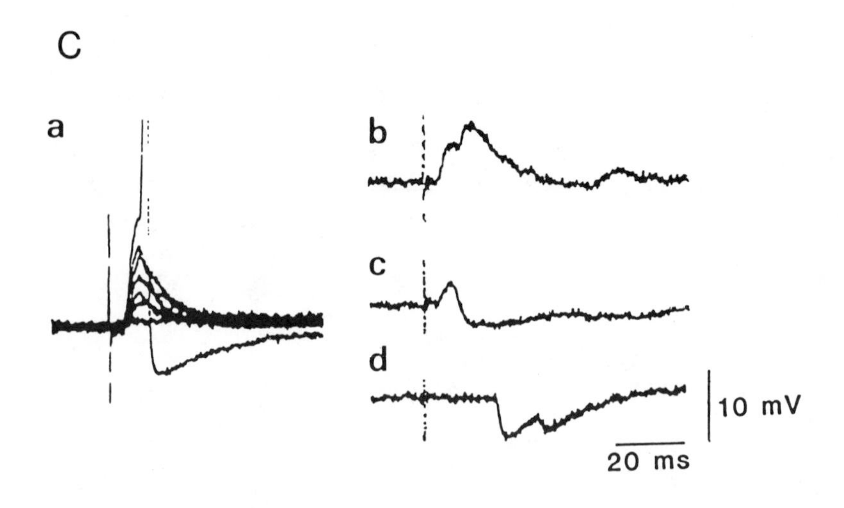

The changes induced in autonomic nervous activity by VMH lesions agree with stimulation experiments. Insulin secretion is increased by stimulation of the VPN and decreased by stimulation of the SPN. The obesity produced by VMH lesion has been prevented by vagotomy or denervation of the pancreas (Powley and Opsahl, 1974; Inoue et al., 1977). Thus, both the increase of VPN and the decrease of SPN activity that follow VMH lesion may facilitate insulin secretion.

It has recently become evident that the VMH has direct projections to the autonomic preganglionic nuclei, the DMV, or the IML (Swanson and Kuypers, 1980), although questions still remain (Saper et al., 1976; Horst et al., 1984). There have been few reports of DMH and PVN involvement in the control of pancreatic hormone secretion. PVN lesion elicits hyperphagia (Leibowitz et al., 1981), and hyperinsulinemia mimicked the effects of VMH lesion (Steffens, 1976). Anatomically the DMH and PVN could be involved in hypothalamic control of pancreatic hormone secretion via the autonomic nervous system, so the role of the PVN in the feeding process should be studied further. Hypothalamic effects on the activity of the vagal and splanchnic efferents to the liver, which control enzymatic activity in

---

Fig. 6. Interconnections within hypothalamus, and with visceral nervous system. (A) Network showing known feeding relations; other connections exist but relations to feeding are not known. DMH, dorsomedial hypothalamus, IML, intermediolateral cell column, CG, central gray, LC, locus coeruleus, PBN, parabrachial nucleus, NTS, nucleus tractus solitarii, symp. N, sympathetic nervous system, Ad. Med., adrenal medulla. (B) Verification of connections. Stimulation of vagus nerve (Vag. N) subdiaphragmatically induced antidromic spikes, and monosynaptic EPSP-IPSPs (upper) and IPSPs (lower) in dorsomotor nucleus of the vagus (DMV). Stimulation of ventromedial nucleus of hypothalamus (VMH) or paraventricular nucleus (PVN) elicited monosynaptic EPSP-IPSPs in DMV. Stimulation of lateral hypothalamus (LHA) elicited monosynaptic EPSPs and EPSP-IPSPs (not shown) in DMV (Oomura, 1985). Calibration: 10 mV for upper 2, 5 mV for lower 3; 20 msec. (C) Intracellular recordings from the DMV neurons by stimulation of tractus solitarius (in vitro slice experiment). a, graded summation of monosynaptic EPSPs by increasing single stimulus intensity; action potential (truncated) evident for supramaximal stimulation. Single stimulus elicited monosynaptic EPSP followed by polysynaptic EPSPs, b; or by single polysynaptic EPSP-IPSP, c; or by multiple polysynaptic IPSPs, d (Mizuno et al., 1985).

the liver (Shimazu, 1982), are similar to effects on efferents to the pancreas (Yoshimatsu et al., unpublished observation).

The VMH, DMH, and PVN, which are all involved in feeding behavior and regulation of the peripheral autonomic nervous system, have intrahypothalamic connections with each other as shown in Fig. 6A, although the connections between the VMH and the LHA are not direct but indirect through the DMH, PVN, and the amygdala (Berk and Finkelstein, 1982; Kita and Oomura, 1982; Luiten and Room, 1980). Thus, these regions might collectively integrate control of pancreatic hormone secretion by balancing parasympathetic and sympathetic neuronal activity.

Stimulation of the subdiaphragmatic vagal nerve (which contains both efferent and afferent fibers) produced an antidromic spike and a monosynaptic EPSP-IPSP or monosynaptic IPSP in the DMV. The same stimulation also produced occasional orthodromic spikes in the NTS. This suggests that afferent fibers of the vagal nerve project directly to the caudal NTS and thence to the DMV. There are some direct connections to the DMV through the solitary tract (ST) (Mizuno et al., 1985). While intracellular recordings in the DMV in slice preparations were being made, stimulation of ST elicited monosynaptic EPSPs and stimulation in the vagal nerve projection path elicited antidromic action potentials (Fig. 6C). An EPSP was elicited in the NTS about 1 msec before independent EPSP or EPSP-IPSP sequences appeared in the DMV when the ST was stimulated while intracellular recordings were made simultaneously in the NTS and DMV (Mizuno et al., 1985). These observations, with appropriate precautions against artifacts, verified monosynaptic excitatory, and multisynaptic excitatory and inhibitory connections between the NTS and DMV. As shown in Figure 6B, stimulation of the VMH or the PVN each produced a monosynaptic EPSP-IPSP sequence indicating direct connections to the DMV from both of these centers. Stimulation of the LHA produced a monosynaptic EPSP-IPSP in the DMV to also show a direct connection.

Lesion of the LHA tended to depress adrenal nerve activity, which is known to be presynaptic, while VMH lesion enhanced adrenal activity (Yoshimatsu et al., 1985). Stimulation of the medial LHA produced frequency-dependent enhancement of adrenal nerve activity. Stimulation of the VMH had no effect. Simultaneous measurement of catecholamine release from the adrenal medulla indicated parallelism between secretion and neuronal activity (Katafuchi et al., 1985b; Yoshimatsu et al., 1985). These results indicate direct connections from the LHA to the IML.

V. OTHER METABOLITES THAT AFFECT GLUCOSE RESPONSIVE UNITS

Neuronal responses in the LHA and VMH are not glucose-specific. The activity of GS and GR neurons in the LHA and VMH is affected by blood-borne metabolites and hormones such as FFA and insulin (Oomura, 1976), glucagon (Oomura and Inokuchi, 1983; Inokuchi et al., 1985a) and other substances which have been characterized as intrinsic feeding or metabolism-related. The systemic levels of these endogenous substances, including the opioids, fluctuate as the nutritional state of an animal varies (Gambert et al., 1980; Margules et al., 1978; Morley, 1980).

A. Opioids

Morphine and met-enkephalin depress food intake under acute conditions (King et al., 1979), but findings are questionable because of addiction and withdrawal effects. Morphine and met-enkephalin, applied electrophoretically, had the same effects as glucose in the LHA and VMH of naive rats (Ono et al., 1980). The effects of morphine and met-enkephalin, but not those of glucose, were blocked by naloxone. Endogenous opioids selectively suppress the LHA GS neurons during noxious stimuli (Sikdar and Oomura, 1985).

B. TRH, TRHA

Electrophoretic application of thyrotropin-releasing hormone (TRH), which inhibits feeding, stimulates GR neurons in the VMH. Its effects in the LHA are not clear (Ishibashi et al., 1979a). TRHA, an analog of TRH found in the urine of anorexia nervosa patients, depresses mouse body weight when injected i.p. It also suppresses GS neurons in the LHA (Oomura, 1981).

C. Calcitonin

Electrophoretically applied calcitonin suppresses GS but not non-GS neurons in the rat LHA. Ouabain blocks suppression by glucose, but not by calcitonin. Phenoxybenzamine and methylsergide block NA and 5-HT effects, respectively, but not calcitonin activity. This indicates that the effect of calcitonin is independent of Na^+-K^+ pump acceleration, and noradrenergic or serotonergic mechanisms. When a phosphodiesterase inhibitor, such as isobutylmethylxanthine or papaverine, is applied simultaneously with calcitonin, the inhibitory effect is augmented, whereas it is blocked by nicotinic acid, an adenylate cyclase inhibitor. These results suggest that inhibition of the GS neuron by calcitonin is mediated by an increase in the intracellular level of cyclic AMP (Shimizu and Oomura, 1985). Dibutyryl cyclic AMP actually inhibits GS neuron activity. Calcitonin hyperpolarizes GS neurons about -15mV and reduces membrane conductivity about 75%. This reduction may be produced by passive decrease of Na^+ ions, possibly due to decreased Na^+ and/or Ca^{++} permeability. However, as yet, there is no indication of how cyclic AMP may affect such changes in ionic conductance. A small amount of calcitonin injected into the third ventricle (0.5 U, corresponding to a concentration of 1.5 U/ml in the CSF, assuming the total volume of cerbrospinal fluid to be 300 μl) reduced the 24-hour food intake by 50%. The blood concentration of calcitonin is about 0.1 U/ml. If we assume that diffusion from the ventricle to the LHA, 20 minutes after injection, does not exceed 1% of the initial amount of injection (de Wildt et al., 1982), the amount of calcitonin (threshold dose for inhibition of food intake, 0.2 U) available to act on the GS neurons might be 2 μU (0.3 ng). This is comparable to the amount that exists in the hypothalamus other than the VMH, approximately 0.21 ng/hypothalamus (Flynn et al., 1981). Thus, the effective dose may be physiological. During food intake calcitonin is released into the blood from the thyroid gland. It precipitates Ca^+ ions, absorbed from food, into bone. This clearly indicates that calcitonin is a satiety-inducing substance and its effect might be mediated through direct inhibition of GS neurons in the LHA.

D. Neurotensin

Neurotensin suppresses feeding (Hoebel et al., 1982), gastric acid secretion (Shiraishi et al., 1982) and insulin secretion, and induces hyperglycemia and hyperglucagonemia (Nagai and Frohman, 1976). Found in various parts of the brain and gut, it is without effect when it alone is applied electrophoretically to GS neurons in the LHA, but it does modulate the effects of insulin or 2DG (Shiraishi et al., 1982).

E. Bombesin

Bombesin, a potent gastric acid secretagogue found in the viscera and brain of mammals is, at relatively high doses, a powerful depressant of feeding (Gibbs et al., 1979) and induces hyperglycemia and hyperglucagonemia (Brown et al., 1979). Electrophoretically, it has no effect on LHA neurons, but does modulate the effects of other agents on GS neurons in the LHA (Shiraishi et al., 1982).

F. CCK

CCK suppresses feeding. After vagotomy, it has no effect on food intake (Smith et al., 1981). Electrophoretically applied CCK was reported to have no effect on LHA or VMH neurons (Ishibashi et al., 1979b), but it does suppress the activity of hepatic vagal afferents (Niijima, 1981). Proglumide, a CCK antagonist, induced feeding when injected into the PVN. Injection of CCK at the same time reduced the proglumide effect (Hoebel et al., 1984).

G. Insulin

The role of insulin in the utilization of glucose is well known. The activity of GR neurons in the VMH is much facilitated by simultaneous electrophoretic application of insulin and glucose, but is slightly suppressed by insulin alone (Oomura, 1973). Activity of GS neurons in the LHA is dose-dependently facilitated by insulin (Oomura et al., 1975). Insulin-binding sites in the brain have been identified (Havrankova et al., 1979; Szabo and Szabo, 1975; Oomura and Kita, 1981). Measurements of insulin in the blood plasma and brain of rats vary widely (Baskin and Dorsa, 1982; Havrankova et al., 1979; Oomura and Kita, 1981; Stein et al., 1982), but two points are consistent: 1) Plasma insulin level varies widely from lean to obese animals and from one strain to another. 2) Although brain insulin content seems to depend on the strain or the investigator, it appears to be independent of plasma level and remains stable within each individual. Specific insulin-binding sites in the brain, most abundant in the hypothalamus and olfactory bulb, are independent of peripheral insulin concentration (Havrankova et al., 1979; Sakamoto et al., 1980).

H. Glucagon

Glucagon-like immunoreactivity (GLI) (measured by glucagon N-terminal specific antiserum, GA-10) has been observed in the LHA and VMH of the rat at concentrations of about 50 ng/mg protein, and there is a very small amount in the pons and olfactory bulb, but immunoreactive pancreatic glucagon (IRG) was not found in any brain region studied. However, many binding sites for both GLI and IRG were found in the hypothalamus. After 48 hours of food deprivation, GLI concentration increased about 1.3 times in the VMH and LHA (Inokuchi et al., 1985b).

Nerve fibers and terminals that reacted for GLI, but not IRG, were found in the periventricular region, PVN, supraoptic nucleus, median eminence, arcuate nucleus, and VMH, but fewer were found in the LHA and DMH (Inokuchi et al., 1985b). Immunoreactive cell bodies were not observed in the hypothalamus, but have been seen in the NTS (our unpublished observation). After 24 hours of food deprivation, immunoreactive cell bodies became visible at the border region between the VMH and arcuate nucleus and ventromedial portion of the VMH, together with an increase in the number of immunostaining fibers and terminals in the PVN, dorsolateral part of the VMH, and ventrolateral portion of the LHA.

Activity of LHA, DMH, and VMH neurons was recorded extracellularly during electrophoretic application of pancreatic glucagon (Oomura and Inokuchi, 1983; Inokuchi et al., 1985a). Glucagon significantly suppressed the activity of GS neurons, but not that of non-GS neurons, in the LHA, and suppressed GR neurons with a higher dose in the VMH. Inhibition of GS neurons by glucagon was blocked by ouabain, so glucagon may regulate cyclic AMP and, consequently, Na^+-K^+ pump activity. Electrophoretically applied glucagon hyperpolarized the GS neuron membrane, but did not affect conductivity; this was the same effect produced by glucose. Glucagon injected into the carotid artery (100 μg) also suppressed glucagon-sensitive neuron activity, indicating the effectiveness of blood glucagon.

Injection of 5 ng of glucagon into the third ventricle (corresponding to 15 ng/ml in the CSF) at 1900 suppressed food intake to 70% of control for the next 3 hours (Inokuchi et al., 1984). The concentration reaching the LHA 20 minutes after injection is assumed to be 150 pg/ml (see Section V,C). This would be within the physiological range of blood glucagon concentration, approximately 100 pg/ml. The concentration of glucagon during feeding reached double the inter-meal concentration, so pancreatic glucagon may hematogenically suppress GS neurons in the LHA to terminate feeding (Geary and Smith, 1982; Langhans et al., 1982), and GLI in the brain may act as an inhibitory neurotransmitter or neuromodulator in the hypothalamus.

VI. ORGANIC ACIDS

A. Physiological Range

The blood levels of the two sugar acids, 3,4-dihydroxy-butanoic acid (2-deoxytetronate, 2-DTA) and 2,4,5-trihydroxy pentanoic acid (3-deoxypentonate, 3-DPA), and the ketone body, 3-hydroxybutric acid (3-HBA), all increase during deprivation, although circadian-related variations are evident. When measured at 1000, the blood concentration of 2-DTA was about 0.1 mM and that of 3 DPA was 0.2 mM. These concentrations rose to 5 and 1.5 times their initial levels, respectively, after 36 hours, which would have been just after the beginning of feeding time the next day. The concentration of 3-HBA increased from an initial level of about 0.2 mM to more than 10 times that level after 96 hours of food deprivation (Hawkins et al., 1971).

The effect of 3-DPA on food intake depended on the time of injection. When injected at 1030, it induced only a transient increase in feeding. Injection at 1900 was most effective; food intake and eating duration measured from 2000 to 2300 increased significantly, although the eating rate did not. Thus, the strongest effect, which lasted for at least 3 hours, was seen when 3-DPA was injected at twilight.

Motor activity, intake of pelleted chow, and drinking were measured for 24 hours after a 2.5 μmol third ventricular injection of 2-DTA at 1630. The mean number of pellets consumed in 24 hours decreased significantly. Prandial drinking was also reduced. The suppression of food intake was apparently not due to general depression, since motor activity was not depressed. In order to clarify the satiety effect of 2-DTA, it was injected into the third ventricle of 72-hour-food-deprived rats at 1900, and cumulative food intake was measured from 2000 to 0100. In the first 5 hours after injection, the deprived rats that were injected with 2-DTA and the non-deprived rats ate the same amount, while the amount eaten by the deprived rats treated with artificial cerebrospinal fluid was about double that eaten by the others. This pattern persisted for 24 hours and then recovered. Thus 2-DTA significantly reduced feeding even in 72-hour-food-deprived rats.

Injection of 2.5 μmol of 3-HBA into the third ventricle at 1030 induced one transient episode of feeding and then depressed further feeding for 24 hours (Arase et al., 1984; Davis et al., 1981). Normal dark-time feeding was almost completely suppressed. Thus, the total amount of food intake was reduced to half that consumed by the controls. Injection of 2.5 μmol of 3-HBA at 1400 increased single LHA neuron activity in a chronic and unanesthetized rat for more than 50 minutes, and one episode of food intake with prandial drinking occurred during this period. Electrophoretic 3-HBA application significantly excited GS neurons in the LHA in acute anesthetized rats (Oomura, 1984).

The brain content of 2-DPA and 3-DPA is not known. The concentration of these sugar acids in the CSF was estimated to be 8.3 mM after a 2.5 μmol injection (rat CSF volume is about 300 μl, see section V,C), disregarding metabolism. Furthermore, if CSF is produced at a rate of 0.5% of the total volume per minute, the concentration of sugar acids would be diluted to 22% of the initial concentration, 1.8 mM, at 5 hours after injection. Both 2-DTA and 3-DPA are effective for 5 hours. The concentration of 3-HBA in the CSF is about 50% of that in the blood (Kadekaro et al., 1980). The 3-HBA concentration in the CSF may thus be estimated to be about 1.6 mM after 96 hours of food deprivation. Therefore, its concentration could be comparable to the 5-hour concentration estimated above for the sugar acids. Approximately 1% of peptides diffuse from the ventricle to the LHA in 20 minutes after application. Assuming a diffusion rate similar to peptides, 2-DTA or 3-DPA concentrations available to act on hypothalamic GS and GR neurons should be no more than 80 μM (18μM, 5 hours after injection). Similar arguments can be used for arriving at 3-HBA levels.

B. Central and Feeding Effects

Changes in single LHA neuron activity were correlated with feeding behavior after injection of 2.5 μmol 2-DTA or 3-DPA into the third ventricle. After presentation of food at 1800, LHA single neuron activity increased concomitantly with the onset of food and water intake. Feeding stopped and neural activity decreased slightly about 5 minutes after the start of 2-DPA injection. Neural activity declined significantly 18 minutes later. Activity of the neuron did not change during motor activity, and there was no feeding for the next 4 hours, although the animal had been deprived of food for 24 hours. An injection of 3-DPA at 1400 in a non-deprived, ad lib-fed rat induced food intake with prandial drinking and increased neural activity that continued for about 40 minutes.

In anesthetized rats, electrophoretic applications of 2-DPA suppressed spontaneous single neuron discharges of GS neurons, and 3-DPA excited GS neurons; both effects were dose-dependent. Opposite effects were observed when 2-DPA was applied to GR neurons in the VMH (Puthuraya et al., 1985). The difference in the effects of 2-DTA and 3-DPA on LHA and VMH neuron activity was statistically significant. Some GS neurons also influence gastric acid secretion. Shiraishi et al. (1985) reported that gastric acid secretion was not affected by peripheral application of either 2-DTA or 3-DPA, nor by injection of 2-DTA alone into the LHA. However, 3-DPA injection into the LHA induced gastric acid secretion, and 2-DTA injection suppressed gastric acid secretion induced by 2DG. Shiraishi et al. also verified previous reports (Shimizu et al., 1984) of 2-DTA and 3-DPA effects on LHA neuron activity.

When 3-DPA was injected into the jugular vein at 1900, it increased cumulative food consumption in a dose-related manner. The concentration was 100 to 500 μmol. The same doses of 2-DTA injected intravenously at the same time had no effect. The smallest effective dose was more than 2.5 mmol. One possible reason for the ineffectiveness of systemic 2-DTA may be that not enough reaches the brain. To test this possibility 2-DTA was encapsulated in liposome vesicles and injected i.p. (Nagai et al., 1985). Injection of 1.25 mmol encapsulated 2-DTA i.p. suppressed feeding. Injections of encapsulating material i.p. or into the third ventricle, or unencapsulated 2-DTA i.p., or saline i.p. were all ineffective. It was estimated that 8% of the liposome-treated 2-DTA was actually encapsulated and that 1% of the encapsulated 2-DTA, or 1 μmol, reached the brain. This dose is comparable to the 2.5 μmol that was injected into the third ventricle. As mentioned above, relatively massive doses of 2-DTA injected into the carotid artery also suppressed feeding (Sakata, personal communication). It is still too early to draw any conclusion, but this might indicate that entrapment of systemically applied 2-DTA within the blood circulatory system may also be a factor in diminishing its potency.

C. Origin of 2-DTA and 3-DPA

Since human urine sometimes contains more 2-DTA than can be accounted for by intake, and carbohydrate loading increases the level of 2-DTA in urine, it is probably a metabolite (Lawson et al., 1976). A metabolic substrate could be 4-hydroxybutyrate (Lee, 1977). Levels of 4-hydroxybutyrate are high in rat brain, and ß-oxidation produces 2-DTA as an intermediate factor (Oomura, 1984).

The origin of 3-DPA is not yet known, but it could be in the metabolic cycle of glucose or acetic acid. The 3-DPA in blood could be produced by continuous carbohydrate and/or lipid metabolism (Oomura, 1984).

It is not yet known which of the two optical isomers of 2-DTA, L or D, exist in the blood, but the synthesized L type used in experiments has the same effect as that extracted from blood. There are four optical isomers of 3-DPA: 2D, 4D; 2L, 4L; 2L, 4D and 2D, 4L, 5-trihydroxypentanoic acid. The 2L, 4L isomer is the only one found to be physiologically active (right chemical structure, below). These exist in the lactone ring form in blood.

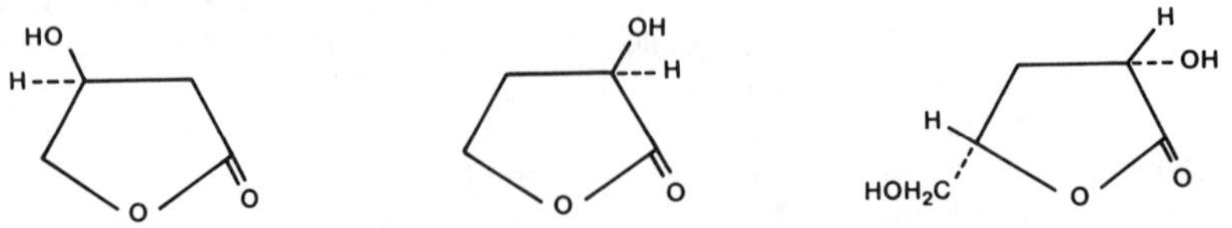

The only difference between the molecules of 2-DTA and 3-DPA is the absence of an OH group from the second carbon of the 2-DTA and from the third carbon of the 3-DPA molecule, but their effects on the activity of GS neurons and on feeding are opposite. Another sugar acid found in the blood of food-deprived rats, 2,4-dihydroxybutanoic acid (3-deoxytetronic acid, middle chemical structure, above), is a 4-carbon chain with OH absent from the 3rd carbon. This sugar acid produced effects similar to those of 3-DPA (Oomura, 1984).

VII. SUMMARY

Control of feeding depends on many things, among which are the levels of glucose and other metabolites in the body, chemoresponsive neurons that can be found in various parts of the nervous system, and networks through which these neurons interact with each other, as well as with effector neurons and effectors. Glucose-responsive elements that have been implicated in feeding control have been reported in the lateral hypothalamus (LHA) and the ventromedial nucleus of the hypothalamus (VMH), among other forebrain sites; in the nucleus tractus solitarii (NTS), the dorsal motor nucleus of the vagus (DMV) and area postrema in the medulla; and in afferent nerves of the liver, pancreas and small intestine. Glucose-responsive neurons in these three regions appear to be morphologically similar to each other, and are morphologically different from non-glucose responsive neurons.

The elements that have been tested for responses to metabolites other than glucose have so far indicated that they are not necessarily glucose-specific. These neurons respond in various ways to opioids, TRH, calcitonin, neurotensin, bombesin, insulin and glucagon, 2-deoxytetronate (2-DTA), 3-deoxypentonate (3-DPA) and other glucose analogues. The reactions to some of these materials, such as 2DG, are rather complex. The feeding response to 2DG is related to the LHA neuronal response in ways that are consistent with the hypothesis that LHA glucose-sensitive neurons exert control over feeding behavior. Some materials, of which neurotensin and bombesin are examples, have no direct effects on neural activity when applied alone topically, but strongly modulate the effects of others. The sugar acids, 2-DTA and 3-DPA, directly affect glucose-responding neurons in the LHA and VMH in ways that agree with their effects on feeding. Secondary feeding-related responses, such as gastric acid secretion, respond to centrally applied 2-DTA and 3-DPA. 2-DTA, 3-DPA, and 3-HBA have effects in chronic recording and behavioral experiments that appear to be mediated by GS neurons in the LHA.

Some brain regions, such as the orbitofrontal cortex (OBF), that are involved in control of feeding behavior and appear to respond to glucose level, do so only through the networks that connect them with glucose-responding chemoresponsive neurons. These networks include direct connections and systems such as the catecholaminergic systems.

Inputs from the tongue to the rostral NTS, and from hepatic, pancreatic and intestinal glucose-responsive units to the caudal NTS, are relayed either directly or through the parabrachial nucleus to the LHA. The LHA, in communication with the VMH, the amygdala, and the motor cortex, the OBF, and dorsolateral prefrontal area, integrates the information it receives and directs effectors, muscular for obtaining food and visceral for maintaining homeostasis, to appropriate responses. Outputs from the LHA and the VMH to the visceral effectors are relayed through the NTS, the DMV, and the intermedio-lateral cell column. Consideration of results indicates that the GS neurons in the LHA consolidate information from internal sources, including neurons that respond to external signals and then project to effectors or their control neurons to initiate or terminate overt feeding behavior.

It appears that feeding can be controlled reflexly by visceral and medullary centers and their networks, but the complex behavior necessary to obtain food involves networks and centers in the forebrain, including the hypothalamus and its associated circuits.

ACKNOWLEDGEMENTS

I thank Professor A. Simpson for help in preparation of this article. This work was partly supported by Grant-in-Aid 57440085 from the Ministry of Education, Science and Culture.

REFEFERENCES

Adachi, A., and M. Kobashi. 1985. Convergence of hepatic gluco- and osmosensitive inputs on chemosensitive units in the medulla oblongata of the rat. In *Neuronal and Endogenous Chemical Control Mechanisms in Emotional Behavior*, ed. Y. Oomura. Japan Scientific Societies Press, Tokyo, and Springer-Verlag, Berlin. In press.

Adachi, A., N. Shimizu, Y. Oomura, and M. Kobashi. 1984. convergence of hepataportal glucose-sensitive afferent signals to glucose-sensitive units within the nucleus of the solitary tract. *Neurosci. Lett.* 46: 215-218.

Aou, S., Y. Oomura, H. Nishino, A. Inokuchi, and Y. Mizuno. 1983. Influence of catecholamines on reward related neuronal activity in monkey orbitofrontal cortex. *Brain Res.* 267: 165-170.

Aou, S., Y. Oomura, L. Lénard, H. Nishino, A. Inokuchi, T. Minami, and H. Misaki. 1984. Behavioral significance of monkey hypothalamic glucose-sensitive neurons. *Brain Res.* 302: 69-70.

Arase, K., T. Sakata, Y. Oomura, M. Fukishima, K. Fujimoto, and K. Terada. 1984. Short-chain polyhydroxymonocarboxylic acids as physiological signals for food intake. *Physiol. Behav.* 33: 261-268.

Baskin, D.G., and D.M. Dorsa. 1982. Brain insulin concentrations in hyperinsulinemic fatty Zucker rats are not elevated. *Soc. Neurosci. Abstr.* 8: 272

Berk, M.L., and J.A. Finkelstein. 1982. Efferent connections of the lateral hypothalamic area of the rat: An autoradiographic investigation. *Brain Res. Bull.* 8: 511-526.

Berthoud, H.R., D.A. Bereiter, and B. Jeanrenaud. 1980. Electrophysiological and neuroanatomical studies of hepatic portal osmo- and sodium-receptive afferent projections within the brain. *J. Auton. Nerv. Syst.* 2: 183-198.

Brown, M., Y. Taché, and D. Fisher. 1979. Central nervous system action of bombesin: Mechanism to induce hyperglycemia. *Endocrinology* 105: 660-665.

Davis, J.D., D. Wirthshafter, K.E. Asin, and D. Brief. 1981. Sustained intracerebroventricular infusion of brain fuels reduces body weight and food intake in rats. *Science* 212:81-83.

De Wildt, D., J. Verhoef, and A. Whitter. 1982. H-Pro-[^{3}H]Leu-Gly-NH_2: Uptake and metabolism in rat brain. *J. Neurochem. 38*: 67-74.

Flynn, F.W., and H.J. Grill. 1983. Insulin elicits ingestion in decerebrate rats. *Science* 221: 188-190.

Flynn, J.J., D.L. Margules, and C.W. Cooper. 1981. Presence of immunoreactive calcitonin in the hypothalamus and pituitary lobes of rats. *Brain Res. Bull.* 6: 547-549.

Gambert, S.R., T.L. Garthwaite, C.H. Pontzer, and T.C. Hagen. 1980. Fasting associated with decrease in hypothalamic ß-endorphin. *Science* 210: 1271-1272.

Geary, N., and G.P. Smith. 1982. Pancreatic glucagon and postprandial satiety in the rat. *Physiol. Behav.* 28: 313-322.

Gibbs, J., D.J. Fauser, E.A. Rowe, B.J. Rolls, E.T. Rolls, and S.P. Maddison. 1979. Bombesin suppresses feeding in rats. *Nature (London)* 282: 208-210.

Havrankova, J., J. Roth, and M.J. Brownstein. 1979. Concentrations of insulin and insulin receptors in the brain are independent of peripheral insulin levels. Studies of obese and streptozotocin-treated rodents. *J. Clin. Invest.* 64: 636-642

Hawkins, R.A., D.H. Williamson, and H.A. Krebs. 1971. Ketone-body utilization by adult and suckling rat brain *in vivo*. *Biochem. J.* 122: 13-18.

Hoebel, B.G., D. Dorfman, and P. Scott. 1984. Effects of CCK in the brain. In *The Neural and Metabolic Bases of Feeding*. Satellite Symposium of Neuroscience Society Meeting.

Hoebel, B.G., L. Hernandez, S. McLean, B.G. Stanley, E.F. Aulissi, F. Glimcher, and D. Margolin. 1982. Catecholamines, enkephalin and neurotensin in feeding and reward. In *The Neural Basis of Feeding and Reward,* ed. B.G. Hoebel and D. Novin, 465-478. Haer Institute for Electrophysiological Research, Brunswick, Maine.

Horst, G.J. ter, P.G.M. Luiten, and F. Kuipers. 1984. Descending pathways from hypothalamus to dorsal motor vagus and ambiguus nuclei in the rat. *J. Auton. Nerv. Syst.* 11: 59-75.

Inokuchi, A., Y. Oomura, and H. Nishimura. 1984. Effect of intracerebroventricular infused glucagon on feeding behavior. *Physiol. Behav.* 33: 397-400.

Inokuchi, A., Y. Oomura, N. Shimizu, and T. Yamamoto. 1985a.

Central action of glucagon in the rat hypothalamus. *Am. J. Physiol.* In press.
Inokuchi, A., Y. Tomita, C. Yanaihara, R. Yui, Y. Oomura, H. Kimura, T. Hase, T. Matsumoto, and N. Yanaihara. 1985b. Glucagon-related peptides in the rat hypothalamus. *Cell Tiss. Res.* In press.
Inoue, M., Y. Oomura, H. Nishino, S. Aou, S.K. Sikdar, M. Hynes, Y. Mizuno, and Y. Katafuchi. 1983. Cholinergic role in monkey dorsolateral prefrontal cortex during bar-press feeding behavior. *Brain Res.* 278: 185-194.
Inoue, M., Y. Oomura, S. Aou, H. Nishino, and S.K. Sikdar. 1985. Reward-related neuronal activity in monkey dorsolateral prefrontal cortex during feeding behavior. *Brain Res.*: 307-312.
Inoue, S., G.A. Bray, and Y.S. Muller. 1977. Effect of transplantation of pancreas on development of hypothalamic obesity. *Nature (London) 266*: 742-744.
Ishibashi, S., Y. Oomura, and T. Okajima. 1979a. Facilitatory and inhibitory effects of TRH on lateral hypothalamic and ventromedial neurons. *Physiol. Behav.* 22: 785-787.
Ishibashi, S., Y. Oomura, T. Okajima, and S. Shibata. 1979b. Cholecystokinin, motilin and secretin effects on the central nervous system. *Physiol. Behav.* 23: 401-403.
Jeanningros, R., and N. Mei. 1980. Vagal and splanchnic effects at the level of the ventromedial nucleus of the hypothalamus in the cat. *Brain Res.* 185: 239-251.
Kadekaro, M., C. Timo-Iaria, and M. deL. Vicentini. 1980. Gastric secretion provoked by functional cytoglycoponenia in the nuclei of the solitary tract in the cat. *J. Physiol. (Lond.)* 299: 397-407.
Katafuchi, T., Y. Oomura, and H. Yoshimatsu. 1985a. Single neuron activity in the rat lateral hypothalamus during 2-deoxy-D-glucose induced and natural feeding behavior. *Brain Res.* In press.
Katafuchi, T., Y. Oomura, A. Niijima, and H. Yoshimatsu. 1985b. Effects of intracerebroventricular 2-DG infusion and subsequent hypothalamic lesion on adrenal nerve activity in the rat. *J. Auton. Nerv. Syst.* In press.
King, M.G., A.J. Kastin, R.D. Olson, and D.H. Coy. 1979. Systemic administration of met-enkephalin, (D-Ala^2)-met-enkephalin, beta-endorphin, and (D-Ala^2)-beta-endorphin: effect on eating, drinking and activity measures in rats. *Pharm. Biochem. Behav.* 11: 407-411.
Kita, H., and Y. Oomura. 1981. Functional synaptic interconnections between the lateral hypothalamus and frontal and gustatory cortices in the rat. In *Brain Mechanisms of Sensation*, ed. Y. Katsuki, R. Norgren and M. Sato, 307-321. John Wiley & Sons, New York.

Kita, H., and Y. Oomura. 1982. An HRP study of the afferent connections to rat medial hypothalamic region. *Brain Res. Bull.* 8: 53-62; 63-71
Langhans, W., V. Zieger, E. Scharrer, and N. Geary. 1982. Stimulation of feeding in rats by intraperitoneal injection of antibodies to glucagon. *Science* 218: 894-896.
Lawson, A.M., R.A. Chalmers, and R.W.E. Watts. 1976. Urinary organic acids in man. I. Normal patterns. *Clin. Chem.* 22: 1283-1287.
Lee, C.R. 1977. Evidence for the ß-oxidation of orally administered 4-hydroxybutyrate in humans. *Biochem. Med.* 17: 284-291.
Leibowitz, S.F., N.J. Hammer, and K. Chang. 1981. Hypothalamic paraventricular nucleus lesions produce overeating and obesity in the rat. *Physiol. Behav.* 27: 1031-1040.
Lénard, L., Y. Oomura, H. Nishino, and S. Aou. 1985. Activity in monkey lateral hypothalamus during operant feeding: modulation by catecholamines and opiate. In *Neuronal and Endogenous Chemical Control Mechanisms in Emotional Behavior,* ed. Y. Oomura. In press. Japan Scientific Societies Press, Tokyo, and Springer-Verlag, Berlin.
Llinas, R., and M. Sugimori. 1980. Electrophysiological properties of *in vitro* Purkinje cell somata in mammalian cerebellar slices. *J. Physiol. (Lond.)* 305: 171-195.
Luiten, P.G., and P. Room. 1980. Interrelations between lateral, dorsomedial and ventromedial hypothalamic nuclei in the rat. An HRP study. *Brain Res.* 190: 321-322.
Margules, D.L., B. Joisset, M.J. Lewio, H. Shibuya, and C.B. Pet. 1978. ß-endorphin is associated with overeating in genetically obese mice (ob/ob) and rats (fa/fa). *Science* 202: 989-991.
Mei, N. 1978. Vagal glucoreceptors in the small intestine of the cat. *J. Physiol. (Lond.)* 282: 485-506.
Minami, T., Y. Oomura, and M. Sugimori. 1985. Electrophysiological properties of guinea-pig ventromedial hypothalamic neurones: an *in vitro* study. *J. Physiol. (Lond.)*. Submitted.
Mizuno, Y., and Y. Oomura. 1984. Glucose-responding neurons in the nucleus tractus solitarii of the rat: *in vivo* study. *Brain. Res.* 307: 109-116.
Mizuno, Y., J. Nabekura, and Y. Oomura. 1985. Electrophysiological study of input-output organization in the dorsal motor nucleus of the vagus *in vitro*. *J. Physiol. (Lond.)*. Submitted.
Morley, J.E. 1980. The neuroendocrine control of appetite: the role of endogenous opiates, cholecystokinin, TRH, gamma-amino-butyric-acid and the diazepam receptor. *Life Sci.* 27: 355-368.

Nagai, K., and L.A. Frohman. 1976. Hyperglycemia and hyperglucagonemia following neurotensin administration. *Life Sci*. 19: 273-279.

Nagai, Y., T. Osanai, T. Sakata, K. Terada, K. Arase, and Y. Oomura. 1985. The use of liposomes for the control of neuronal activity and its disorders. International Symposium of Liposome.

Nakano, Y., Y. Oomura, H. Nishino, S. Aou, T. Yamamoto, and S. Nemoto. 1984. Neuronal activity in the medial orbitofrontal cortex of the behaving monkey: modulation by glucose and satiety. *Brain Res. Bull*. 12: 381-385.

Niijima, A. 1980. Glucose-sensitive afferent nerve fibers in the liver and regulation of blood glucose. *Brain Res. Bull*. 5, Suppl. 4: 175-179.

Niijima, A. 1981. Visceral afferents and metabolic function. *Diabetologia*, Suppl. 20: 325-330.

Nishino, H., T. Ono, M. Fukuda, and K. Sasaki. 1982. Lateral hypothalamic neuron activity during monkey bar press feeding behavior: modulation by glucose, morphine and naloxone. In *The Neural Basis of Feeding and Reward,* ed. B.G. Hoebel and D. Novin, 355-372. Haer Institute for Electrophysiological Research, Brunswick, Maine.

Nishino, H., Y. Oomura, L. Lénard, and S. Aou. 1985. Catecholaminergic contributions to feeding-related lateral hypothalamic activity in the monkey. *Brain Res*. Submitted.

Norgren, R. 1978. Projections from the nucleus of the solitary tract in the rat. *Neuroscience* 3: 207-218.

Novin, D., D.A. Vanderweele, and M. Rezek. 1973. Infusion of 2-deoxy-D-glucose into the haepatic-portal system causes eating: evidence for peripheral glucoreceptors. *Science* 181: 858-860.

Ono, T., Y. Oomura, H. Nishino, K. Sasaki, K. Muramoto, and I. Yano. 1980. Morphine and enkephalin effects on hypothalamic glucoresponsive neurons. *Brain Res*. 185: 208-212.

Ono, T., H. Nishino, M. Fukuda, K. Sasaki, K. Muramoto, and Y. Oomura. 1982. Glucoresponsive neurons in rat ventromedial hypothalamic tissue slices *in vitro*. *Brain Res*. 232: 494-499.

Ono, T., Y. Oomura, H. Nishino, K. Sasaki, M. Fukuda, and K. Muramoto. 1981. Neural mechanisms of feeding behavior. In *Brain Mechanisms of Sensation,* ed. Y. Katsuki, R. Norgren, and M. Sato, 272-286. John Wiley & Sons, New York.

Ono, T., H. Nishino, M. Fukuda, and K. Sasaki. 1984a. Monkey amygdala, lateral hypothalamus and prefrontal cortex roles in food discrimination, motivation to bar press, and ingestion reward. In *Modulation of Sensorimotor Activity*

During Alterations in Behavioral States, ed. R. Bandler, 251-268. Alan R. Liss, Inc., New York.
Ono, T., H. Nishino, M. Fukuda, K. Sasaki, and M. Nishijo. 1984b. Single neuron activity in dorsolateral prefrontal cortex of monkey during operant behavior sustained by food reward. *Brain Res.* 311: 323-332.
Oomura, Y. 1973. Central mechanism of feeding. In *Advances in Biophysics,* ed. M. Kotani, 65-142. University of Tokyo Press, Tokyo.
Oomura, Y. 1976. Significance of glucose, insulin, and free fatty acid on the hypothalamic feeding and satiety neurons. In *Hunger: Basic Mechanisms and Clinical Implications,* ed. D. Novin, W. Wrywicka, and A. Bray, 145-157. Raven Press, New York.
Oomura, Y. 1980. Input-output organization in the hypothalamus relating to food intake behavior. In *Handbook of the Hypothalamus,* Vol. II, ed. P. J. Morgane and J. Panksepp, 557-620. Marcel Dekker, New York and Basel.
Oomura, Y. 1981. Chemosensitive neuron in the hypothalamus related to food intake behavior. *Jpn. J. Pharmacol.* 31, Suppl. 1: 12.
Oomura, Y. 1983. Glucose as a regulator of neuronal activity. In *Advances in Metabolic Disorders,* Vol. 10, ed. A.J. Szabo, 31-63. Academic Press, New York.
Oomura, Y. 1984. Newly characterized C-chain organic acids for neuronal control of food intake. *Biomed. Res. 5,* Suppl.: 91-104.
Oomura, Y. 1985. Feeding control through bioassay of body chemistry. *Jpn. J. Physiol.* 35: 1-19.
Oomura, Y., and A. Inokuchi. 1983. Effect of glucagon on feeding behavior. *Biomed. Res. 4,* Suppl.: 209-215.
Oomura, Y., and H. Kita. 1981. Insulin acting as a modulator of feeding through the hypothalamus. *Diabetologia,* Suppl. 20: 290-298.
Oomura, Y., and H. Yoshimatsu. 1984. Neuronal network of glucose monitoring system. *J. Auton. Nerv. Syst.* 10: 359-372.
Oomura, Y., H. Ooyama, M. Sugimori, T. Nakamura, and Y. Yamada. 1974. Glucose inhibition of the glucose-sensitive neurone in the rat lateral hypothalamus. *Nature (London)* 247: 284-286.
Oomura, Y., M. Sugimori, T. Nakamura, and Y. Yamada. 1975. Contribution of electrophysiological techniques to the understanding of central control systems. In *Neural Integration of Physiological Mechanisms and Behaviour,* ed. G.J. Mogenson and F.R. Calaresu, 375-395. University of Toronto Press, Toronto.
Oomura, Y., T. Ono, N. Shimizu, H. Kita, S. Ishizuka, S. Aou,

H. Yoshimatsu, and K. Yamabe. 1980. Monkey cortical neuron discharge variations in high fixed ratio schedules for food. *Brain Res. Bull.* 5, Suppl. 4: 151-161.

Perrin, J., J. Crousillat, and N. Mei. 1981. Assessment of true splanchnic glucoreceptors in the jejuno-ileum of the cat. *Brain Res. Bull.* 7: 625-628.

Powley, T.L., and C.A. Opsahl. 1974. Ventromedial hypothalamic obesity abolished by subdiaphragmatic vagotomy. *Am. J. Physiol.* 226: 25-33.

Puthuraya, K.P., Y. Oomura, and N. Shimizu. 1985. Effects of endogenous sugar acids on the ventromedial hypothalamic nucleus of the rat. *Brain Res.* *332*: 165-168.

Ritter, R.C., and G.L. Edwards. 1984. Area postrema lesions cause overconsumption of palatable foods but not calories. *Physiol. Behav.* 32: 923-927.

Ritter, R.C., G.E. Edwards, and V.K. Nonovinakere. 1985. Hindbrain control of food intake: behavioral evidence for metabolic sensing and modulation of orosensory responsiveness. In *Neuronal and Endogenous Chemical Control Mechanisms in Emotional Behavior*, ed. Y. Oomura. In press.

Russek, M. 1970. Demonstration of the influence of an hepatic glucosensitive mechanism of food-intake. *Physiol. Behav.* 5: 1207-1209.

Sakamoto, Y., Y. Oomura, H. Kita, S. Shibata, S. Suzuki, T. Kuzuya, and S. Yoshida. 1980. Insulin content and insulin receptors in the rat brain. Effect of fasting and streptozotocin treatment. *Biomedical Res.* 1: 334-340.

Saper, C.B., L.W. Swanson, and W.M. Cowan. 1976. The efferent connections of the ventromedial nucleus of the hypothalamus of the rat. *J. Comp. Neurol.* 169: 409-442.

Shimazu, T. 1981. Central nervous system regulation of liver and adipose tissue metabolism. *Diabetologia* 20, Suppl.: 343-356.

Shimizu, N., and Y. Oomura. 1985. Calcitonin-induced anorexia in rats: evidence for its inhibitory action on lateral hypothalamic chemosensitive neurons. *Brain Res.* In press.

Shimizu, N., Y. Oomura, D. Novin, C. Grijalva, and P. Cooper. 1983. Functional correlations between lateral hypothalamic glucose-sensitive neurons and hepatic portal glucose-sensitive units in rat. *Brain Res.* 265: 49-54.

Shimizu, N., Y. Oomura, and T. Sakata. 1984. Modulation of feeding by endogenous sugar acids acting as hunger or satiety factors. *Am. J. Physiol.* 246: R542-R550.

Shiraishi, T., Tsutsui, K., Sakata, T., and A. Simpson. 1982. 2-deoxy-D-glucose effects on hypothalamic neurons, and on short- and long-term feeding. In *The Neural Basis of Feeding and Reward*, ed. B.G. Hoebel and D. Novin,

373-390. Haer Institute for Electrophysiological Research, Brunswick, Maine.

Shiraishi, T., M. Kawashima, and Y. Oomura. 1985. Endogenous sugar acid control of hypothalamic neuron activity and gastric acid secretion in rats. *Brain Res. Bull.* 14. In press.

Sikdar, S.K., and Y. Oomura. 1985. Selective inhibition of glucose-sensitive neurons in rat lateral hypothalamus by noxious stimuli and morphine. *J. Neurophysiol.* 53: 17-31.

Smith, G.P., C. Jerome, B.J. Cushin, R. Eterno, and K.J. Simansky. 1981. Abdominal vagotomy blocks the satiety effect of cholecystokinin in the rat. *Science* 213: 1036-1037.

Steffens, A.B. 1976. Influence of the oral cavity on insulin release in the rat. *Am. J. Physiol.* 230: 1411-1415.

Stein, L.J., D.L. Hjeresen, D. Porte, Jr., and S.C. Woods. 1982. Genetically obese Zucker rats have inappropriately low immunoreactive insulin levels in cerebrospinal fluid. *Soc. Neurosci. Abstr.* 8: 273.

Swanson, L.W., and H.G.J.M. Kuypers. 1980. A direct projection from the ventromedial nucleus and ventrochiasmatic area of the hypothalamus to the medulla and spinal cord of the rat. *Neurosci. Lett.* 17: 307-312

Szabo, O., and A.J. Szabo. 1975. Studies on the nature and mode of action of the insulin-sensitive glucoregulator receptor in the central nervous system. *Diabetes* 24: 328-336.

Takagi, S.F. 1984. The olfactory nervous system of the old world monkey. *Jpn. J. Physiol.* 34: 561-573.

Thorpe, S.J., E.T. Rolls, and S. Maddison. 1983. The orbitofrontal cortex: neuronal activity in the behaving monkey. *Exp. Brain Res.* 49: 93-115.

Yamamoto, T., Y. Oomura, H. Nishino, S. Aou, Y. Nakano, and S. Nemoto. 1984. Monkey orbitofrontal activity during emotional and feeding behavior. *Brain Res. Bull.* 12: 441-443.

Yoshimatsu, H., A. Niijima, Y. Oomura, K. Yamabe, and T. Katafuchi. 1984. Effects of hypothalamic lesion on pancreatic nerve activity in the rat. *Brain Res.* 303: 147-152.

Yoshimatsu, H., Y. Oomura, T. Katafuchi, A. Niijima, and A. Sato. 1985. Lesions of the ventromedial hypothalamic nucleus enhanced sympatho-adrenal function. *Brain Res.* In press.

Chapter 9

GLUCOPRIVATION AND THE GLUCOPRIVIC CONTROL OF FOOD INTAKE

Sue Ritter

FEEDING BEHAVIOR
Neural and Humoral Controls

I. INTRODUCTION

Smith and Epstein (1969) were the first to observe the dramatic increase in food intake associated with injection of 2-deoxy-D-glucose (2DG), a glucose analogue that competitively inhibits intracellular glucose metabolism (Brown, 1962; Horton et al., 1973; Tower, 1958; Wick et al., 1957). Since glucose deprivation ("glucoprivation") brought about by 2DG administration appeared to be the stimulus for increased food intake, Smith and Epstein asserted the existence of a "glucoprivic control." Subsequent experiments have demonstrated the operation of the glucoprivic control in many mammalian species, including humans (Houpt et al., 1977; Houpt, 1974; Houpt and Hance, 1977; Smith and Epstein, 1969; Thompson and Campbell, 1977; Mackay et al., 1940). Although the glucoprivic control is the only control of feeding so far identified that can be activated by blockade of a specfic metabolic pathway, it is nevertheless only one of many controls of feeding that together contribute to metabolic homeostasis. The relative importance of glucoprivation in the physiological control of feeding is a compelling issue that has not yet been resolved.

The glucoprivic control of feeding was reviewed in 1975 by Epstein, Nicolaidis, and Miselis. Their insightful review crystallized the most salient questions concerning this control and gave impetus to a decade of research. It is required reading for anyone interested in this control. The present review is not comprehensive, but focuses on issues guiding the work in our laboratory over the past few years.

II. GLUCOPRIVIC CONTROL VERSUS GLUCOSTATIC REGULATION OF FEEDING

In their review of glucoprivic feeding, Epstein et al. (1975) articulated the relationship of the glucoprivic control of food intake to the glucostatic theory of feeding behavior, proposed by Jean Mayer (1955, 1972). The glucostatic theory proposes that feeding is controlled in defense of glucose homeostasis and predicts that changes in glucose availability will produce alterations in food intake appropriate for restoration of glucose homeostasis. In contrast to this comprehensive theory, the glucoprivic control simply describes an event, glucoprivation, leading to increased food intake.

It is important to make the distinction between these two terms, in part because the predictions of the glucostatic theory of food intake have not been fully supported by experimental data, whereas the glucoprivic control of feeding is well established. The glucostatic theory makes two major predictions about food intake: that decreased glucose availability will stimulate food intake and that increased glucose availability will inhibit food intake. Obviously, the glucoprivic control might be one component of a glucostatic mechanism, and to the extent that this is true, evidence demonstrating this control would provide support for the first prediction of the glucostatic theory. The second prediction, however, has not found strong experimental support. An ever-growing body of evidence indicates that elevated blood glucose levels do not directly inhibit food intake (for example see Smyth et al., 1947; Baile et al., 1971; Bellinger et al., 1977) and that the conditions under which glucose does reduce intake are indeed limited (for example see Yin et al., 1970; VanderWeele et al., 1974). Several investigators have even demonstrated that experimentally produced increases in glucose availability do not necessarily block feeding aroused by glucoprivation (Ritter et al., 1978; Friedman et al., 1983; Bellin and Ritter, 1981a). Future research may yet reveal that increased glucose *utilization* inhibits feeding, but present data do not provide compelling support for this prediction.

III. PHYSIOLOGICAL RESPONSES TO GLUCOPRIVATION

Glucoprivation elicits various compensatory physiological responses in addition to the stimulation of food intake. Noteworthy among these is an enhancement of sympathoadrenal catecholamine release (Cannon et al., 1924; Hokfelt and Bydgeman, 1961; Himsworth, 1968), which stimulates hepatic glycogenolysis and inhibits insulin release (Frohman et al., 1973; Yamamoto et al., 1983). By virtue of these two effects, glucose utilization by peripheral insulin-sensitive tissues is diminished and glucose availability to insulin-insensitive brain tissue is enhanced. The function of brain tissue, which under usual physiological circumstances depends entirely upon blood-borne glucose for its energy metabolism, is thereby conserved. Thus, in the normal animal, the sympathoadrenal response subserves the unceasing metabolic requirements of the brain over the short-term, until the animal can repair its energy deficit by eating, or until longer-term physiological support can be generated by other mechanisms, such as gluconeogenesis and ketogenesis. It is important to note that when cellular glucoprivation is induced experimentally by administration of a glucose antimetabolite, glucoprivation actually occurs in spite of the sympathoadrenal response. Under these conditions, cells do not fully benefit from the hyperglycemia because intracellular glucose utilization itself is diminished by the drug.

Other responses elicited by glucoprivation include increased gastric acid secretion (Colin-Jones and Himsworth, 1969) and rate of gastric emptying (Friedman et al., 1982; but see also Granneman, 1983), release of glucagon (Unger, 1976), release of growth hormone and 17-hydroxycorticosteroids (Smith and Root, 1969) and mobilization of free fatty acids (Laszlo et al., 1969; Larue-Achagiotis and LeMagnen, 1980). Future research may reveal an important role for these responses in the physiological expression of the glucoprivic control of food intake (for example see, Stricker and McCann, 1985). However, they will be mentioned only briefly in this review.

IV. WHAT IS THE STIMULUS FOR GLUCOPRIVIC FEEDING?

A. Pharmacological Activation of the Glucoprivic Control

The glucoprivic control of food intake can be experimentally activated by injection of high doses of insulin or by administration of a glucose antimetabolite, such as 2DG (Smith and Epstein, 1969), 1,5-anhydroglucitol (lDG) (Sakata et al., 1981), 5-thioglucose (5TG) (Ritter and Slusser, 1980), or 3-0-methylglucose (Booth, 1972) (Fig. 1). These drugs have in common the ability to inhibit cerebral glucose utilization, and this appears to be the critical event required for the stimulation of feeding. Insulin produces cerebral glucoprivation indirectly by facilitating glucose uptake by peripheral tissues. As glucose is removed from the blood, its availability to insulin-insensitive brain tissue is diminished. Thus, cerebral glucoprivation is a consequence of the insulin-induced hypoglycemia. Glucose antimetabolites, on the other hand, produce glucoprivation directly, by competitive blockade of glycolysis (Wick et al., 1957; Chen and Whistle, 1975; Pazur and Kleppe, 1964; Sols and Crane, 1954). Cells are deprived of glucose as an energy substrate because their ability to metabolize available glucose is diminished by the drug.

B. Cerebral Glucoprivation Is an Adequate Stimulus for Feeding

The discovery by Smith and Epstein (1969) that 2DG stimulates feeding was critical for the conceptualization of the glucoprivic control. Since feeding induced by 2DG actually occurs in the presence of elevated blood glucose levels, their discovery established that cellular glucoprivation, not hyperinsulinemia or hypoglycemia per se, is the essential stimulus for activation of this control. Smith and Epstein speculated that the glucoprivic control is related to glucose metabolism in the brain, rather than in peripheral tissues. Miselis and Epstein (1975) later showed that injection of 2DG into the lateral ventricle at systemically ineffective doses stimulated a vigorous feeding response, thus demonstrating that cerebral glucoprivation is an adequate stimulus for activation of the glucoprivic control.

C. Delayed Glucoprivic Feeding and the Stimulus to Eat

Although it has been presumed that feeding in response to glucoprivic agents is a consequence of ongoing reduction in glucose utilization, R.C. Ritter and his colleagues made a discovery that forces us to consider alternative hypotheses regarding the actual stimulus (or stimuli) governing the

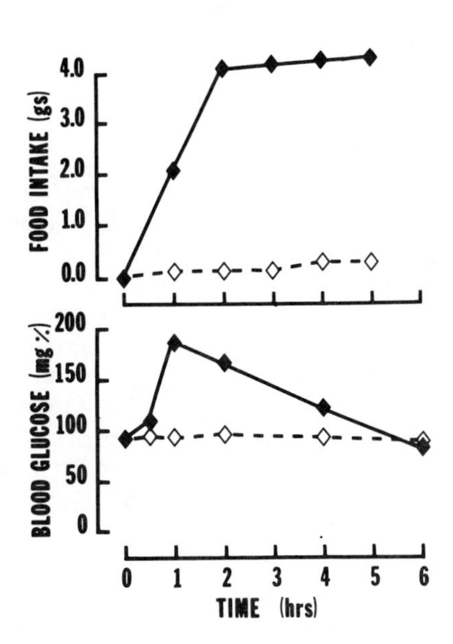

Fig. 1. Increased feeding and increased blood glucose (solid line) after 2DG (200 mg/kg) given at time 0. Feeding and blood glucose after an isotonic saline control injection (broken line). From R.C. Ritter et al. (1978), *American Journal of Physiology* 234.

feeding response. They observed that the feeding response to either systemic (Ritter et al, 1978) or central (Engeset and Ritter, 1980) glucoprivation is still present even when access to food is delayed until other signs of glucoprivation have abated. This finding raises the possibility that glucoprivic feeding is not controlled exclusively by the glucoprivation itself. Rather, some consequence of glucoprivation that outlasts the glucoprivic episode may also contribute to the response. Several possibilities have been considered and will be discussed further in Section VIII.

D. Do Insulin and 2DG Provide Equivalent Stimuli for Food Intake?

Several lines of evidence suggest that insulin and 2DG are not equivalent in their effects on food intake: in addition to activating the glucoprivic control, insulin may also be capable of stimulating feeding by nonglucoprivic mechanisms. First, certain lesions differentially impair 2DG-induced feeding. For example, zona incerta lesions (McDermott and Grossman, 1980), ventromedial hypothalamic lesions (King et al., 1978), brainstem transections anterior to the solitary nucleus (McDermott and Grossman, 1979), and coronal knife cuts in the midbrain tegmentum (Alheid et al., 1977; Grossman and Grossman, 1977), all impair the feeding response to 2DG while leaving the feeding response to insulin largely intact. Intraventricular (i.c.v.) alloxan injections (Murnane and Ritter, 1982) and subdiaphragmatic vagotomy (Booth, 1972) also seem to produce greater deficits in 2DG- than in insulin-induced feeding. Second, it has been demonstrated that hamsters and gerbils do not eat in response to 2DG or 5TG, but do increase their food intake in response to insulin (Ritter and Balch, 1978; Rowland, 1978; Silverman, 1978; Sclafani and Eisenstadt, 1980; DiBattista, 1982). Thus, it appears that some animals may normally lack the glucoprivic control, activated by direct and selective blockade of glucose metabolism by antimetabolic glucose analogues, and still increase their food intake in response to insulin.

The idea that insulin may activate more than one control of feeding should not be surprising since insulin is a complex hormone with multiple metabolic and physiological actions. Before this explanation of the data is accepted, however, due consideration should be given to the possibility that 2DG may produce a more profound depressive effect on behavior than insulin because of its rapidly developing, simultaneous blockade of both central and systemic glucose utilization. In contrast, physical impairment after insulin may be less profound because systemic glucose utilization is actually facilitated during the initial phase of insulin action. Behavioral responses to glucoprivation may therefore survive a larger variety of experimental manipulations when insulin is used as the glucoprivic agent than when 2DG is used. In support of this possibility, McDermott and Grossman (1980) have found that in zona incerta lesioned rats, which increase their food intake in response to insulin, but not to 2DG, 2DG-induced feeding can be partially restored by administration of mild stimulant drugs prior to the 2DG feeding test.

V. BRAIN GLUCORECEPTORS AND METABOLIC MONITORING

The term "glucoreceptors" is used in this review to refer to putative metabolic receptor cells that arouse food intake in response to glucoprivation. The existence of these glucoreceptors is still hypothetical. However, glucose-responsive neurons--neurons which increase or decrease their firing rate in response to changes in glucose concentration--have been identified by electrophysiological studies in many central and peripheral locations (Oomura, 1976; Mei, 1978; Adachi, 1985; Mizuno et al., 1983; Niijima, 1980). It seems reasonable to speculate that a specific population of neurons with responses such as these may control glucoprivic feeding.

Several ideas have been put forth regarding the way in which metabolic monitoring might be accomplished by glucoreceptor cells. One idea is that glucose availability may be monitored at the cell membrane by surface receptors that bind specifically with glucose (Woods and McKay, 1978). An alternative proposal is that intracellular metabolism of glucose or other energy substrates may be monitored by the receptor cell.

Convincing support for the membrane hypothesis is lacking. Although 3-0-methylglucose and phlorizin, which antagonize glucose transport into some cells (Crofford and Renold, 1965; Betz et al., 1975), stimulate glucoprivic feeding (Glick and Mayer, 1968; Flynn and Grill, 1985), blockade of glucose entry also reduces intracellular glucose utilization. Thus, this particular evidence does not distinguish the membrane from the metabolic hypothesis. Evidence against the membrane hypothesis and in favor of the metabolic hypothesis comes from the fact that after administration of hyperphagic doses of 2DG, blood glucose concentration--and hence, the number of glucose molecules in contact with surface receptors--is actually increased by virtue of the sympathoadrenal response. In addition, the molar ratio between 2DG and glucose under such conditions would greatly favor glucose. Moreover, membrane receptors probably cannot distinguish between glucose and 2DG (Sokoloff, 1977). Studies with 2DG thus demonstrate that glucoreceptor cells controlling glucoprivic feeding are influenced by intracellular glucose utilization, not extracellular glucose concentration.

The term glucoreceptor has been criticized as being metabolically restrictive. The fact that glucoprivic feeding can be blocked by ß-hydroxybutyrate (Stricker et al., 1977; Stricker and Rowland, 1978; Bellin and Ritter, 1981) suggests that receptors governing glucoprivic feeding might be influenced not only by glucose, but by any substrate they can utilize for energy metabolism. This criticism can be countered with two arguments. First, evidence that ketone bodies influence food intake by acting on glucoreceptors is presumptive at best, and the possibility that they inhibit feeding by some other mechanism cannot be disregarded. Second, although glucoreceptors may in fact be capable of utilizing ß-hydroxybutyrate for energy metabolism, glucose is the only substrate available to brain cells under most normal physiological conditions (Sokoloff, 1972; Owen et al., 1967; Krebs et al., 1971; Hawkins and Biebuyck, 1979). For these reasons, the term glucoreceptor seems appropriate, though possibly somewhat restrictive.

How deficient substrate availability or diminished energy production influences cellular function in glucoreceptor cells and leads to the feeding response is not yet known. Localization and electrophysiological studies of glucoreceptor cells controlling the glucoprivic feeding response will be required to answer these questions.

VI. LOCATION OF BRAIN GLUCORECEPTORS CONTROLLING GLUCOPRIVIC FEEDING

A. The Search for Glucoreceptors in the Forebrain

A substantial body of evidence now supports the hypothesis that glucoreceptors controlling glucoprivic feeding reside within the brain (Epstein and Miselis, 1975; Slusser and Ritter, 1980), but their precise location is not yet known. Three criteria must be satisfied to qualify any tissue for this particular glucoreceptive function: destruction of the cells in question should impair or abolish glucoprivic feeding, local glucoprivation should stimulate the feeding response, and the cells involved must be capable of altering their electrical activity or synaptic function in response to changes in glucose availability. The likelihood that these criteria can be satisfied for glucoreceptors controlling glucoprivic feeding depends in part on the distribution of the receptors within the brain. If the glucoreceptor cells are widely distributed within the brain, proof of their participation in the glucoprivic control will be more difficult to obtain using conventional lesion, injection, and elctrophysiological approaches than if they are highly localized.

Although a number of structures have been proposed to control glucoprivic feeding, the above criteria for glucoreceptors have not yet been satisfied for any specific site. The central role of the hypothalamus in food intake and the electrophysiological detection of glucose-responsive neurons there (Oomura, 1976) have fostered the notion that glucoreceptors supporting glucoprivic feeding are located within the hypothalamus (for example see Mayer, 1955; Panksepp, 1974), perhaps in the lateral (LH) or ventromedial (VMH) regions. However, this widely held belief has been called into question. Whether animals with VMH lesions eat in response to glucoprivation remains controversial (Miselis and Epstein, 1971; King et al., 1978; Larue-Achagiotis and LeMagnen, 1981). Nevertheless, the participation of VMH glucoreceptors in glucoprivic feeding seems unlikely because injections of 2DG directly in the VMH fail to elicit feeding (Miselis and Epstein, 1975).

The lateral hypothalamus also seems unlikely as a major site for the glucoreceptors. Although LH lesions greatly impair the response to glucoprivation (Epstein and Teitelbaum, 1967; Wayner et al., 1971), the response can still be demonstrated under the appropriate experimental conditions (Kanarek et al., 1981; Stricker et al., 1975). The report of Balagura and Kanner (1971) claimed that implantation of crystalline 2DG into the lateral or medial hypothalamus stimulates feeding, but in the same study, implantation of crystalline glucose also had a stimulatory effect. Other investigators have failed to stimulate feeding with hypothalamic injections of 2DG (Miselis and Epstein, 1975), and the small effects reported in the original study were most certainly due to nonspecific osmotic stimulation of hypothalamic tissue. Such osmotic stimulation of feeding by hypothalamic injections was recognized prior to the report of Balagura and Kanner (Epstein, 1960).

The paraventricular nucleus of the hypothalamus (PVN), which appears to be extensively involved in feeding behavior, has also been examined for possible involvement in glucoprivic feeding. However, lesions of this region do not impair glucoprivic feeding. Thus, the hypothalamus may contain circuitry necessary for the normal expression of the glucoprivic feeding response, but failure to elicit feeding by localized glucoprivation indicates that the glucoreceptors themselves are located elsewhere.

Numerous other brain lesions have been reported to impair glucoprivic feeding, but it has not yet been demonstrated that any of these sites have glucoreceptive functions (Bellinger et al., 1978; Brandes and Johnson, 1978; McDermott and Grossman, 1980; Grossman and Grossman, 1977; McDermott et al., 1979). In some cases it has been difficult to establish whether lesioned animals actually lack the glucoprivic control, rather than the capacity to respond during the glucoprivic emergency (see for example Marshall and Richardson, 1974; Kanarek et al., 1981; Stricker et al., 1975).

B. Glucoprivic Feeding Appears To Be Controlled by Glucoreceptors in the Caudal Hindbrain

Recent evidence strongly indicates that glucoreceptors mediating both the feeding and the sympathoadrenal hyperglycemic responses to i.c.v. injections of glucose antimetabolites are not located in the forebrain, but reside in the caudal hindbrain. Ritter et al. (1981) demonstrated that feeding and hyperglycemia could no longer be elicited by lateral ventricle injections of the glucoprivic agent 5TG if the distribution of 5TG was restricted to the forebrain by the insertion of a silicone grease plug into the mesencephalic aqueduct. Feeding and hyperglycemic responses to fourth ventricular 5TG injections, however, occurred regardless of whether the plug was in place and were not diminished by the plug (Fig. 2).

Grill and his colleagues have provided additional evidence for the caudal hindbrain location of the critical glucoreceptor cells. They have demonstrated that both feeding (Flynn and Grill, 1982) and hyperglycemic (DiRocco and Grill, 1979) responses to systemic glucoprivation are present in chronic decerebrate rats. Thus, the caudal hindbrain contains neural machinery adequate for both the arousal and the reflexive expression of the glucoprivic control; forebrain structures are not essential. This is not to say that forebrain structures are unimportant in the elaboration of normal appetitive responses to glucoprivation, but the basic circuitry for detection of glucoprivation and arousal of the feeding response must be present in the caudal hindbrain. Feeding has been reported in decerebrate rats only in response to insulin-induced hypoglycemia, which may qualify these results. It will also be important to know whether intake is stimulated by 2DG or 5TG in these rats. If insulin is capable of stimulating feeding by both glucoprivic and nonglucoprivic mechanisms, as discussed above, specific loss of the glucoprivic control might not be detected when insulin is used as the glucoprivic agent.

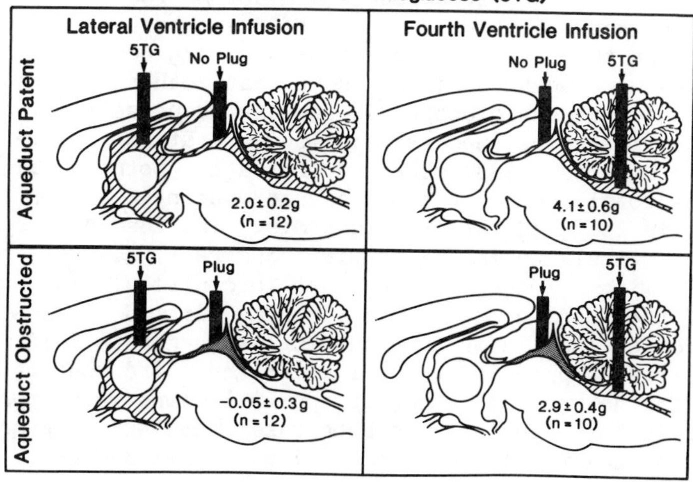

Fig. 2. Increased food intake following lateral or 4th i.c.v. 5TG prior to and subsequent to obstruction of the cerebral aqueduct. The values shown represent the change in food intake (mean ± SEM) following 5TG minus intake following artificial cerebrospinal fluid. Note that lateral ventricle 5TG injection increased feeding only when the aqueduct was patent. In contrast, feeding stimulated by 4th i.c.v. 5TG was not significantly reduced by aqueduct obstruction. From Ritter, R.C., in *Emotions: Neuronal and Chemical Control*, ed. Y. Oomura, 1986. With permission from Japan Scientific Societies Press.

C. The Search for Glucoreceptors in the Area Postrema

Contreras and co-workers have reported that area postrema (AP) lesions impair feeding in response to either systemic or i.c.v. injection of antimetabolic glucose analogues (Contreras et al., 1982; Bird et al., 1983), and they have proposed that glucoreceptors controlling the feeding response may be localized in the AP. The AP would seem to be a likely site for glucoreceptors because of its recently discovered involvement in food intake (see R.C. Ritter, this volume), the demonstration of glucose responsive cells in both the AP (Adachi and Kobashi, 1985) and the adjacent nucleus of the solitary tract (NST) (Mizuno et al., 1983), and the absence of a blood-brain barrier in the AP. (Roth and Yamamoto, 1968). These characteristics equip this structure for a pivotal role in monitoring both central and peripheral nutrient flux. Moreover, the AP is located in the caudal hindbrain where glucoreceptors appear to reside.

Despite the appeal of this evidence, locating the glucoreceptors for glucoprivic feeding in the AP requires further proof. Although Contreras et al. (1982) observed profound deficits in their AP-lesioned animals when glucoprivic feeding responses were measured in a 2-hour test, other investigators have found that the glucoprivic feeding response is clearly expressed in AP-lesioned rats if longer testing times are used. Hyde and Miselis (1983) observed that after both insulin and 2DG, the total amount of food consumed during a 6-hour test was similar in lesioned rats and controls. The feeding response was not entirely normal in the lesioned rats, however, since it occurred with a longer latency than in controls. In controls, most of the feeding occurred during the first 3 hours after injection of insulin or 2DG, but in lesioned rats feeding was not increased appreciably until after the third hour of the test. The nature and extent of the AP lesions were apparently similar in both studies. In both cases the lesions produced near-total destruction of the AP with minimal damage to the adjacent NST.

Nonavinakere and Ritter (1984) studied 2DG-induced feeding in rats with intentionally large lesions in this region. These lesions totally destroyed the AP as well as 63% of the adjacent NST. In spite of their large lesions, 8 out of 12 rats increased their food intake in response to subcutaneous 2DG in doses of 200 mg/kg or more, but the total amount consumed in a 6-hour test was less than in controls. As in the Hyde and Miselis study, the response occurred with a longer latency in the lesioned rats than in controls. These investigators also found that 61% of the AP-lesioned rats increased their food intake in response to intraventricular 5TG in a 3-hour test.

Several conclusions can be made from these results. (1) Area postrema lesioned rats are clearly capable of increasing their food intake in short-term tests in response to other stimuli (Edwards and Ritter, 1981). Therefore, their abnormal glucoprivic response may indicate that some glucoreceptor cells have been destroyed by the lesion. However, this conclusion is tenuous until it can be shown that feeding can be stimulated by injection of a glucoprivic agent directly into the AP. (2) The failure of AP lesions to totally eliminate glucoprivic feeding suggests that if the AP contains glucoreceptors controlling this response, it must not be the only hindbrain location for such cells. (3) These data suggest that more than one type of receptor contributes to the control of feeding during glucoprivation: one type located in the AP may mediate short latency feeding responses and others outside the AP may mediate longer latency responses or responses to the more delayed physiological consequences of glucoprivation.

VII. ARE EXTRACEREBRAL GLUCORECEPTORS INVOLVED IN THE GLUCOPRIVIC CONTROL OF FEEDING?

The demonstration that brain glucoreceptors are independently capable of eliciting feeding in response to glucoprivation does not necessarily preclude the participation of extracerebral receptors in the glucoprivic control. In fact, the importance of peripheral signals in other aspects of feeding and the identification of glucose-responsive neurons in the peripheral vagus nerve have made this idea particularly enticing. Several investigators have explored this possibility, but a convincing demonstration of the existence and role of peripheral glucoreceptors in the *initiation* of glucoprivic feeding is still lacking.

Novin and his colleagues have proposed that peripheral receptors involved in the initiation of glucoprivic feeding may be situated in the liver (Novin et al., 1973). They found that feeding in rabbits was stimulated to a greater degree and with a shorter latency by hepatic portal injections of 2DG than by jugular injections. Furthermore, this effect was abolished by vagotomy. However, later work by other investigators failed to identify differences between the hepatic and jugular routes of administration (Russell and Mogenson, 1975; Stricker and Rowland, 1978) in rats. This inconsistency has not been resolved, but a primary role for the hepatic vagal innervation seems dubious because neither selective hepatic vagotomy (Tordoff et al., 1982; Bellinger and Williams, 1983) nor celiac vagotomy (Tordoff and Novin, 1982) impairs glucoprivic feeding.

Stricker and his colleagues (Stricker and Rowland, 1978; Stricker et al., 1977) have provided additional evidence against the location of glucoreceptors in the liver. They found that both insulin- and 2DG-induced feeding could be blocked by co-administration of ß-hydroxybutyrate or glucose with the glucoprivic agent. Because hepatocytes cannot utilize ketone bodies (Owen et al., 1967; Krebs et al., 1971), the effectiveness of ß-hydroxybutyrate in blocking the response suggests that the liver parenchyma is probably not the receptive tissue. Furthermore, they found that 2DG-induced feeding could not be blocked by co-administration of fructose with the 2DG. Because 2DG does not impair the transport or utilization of fructose in liver cells or the formation of hepatic glycogen from this sugar (Woodward and Hudson, 1954; Brown, 1962), the ineffectiveness of fructose in blocking 2DG-induced feeding can be viewed as further evidence that the receptive tissue is not the liver itself.

Our work with alloxan also casts doubt on the existence of peripheral receptors that are independently capable of stimulating glucoprivic feeding. This work shows that fourth ventricular injection of alloxan severely impaired the glucoprivic feeding response induced by both central and systemic administration of glucoprivic agents (Murnane and Ritter, 1983). This result suggests that feeding in response to both systemic and central glucoprivation is mediated by the same or by similar central neural substrates. If alloxan impairs glucoprivic feeding by destroying glucoreceptor cells, an idea discussed elsewhere in this review, then it seems likely from our results that these glucoreceptors are located solely in the caudal hindbrain and that glucose-sensitive neurons located elsewhere are not capable of eliciting glucoprivic feeding. Another interesting possibility, however, is that the caudal hindbrain glucoreceptors which are activated by centrally administered 2DG or 5TG are the central projections of primary afferent glucoreceptive neurons arising in the periphery. If so, glucoprivic feeding might be activated by either central or systemic glucoprivation, but destruction of the central projection alone would block the response.

Overall, it has been difficult to demonstrate the existence of receptors that are independently capable of stimulating feeding in response to glucoprivation. In contrast, evidence that the glucoprivic feeding response, once aroused, can be modulated by peripheral factors is more convincing and will be discussed in the following section.

One additional point deserves comment before leaving the subject of peripheral glucoreceptors. This point pertains to experiments using various sugars to distinguish between central and peripheral glucoreceptors. It has been noted that glucose, but not fructose, is capable of blocking 2DG-induced feeding. Although both sugars are potentially capable of nourishing both peripheral tissues and brain, only glucose passes the blood-brain barrier. Therefore, the failure of fructose to block 2DG-induced feeding has been cited as evidence for the central location of glucoreceptors and against their localization outside the blood brain barrier. This is a plausible interpretation of the data. However, due to the characteristics of its metabolism, the availability of fructose to peripheral as well as central tissues may also be limited in comparison with glucose. In vitro studies indicate that in brain tissue 2DG may be many times more potent in blocking fructose utilization than in blocking glucose utilization (Woodward and Hudson, 1954). Thus, if glucoreceptors controlling glucoprivic feeding are located outside the blood-brain barrier in circumventricular sites, the ability of these receptors to utilize fructose would be less than their ability to utilize glucose during 2DG-induced metabolic blockade. Furthermore, whole body fructose oxidation is reduced to approximately the same extent as glucose oxidation by doses of 2DG similar to those used in feeding tests (Brown, 1962). These facts of fructose metabolism should be considered, together with the experimental protocols utilized in sugar-induced prevention of glucoprivic feeding (Stricker and Rowland, 1978). In these protocols, sugars were infused intravenously (i.v.) beginning 15 min prior to 2DG administration. Under these circumstances, insulin-facilitated uptake and utilization of glucose by peripheral tissues would provide a competitive advantage to glucose in antagonizing 2DG- induced glucoprivation. Fructose, in contrast, lacks this advantage because it provides a much weaker stimulus for insulin release and because the uptake and utilization of fructose are not greatly facilitated by insulin. Thus, the accumulation of fructose in peripheral cells is apt to be lower than glucose, and fructose utilization would be even more severely compromised by the 2DG. Finally, the fact that fructose is partially effective in blocking glucoprivic feeding in the absence of metabolic blockade (that is, when insulin is used as the glucoprivic agent) is consistent with a role for peripheral or circumventricular glucoreceptors in the feeding response (Stricker et al., 1977; Bellin and Ritter, 1981). For these reasons, the peripheral or circumventricular location for glucoreceptors cannot be ruled out on the basis of the fructose evidence alone. Additional evidence is required.

VIII. DELAYED GLUCOPRIVIC FEEDING

A. Delayed Glucoprivic Feeding and the Metabolic Stimulus for Feeding

It has been presumed that feeding occuring after administration of glucoprivic agents is a consequence of ongoing reduction in glucose utilization. However, Ritter et al. (1978) discovered that insulin- and 2DG-induced feeding still occur even if food is withheld until other physiological signs of glucoprivation have spontaneously abated. In their experiments, rats were denied access to food for 6 to 8 hours after subcutaneous injection of 2DG or insulin, at which time the return of blood glucose levels to normal indicated that the glucoprivic stimulus had dissipated. When food was returned, the animals consumed amounts similar to those consumed when food was returned immediately after the injection of the glucoprivic agents (Fig. 3A,B). The delayed feeding response does not depend upon systemic administration of the glucoprivic

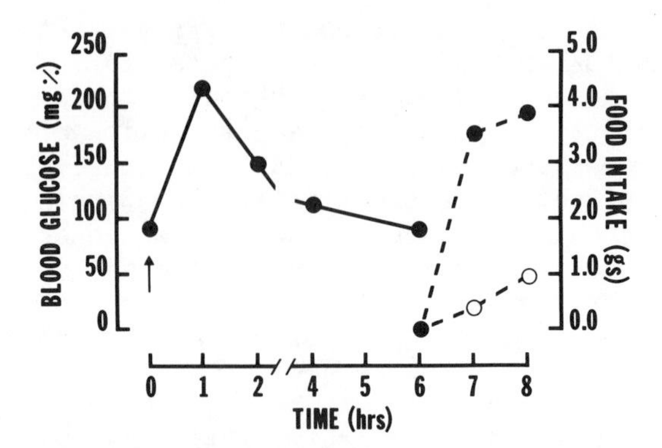

Fig. 3A. Feeding in response to 2DG (200 mg/ kg) (broken line, solid circles). Food was withheld for 6 hr after 2DG (given at arrow). Sympathoadrenal hyperglycemia (solid line) has abated by 6 hr post-2DG. Feeding 6 hr after a saline control injection (broken line, open circles). From R.C. Ritter et al. (1978), *American Journal of Physiology* 234.

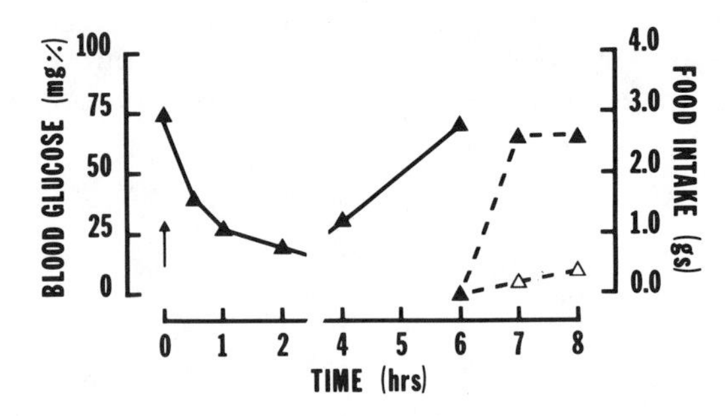

Fig. 3B. Increased feeding (broken line, solid triangles) after a single injection of regular insulin (2 U/kg) at time 0 (arrow). Food was withheld for 6 hr after insulin injection at which time blood glucose levels (solid line, solid triangles) were normal. Feeding when food is returned 6 hr after a saline control injection shown by broken line, open tri angles. From R.C. Ritter et al. (1978), *American Journal of Physiology* 234.

agent since it also occurs after i.c.v. administration of 2DG (Engeset and Ritter, 1980; Granneman and Friedman, 1983).

The delayed or "postglucoprivic" feeding response has stimulated a considerable amount of research, partly because this response may be another manifestation of the glucoprivic feeding response and, as such, may be controlled by the same stimuli governing the occurrence of feeding during ongoing glucoprivation. The opportunity to study the same response under differing sets of physiological conditions increases the possibilities for factoring out those conditions essential to the control of food intake. The delayed response is also of theoretical interest: since many studies have failed to reveal a close association between blood glucose levels and spontaneous feeding, the expression of the delayed feeding response in the presence of normoglycemia lends a new dimension to the possible role of glucoprivic control in the overall regulation of "spontaneous" feeding. If neither of the above assumptions about the delayed response is supported, this robust feeding response must still be accounted for in its own right.

There are several possible explanations for the delayed feeding response. One possibility is that reduced cellular glucose oxidation might still prevail during the postglucoprivic period, even though blood glucose levels have returned to normal. Ritter and his colleagues have provided evidence to contradict this hypothesis: they have shown that providing animals with an excess of glucose during the postglucoprivic period (5-8 hours after insulin) by glucose injection does not block the delayed feeding response (Ritter et al., 1978; see also Bellin and Ritter, 1981, and Friedman and Granneman, 1983). In addition, they have shown that 2DG itself clears almost entirely from the brain within 6 hours of the 2DG injection and that residual levels are insufficient to stimulate feeding (Engeset and Ritter, 1980). Furthermore, both systemic glucose oxidation, as indicated by CO_2 evolution, and brain glucose availability, as measured by brain uptake of tracer amounts of labeled 2DG, are normal during the postglucoprivic period (Nonavinakere and Ritter, 1983). Perhaps more sensitive studies may yet find impaired glucose oxidation regionally within the brain during the postglucoprivic period. However, the evidence now available suggests that some other explanation accounts for the delayed feeding response.

Another possibility is that feeding in response to prevailing or prior glucoprivation is driven by some neural or metabolic consequence of glucoprivation that persists longer than the glucoprivic episode itself. Such an arrangement would be adaptive since it would permit rapidly responding homeostatic mechanisms to restore blood glucose to levels necessary for normal brain function without interfering with the search for and ingestion of food needed for repletion of energy deficits. Friedman and Granneman have speculated that activation of systemic counterregulatory responses to glucoprivation may play such a role (Friedman and Granneman, 1983; Friedman et al., 1982; Granneman and Friedman, 1983). At present, this hypothesis seems implausible, however, since none of the humoral counterregulatory responses have been shown to elicit feeding. An hypothesis that receives stronger support from their data is that the stimulus to eat in the aftermath of a glucoprivic episode might arise from or be modulated by signals of hepatic origin. They found that injection of fructose during insulin-induced hypoglycemia attenuated postglucoprivic feeding and also prevented hepatic glycogen depletion. Hepatic vagotomy blocked the effect of fructose on both feeding and hepatic glycogen content. Overall, their work indicates, however, that delayed glucoprivic feeding is not contingent upon hepatic glycogen depletion and that signals important for feeding are more apt to arise from some aspect of hepatic metabolism other than glycogen content per se.

B. Delayed Glucoprivic Feeding: Modulation by Specific Nutrients

In examining the mechanisms of delayed glucoprivic feeding, Steve Bellin, Nancy Pelzer, and I tested various nutrients, ingested or infused during the glucoprivic episode, for their effectiveness in blocking the delayed glucoprivic feeding response (Bellin and Ritter, 1981a,b; Ritter et al., 1981). We infused glucose, fructose and ß-hydroxybutyrate i.v. 1.5-2 hours post-insulin, after the glucoprivic stimulus had fully developed, but several hours prior to the delayed feeding test. In other tests the rats were allowed to consume glucose, fructose, or saccharin solutions, or lard. The caloric value (7.9 kcal) of both the ingested and infused nutrients was equated with an amount of pelleted food which, if ingested during the same 30-minute period, would block the delayed feeding response.

We found that the delayed feeding response was suppressed only by infused ß-hydroxybutyrate, a substance distinguished by its ability to penetrate the brain and serve as an energy substrate for both brain and peripheral tissues (Fig. 4), and by ingested substances capable of elevating blood glucose levels (i.e., chow and glucose) (Fig. 5). Nutrients capable of nourishing peripheral tissues, but not brain (i.e., fructose and lard), were partially effective in blocking delayed feeding. Nonnutritive orogastric stimulation brought about by saccharin ingestion was ineffective. In fact, saccharin actually increased delayed feeding. Infused glucose was also ineffective. Surprisingly, delayed feeding was eliminated if animals were allowed to drink saccharin during the infusion of glucose (Fig. 6). These results suggested to us that restoration of nutrients to the brain is necessary for the termination of delayed glucoprivic feeding but that a signal of peripheral origin, brought about by peripheral nutrient metabolism, also contributes to the termination of the response.

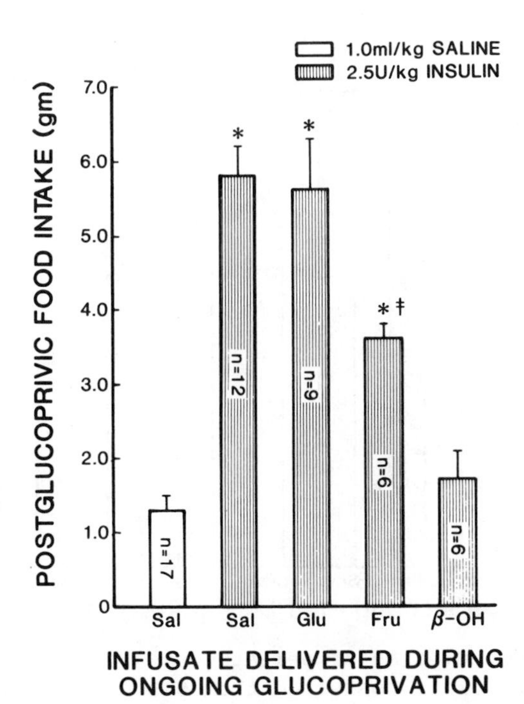

Fig. 4. Effect of infused nutrients on food intake in a delayed feeding test conducted during the postglucoprivic period 6 to 8 hr after insulin (2.5 U/kg, s.c.) or saline (1.0 ml/kg) injection. Nutrients or saline were delivered via intra-atrial cathethers from 1.5 to 2 hr postinjection. Nutrient infusions were calorically equivalent to 2.5 g of pelleted food (7.9 kcal), an amount of food which abolished delayed glucoprivic feeding ($P \geq 0.50$ versus insulin-Sal). Fructose significantly attenuated († = $P \leq 0.01$, insulin-Fru vs. insulin-Sal) but did not abolish postglucoprivic feeding (* = $P \leq 0.01$, insulin-Fru vs. saline-Sal). ß-Hydroxybutyrate abolished delayed glucoprivic feeding ($P \geq 0.50$ vs. saline-Sal). The values presented are the means ± SEM. Abbreviations: Fru = fructose, Glu = glucose, ß-OH = ß-hydroxybutyrate, Sal = saline. From Bellin and Ritter (1981).

The ineffectiveness of infused glucose in blocking the feeding response at first seemed puzzling, since both ingested glucose and infused ß-hydroxybutyrate were effective. However, in combination with other results, this finding suggested to us that perhaps disposal of glucose in peripheral tissues might be impaired as a result of prior or ongoing glucoprivation, and that this might reduce the effectiveness of infused glucose in

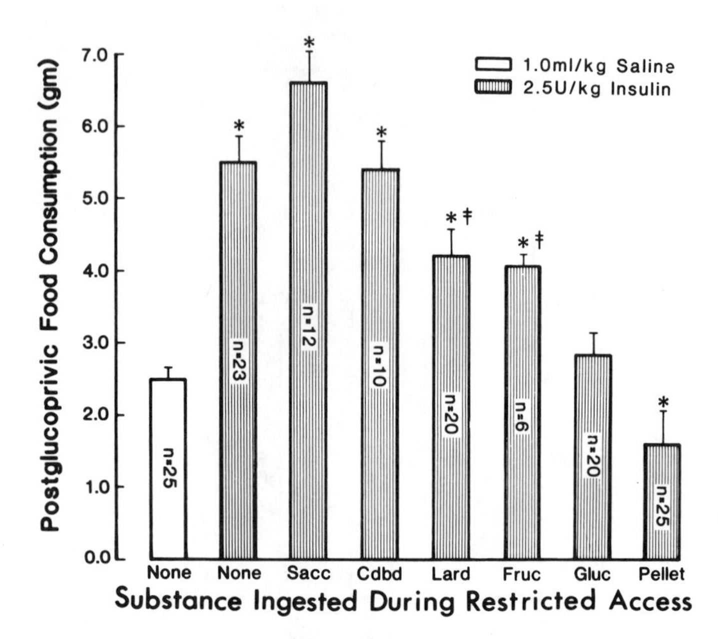

Fig. 5. Consumption of pelleted food during delayed glucoprivic feeding trials conducted from 6 to 8 hr after subcutaneous insulin administration. Animals were allowed to ingest a variety of noncaloric or qualitatively different isocaloric substances from 1.5 to 2 hr after insulin injections. Noncaloric substances failed to alter the amount of delayed glucoprivic feeding significantly. The ingestion of lard or fructose attenuated delayed glucoprivic feeding. († = $P \leq 0.05$ versus insulin, no food), but the amount eaten was significantly greater than the intake observed in saline-treated animals (* = $P \leq 0.05$). Ingestion of either glucose or pelleted food during the restricted access period totally abolished delayed glucoprivic feeding. The data presented are the means ± SEM. Abbreviations: Cdbd = flavored cardboard, Fruc = fructose, Gluc = glucose, Sacc = saccharin. From S. Ritter et al. (1981).

blocking the delayed response. As a partial test of this hypothesis, we generated tolerance curves for ingested and infused glucose in normal rats and in rats exposed to insulin-induced glucoprivation (Ritter and Pelzer, 1981). Glucose was administered 3.5-4 hours after insulin injection when exgenous insulin had disappeared from the blood, as determined by radioimmunoassay. We found that after glucoprivation, clearance of both ingested and infused glucose was significantly impaired, as compared to clearance rates in saline-injected controls

(Fig. 7). After glucose infusion, however, the impairment was more pronounced than after glucose ingestion.

The differential clearance of glucose from the blood after ingestion and infusion would be consistent with a differential effect of these two routes of glucose administration on release of endogenous insulin. Previous work in monkeys has shown that during 2DG-induced glucoprivation, endogenous insulin is not secreted in response to infused glucose (Smith et al., 1973), but is secreted in response to ingestion of food (Smith and

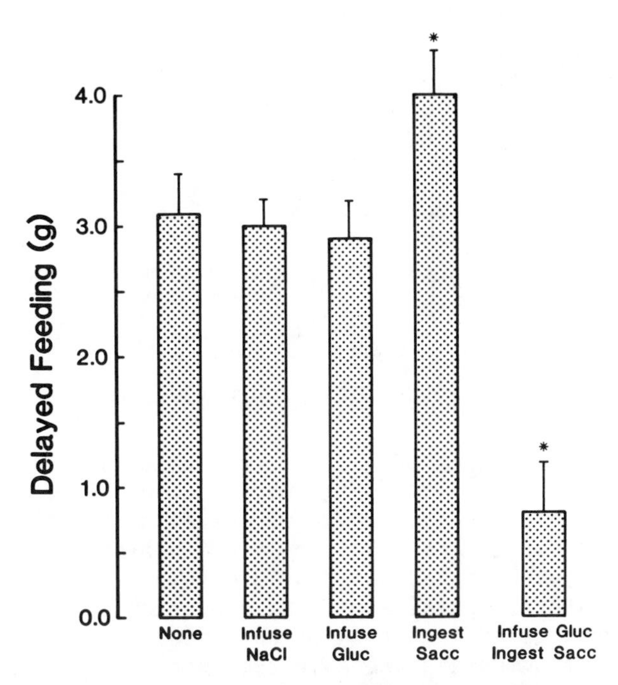

Fig. 6. Intake of pelleted rat chow during delayed feeding tests conducted 6-8 hr after insulin administration (2.5 U/kg, s.c.). Insulin-treated rats were infused remotely through a jugular catheter with 0.9% saline (NaCl), glucose solution (Gluc, 7.9 kcal), or given no treatment (None) during a 30 min period 2-2.5 hr following insulin injection. In other tests, rats were allowed, during this same time period, to ingest a nonnutritive saccharin solution (Ingest Sacc) alone or in combination with an i.v. glucose infusion (Infuse Gluc, Ingest Sacc). Saccharin ingestion and glucose infusion in combination reduced the delayed feeding response, but neither treatment was effective when administered alone.

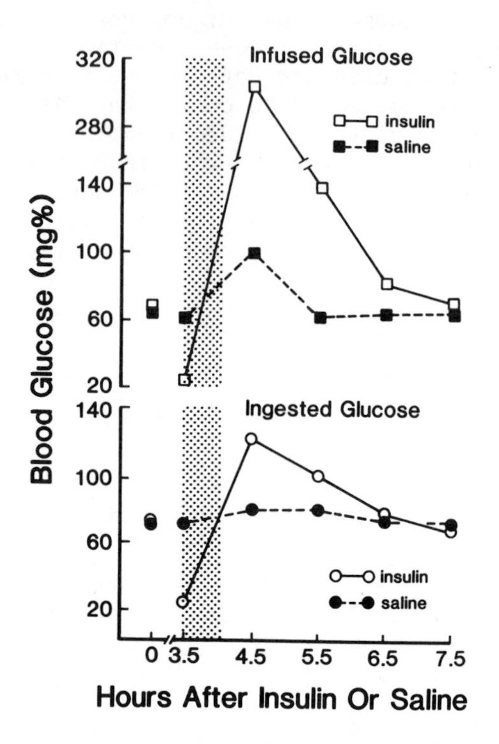

Fig. 7. Clearance of glucose from blood in rats injected at 0 hr with insulin (2.5 U/kg, s.c., solid symbols) or physiological saline (open symbols). Top panel shows the elevation and clearance of glucose (7.9 kcal) administered remotely through an i.v. catheter 3.5-4 hr after insulin or saline injection. Bottom panel shows the elevation and clearance of blood glucose after ingestion of an equicaloric amount of glucose solution during the same time period after insulin or saline injection. Blood glucose was sampled from the tail vein. Blood glucose reached higher levels and clearance time was prolonged in animals subjected to prior glucoprivation, compared to controls, and these differences were greater when the glucose was infused than when it was ingested.

Root, 1969). Presumably, food ingestion is capable of stimulating insulin release by preabsorptive neurally-mediated mechanisms (Woods et al., 1975) that, unlike the insulin secretory response to i.v. glucose, are not impaired by the sympathoadrenal response.

Several findings suggest that our results may be attributable to a stimulation of insulin release that is greater for ingested than for infused glucose. First, the infusion of glucose in combination with epinephrine mimicked the effect of prior glucoprivation. Epinephrine is known to be elevated by glucoprivation and to inhibit glucose-stimulated insulin release (Smith et al., 1973). Second, when we measured release of endogenous insulin in our paradigm, we found that ingestion of glucose was more effective than an equicaloric glucose infusion in stimulating insulin release (Fig. 8).

Our data are consistent with the hypothesis that restoration of nutrients to both insulin-independent brain cells and some insulin-dependent tissue may cooperatively terminate the stimulus for glucoprivic feeding. For glucose to be fully effective in blocking delayed feeding, it must be administered by a route or at a time that ensures its coexistence in the circulation with the insulin required for its uptake and utilization by critical insulin-dependent tissues. In contrast to glucose, ß-hydroxybutyrate (Williamson et al., 1975; Robinson and Williamson, 1980) and fructose (Park et al., 1957) do not require insulin for uptake or utilization and so may reduce glucoprivic feeding even when insulin release is impaired. Saccharin ingestion, although itself ineffective in blocking the delayed response, may enhance the peripheral utilization of infused glucose by stimulation of cephalically-mediated insulin release (Woods et al., 1975).

IX. NEUROCHEMISTRY AND NEUROPHARMACOLOGY OF GLUCOPRIVIC FEEDING

A. Noradrenergic Contributions to Glucoprivic Feeding

Many studies suggest a role for brain catecholamine (CA) neurons in feeding behavior (see Leibowitz and Stanley, this volume). Norepinephrine (NE) may participate more directly in food intake than dopamine (DA), since i.c.v. and intrahypothalamic injections of NE, but not DA, stimulate feeding (Ritter et al., 1975; Ritter and Epstein, 1975). Several lines of evidence suggest that CA neurons participate in the glucoprivic control. Electrolytic lesions of the lateral hypothalamus, as well as more selective chemical lesions with 6-hydroxydopamine of CA pathways in that region, greatly impair the feeding response to both 2DG and insulin (Marshall, 1974; Wayner et al., 1971; Stricker et al., 1975). Blockade of α-noradrenergic receptors, but not ß-receptors (Muller et al., 1972) or DA receptors (Anderson et al., 1979), and blockade of NE synthesis with α-methyl-p-tyrosine or diethyl-dithiocarbamate (Balch and Ritter, 1976) significantly impair glucoprivic feeding.

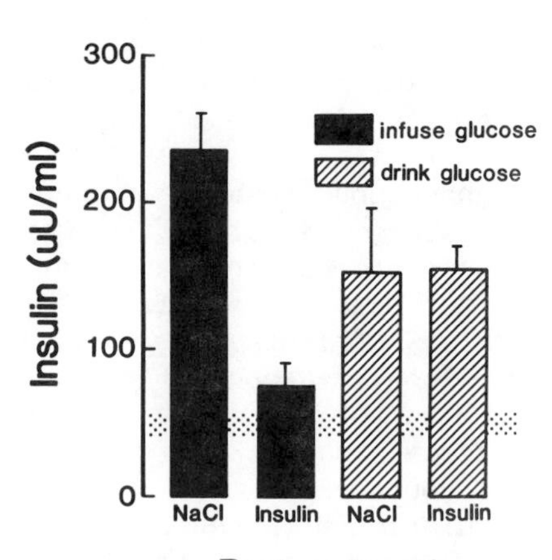

Fig. 8. Plasma levels of endogenous insulin after ingestion or infusion of glucose following insulin-induced glucoprivation. Nondeprived rats were injected with insulin (2.5 U/kg, s.c.) or physiological saline at 0 hr. Food was removed at the time of the injection. Four hr later, rats were allowed to drink a limited quantity of glucose solution over a 20-min period or received an equicaloric glucose infusion (7.9 kcal), administered remotely over a 20-min period through a jugular catheter. Blood was sampled for determination of insulin at 0 and 4 hr after insulin injection and again immediately after glucose administration. Insulin levels at 0 and 4 hr did not differ, indicating that injected insulin had cleared from the blood before glucose administration. Results show that insulin secretion in response to infused glucose (solid bars) was impaired following the glucoprivic episode, whereas insulin secretion in response to ingested glucose was not impaired.

Furthermore, brain NE depletion induced "physiologically" by stress exposure impairs 2DG-induced feeding. Drinking in response to hypertonic saline, a behavioral response not mediated by brain CAs (Ritter et al., 1977), is not impaired by stress. Return of the feeding response after stress parallels the recovery of NE concentrations in the hypothalamus. Finally, hypothalamic NE turnover is enhanced during glucoprivation (Ritter and Neville, 1976), and local perfusion of 2DG through a push-pull cannula increases the efflux of CAs from the perfused brain tissue (McCaleb et al., 1979).

We attempted to assess in greater detail the possible involvement of brain CA neurons in glucoprivic feeding (Bellin and Ritter, 1981a,b; Ritter et al., 1981). Our approach was to assess regional changes in turnover of CAs during glucoprivation by measuring the rate of CA depletion after synthesis inhibition with α-methyl-para-tyrosine (AMPT). We used the delayed glucoprivic feeding paradigm for our studies, which was advantageous for several reasons. First, with the delay paradigm, glucoprivic feeding and associated changes in turnover could be assessed after the stress of glucoprivation itself had abated, thus avoiding to some degree the potentially confounding influence of stress on behavior and CA turnover (Stone, 1973; Ritter and Pelzer, 1980; Ritter and Ritter, 1977). Second, the delay between the administration of the glucoprivic agent and the feeding test would permit us to evaluate the effects of various treatments in blocking the feeding response without interfering with the development of the glucoprivic stimulus. That is, nutrient infusions could be administered after the glucoprivic episode was fully developed. Our results could thus be interpreted in terms of blocking the response, not the stimulus. Finally, infusion of some substances, such as ketone bodies and hypertonic glucose solutions, may nonspecifically suppress behavior. By administering these treatments during the delay period, 3-4 hours prior to the feeding test, we hoped to minimize such effects.

For these experiments, insulin was administered in the morning and food was withheld for 6 hours to allow blood glucose to return to normal levels (about 5 hours after insulin). Food was then returned and feeding was measured in a 2-hour test, 6-8 hours after the insulin injection. In some experiments, animals were infused with specific nutrient or control substances or were allowed to eat a limited quantity of food during the delay. Turnover was measured beginning 5.5 hours after insulin.

Regional CA analyses revealed that rates of turnover of NE were inversely related to glucose availability in all brain regions analyzed (hypothalamus, striatum, and neocortex). In the hypothalamus, however, NE turnover was more consistently enhanced by glucoprivation than in other regions. In addition, hypothalamic NE turnover, but not NE or DA turnover in other brain regions, remained elevated even after the spontaneous restoration of normoglycemia in the absence of food.

Additional studies showed that glucoprivation-induced increases in hypothalamic NE turnover could be normalized by ingestion or infusion of substances providing substrates for brain energy metabolism (ingested glucose or rat chow, infused glucose, or β-hydroxybutyrate), but not by ingestion of lard or saccharin, by ingestion or infusion of fructose, or by nonnutritive orogastric stimulation. The entry of nutrients into the brain appeared to be both necessary and sufficient to reverse the effects of glucoprivation on hypothalamic NE neurons. Substrates for brain energy metabolism, whether ingested or infused, normalized NE turnover.

As with NE turnover, enhanced feeding was abolished by ingestion of glucose or rat chow or by infusion of β-hydroxybutyrate, as discussed in Section VIII,B. However, the influence of nutrients on the feeding response differed from their influence on NE turnover in two important respects. First, the feeding response was significantly attenuated (although clearly not abolished) by ingestion or infusion of fructose and by ingestion of lard, even though these substances had no effect on hypothalamic NE turnover. Second, ingested glucose blocked the feeding response, but glucose infusions, which reduced NE turnover to normal or even subnormal rates, did not attenuate it.

Taken together with the previously described pharmacological results, the close correlation between feeding and the activity of hypothalamic NE neurons strengthens the hypothesis that NE neurons participate in initiation of glucoprivic feeding. However, although impaired NE neuron function inhibits glucoprivic feeding, our data also suggest that enhanced activity of these neurons is not necessary for the delayed response. Consequently, the specific role of hypothalamic NE neurons in glucoprivic feeding requires additional investigation. They may mediate a specific component of the glucoprivic feeding response, for example, the enhanced carbohydrate appetite stimulated by glucoprivation (Kanarek et al., 1980, 1983; Leibowitz, this volume). Or they may play a more general role in motivating or integrating aspects of the behavioral response (Hoebel, 1975; Rolls, this volume).

By what mechanism is the activity of brain norepinephrine neurons increased during glucoprivation? Possibly these neurons are innervated by glucoreceptors. However, a number of studies support the intriguing possibility that the activity of some NE neurons, or synaptic release of NE, may be modulated directly by availability of energy substrate (see also O'Fallon and Ritter, 1982; McCaleb et al., 1979; Smythe et al., 1984; Kow and Pfaff, 1985; Malik and McGiff, 1974).

It is encouraging that several recent studies have begun to investigate neurochemical mediators other than the CAs as possible participants in the glucoprivic control (Yim et al., 1982; Sewell, 1980; Foley et al., 1980; Rowland and Bartness, 1982; Ipp et al., 1984). Some studies implicate endogenous opiate peptides, but the results are preliminary, and contradictory evidence has not yet been reconciled.

B. Alloxan: A Neurotoxic Probe of Glucoprivic Feeding

Alloxan is best known for its ability to produce experimental diabetes by selective destruction of the pancreatic ß-cell (Dunn et al., 1943). It is also known, however, that alloxan produces a selective but reversible blockade of lingual glucoreception (Zawalich, 1973). With these facts in mind, Woods and McKay (1978) speculated that i.c.v. administration of alloxan might impair glucoprivic feeding. Since the cells controlling this response, like the pancreatic ß-cell and lingual glucoreceptors, must be glucoreceptive, they might share cellular characteristics that would render them vulnerable to alloxan's toxic effect. Indeed, Woods and McKay found that lateral ventricular injection of alloxan did impair the glucoprivic feeding response without producing other obvious neurological deficits or illness.

Joan Murnane and I also injected alloxan into the lateral ventricle and obtained results that were very similar to those reported by Woods and McKay (Ritter et al., 1982; see also Murnane and Ritter, 1985a,b). We also administered alloxan into the fourth ventricle. Since previous evidence suggested that glucoreceptors controlling glucoprivic feeding are located in the caudal hindbrain (Ritter et al., 1981; Flynn and Grill, 1982), we hypothesized that the fourth ventricle, like the lateral ventricle, might be an effective site for alloxan administration. Indeed, we found that impairment of the feeding responses to both insulin and 2DG was somewhat more severe after fourth ventricular alloxan injections than after lateral ventricular injections (Fig. 9). In contrast, the hyperglycemic response to 2DG was not impaired by injection of alloxan at either site.

Several known neurotoxins (for example, capsaicin and glutamate) stimulate the cells for which they are toxic (Monsereenusorn et al., 1982; Olney, 1983). Alloxan also appears to produce both excitatory and toxic effects in the pancreatic ß-cell since systemic administration stimulates insulin release from the ß-cell prior to its destruction (Fischer and Rickert, 1975; Rerup, 1970). Therefore, we speculated that i.c.v. alloxan might also have excitatory, as well as toxic, effects on neurons involved in glucoprivic feeding. If so, i.c.v. injection of low alloxan doses might actually stimulate food intake.

We faced several problems with this experiment. First, alloxan is degraded very quickly under physiological conditions (Patterson et al., 1949) and therefore might produce only a brief stimulatory effect. Second, the dose range that could be tested would be limited at the upper end by the toxicity of the compound. Despite these experimental limitations, we found that fourth ventricular alloxan did indeed stimulate food intake (Ritter and Strang, 1982). Although the magnitude of the feeding was modest--only about 2.5 g in one hour--the response occurred immediately after the injection and was dose-related over the limited range of testable doses (Fig. 10).

We also found that alloxan did not produce a hyperglycemic response, despite the vigorous response obtained when 5TG was injected subsequently through the same cannula. Therefore, the stimulation and lesion data are consistent in showing a selectivity of alloxan for cells involved in the feeding response. A similar distinction between receptors mediating these two

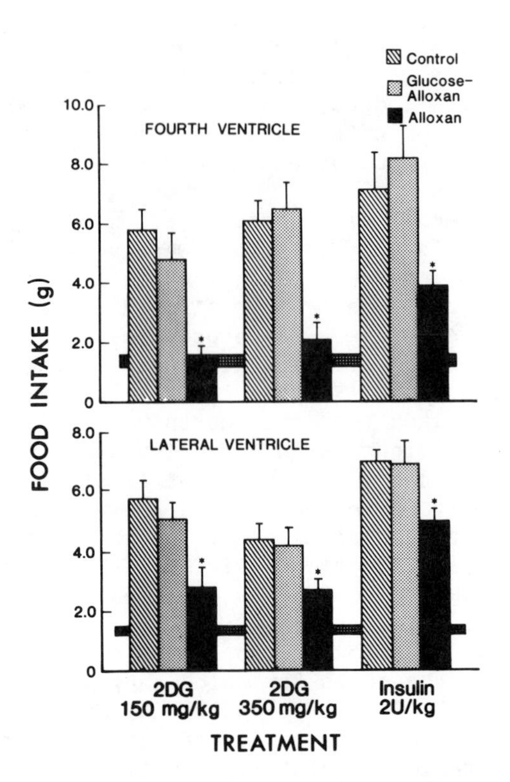

Fig. 9. Impairment of glucoprivic feeding by lateral or fourth ventricle alloxan injection (bottom and top, respectively). Alloxan (40μg in 5 μl saline, pH 3.0), glucose plus alloxan (40μg alloxan in 5 μl of 3 M D-glucose, pH 3.0), or sterile physiological saline (5 μl, pH 3.0) was injected into brain via fourth or lateral ventricular cannulas. Subsequently, food intake was measured for 6 hr following administration of 2-deoxyglucose (2DG, 150, and 350 mg/kg s.c.), insulin (2 U/kg s.c.), or saline. Figure shows means ± SE. Baseline feeding (mean of 3 tests for each rat) was statistically indistinguishable across treatment groups for each injection site. Horizontal stippling, therefore, represents mean of pooled baselines ± SE for each cannula site. * $P < 0.01$ vs. control and vs. glucose-alloxan. From S. Ritter et al. (1982), American Journal of Physiology 243.

responses has been reported after the administration of the drug phlorizin (Flynn and Grill, 1985), which blocks glucose transport into some cells (Betz, 1975). Like alloxan, phlorizin stimulates the feeding response, but does not produce

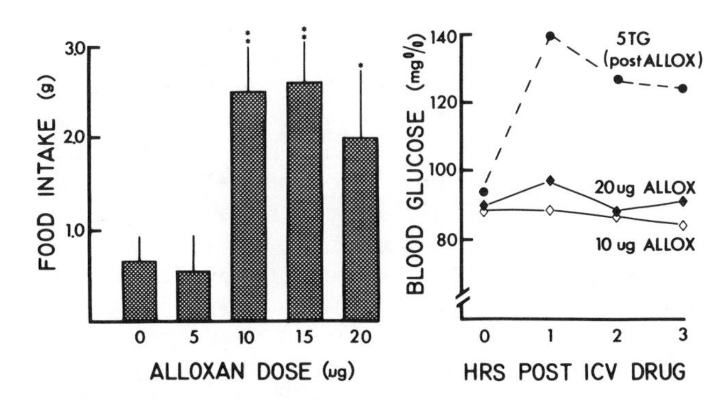

Fig 10. Effects of fourth ventricular (i.c.v.) alloxan injection on food intake and blood glucose concentration. Food intake was measured in a 4 hr test immediately following alloxan or saline injections. Blood glucose was measured from tail blood in the same subjects in separate tests after i.c.v. alloxan injection. At the conclusion of the study, feeding (see text) and blood glucose were measured in a similar manner after i.c.v. 5-thioglucose injections (5TG postALLOX). Statistical comparisons were made using the paired t test. ** $P < 0.01$ vs 0 μg; * $P < 0.05$ vs 0 μg. From Ritter and Strang (1982).

hyperglycemia. These data suggest that these two receptor types are chemically, as well as anatomically, distinct.

The stimulation of feeding by alloxan is of great interest to us because it suggests that alloxan affects brain cells normally involved in food intake. The impairment of the glucoprivic feeding response by higher alloxan doses suggests that the cells on which alloxan acts to stimulate feeding may be those normally activated by glucoprivation: that is, glucoreceptor cells. If so, we hope that this drug might be applied as an anatomical probe to facilitate the precise localization of glucoreceptor cells. In addition, a fine-grained analysis of feeding behavior in alloxan-treated rats, which appear to have a specific loss of the glucoprivic control, may allow us to discover a role for this control in normal feeding.

Although we have proposed that alloxan impairs glucoprivic feeding by a relatively selective effect on brain tissue, the data do not yet permit a convincing defense of this hypothesis. Several alternative hypotheses are possible. For example, alloxan may cause nonselective damage to neurons surrounding the brain ventricles. If glucoreceptors are located within the area of destruction, they may be coincidentally lesioned. However, the fact that lateral ventricle alloxan does not impair angiotensin-induced drinking, which is mediated by neurons in close proximity to the ventricle, or reduce the CA concentrations in forebrain sites, is evidence against a totally nonspecific destruction of brain tissue. The possibility must also be considered that alloxan does not damage glucoreceptors at all, but impairs the function of neural systems required more generally for expression of the behavioral response. However, other ingestive responses do not appear to be impaired by alloxan. The stimulation of feeding by low alloxan doses indicates that alloxan interacts with cells normally involved in food intake, as discussed above. In addition, the failure of alloxan to either stimulate or impair the sympathoadrenal response to glucoprivation provides further evidence for a selective effect, since the ability to stimulate this response by i.c.v. 5TG injections indicates that the cells controlling this response must lie in the diffusion path of ventricular infusates. Therefore, although we remain cautious regarding alloxan's specificity as a glucoreceptor cytotoxin, the data at least encourage us to continue to explore its potential as a tool to study glucoprivic feeding.

X. THE ROLE OF GLUCOPRIVATION IN SPONTANEOUS FEEDING

The existence of the glucoprivic control of food intake is clearly supported by the experimental evidence. Furthermore, the importance of the glucoprivic control as an emergency control to preserve brain function in the face of large glucose deficits and aid in the restoration of glucose homeostasis is well accepted. However, the participation of the glucoprivic control in the initiation or maintenance of spontaneous ingestion is uncertain. Many investigators have assumed that this control operates only during dire metabolic emergency, an assumption based primarily on two experimental observations. First, lesions that impair the glucoprivic control do not necessarily interfere with body weight regulation or produce noticeable disruptions of spontaneous feeding (e.g., Woods and McKay, 1978; McDermott and Grossman, 1980; Smith et al., 1972). This is not a convincing argument, since spontaneous feeding is under multiple controls that may provide for compensatory adjustments of intake in the absence of any single control. Second, glucoprivation of the magnitude occurring during 2DG treatment or insulin hypoglycemia is not a normal physiological event and does not occur in the freely feeding rat. Steffens (1964) found, for example, that feeding in response to insulin-induced glucoprivation did not occur until blood glucose levels had fallen below 50 mg%. Certainly, studies examining spontaneous preprandial glucose levels have not identified such large drops in blood glucose prior to meal onset (Strubbe et al., 1977). This evidence seems to justify the conclusion that glucoprivic control does not contribute to normal feeding, but such a conclusion is premature. It is equally reasonable to assume that feeding elicited by profound pharmacologically-induced glucoprivation is the manifestation of a control that normally responds to more subtle variation in glucose availability.

If glucoprivic feeding is the expression of a control operating when glucose availability is within physiological limits, then it should be possible (1) to demonstrate that animals that have been exposed to a glucoprivic stimulus will continue to feed even after the glucoprivic emergency has abated or (2) to demonstrate that small physiological changes in glucose availability are associated with feeding. Recently, such evidence has been reported.

The increased food intake in animals whose blood glucose levels have returned to normal following a glucoprivic emergency has been discussed previously in this review (Ritter et al., 1978). This finding suggests that the feeding response is not coupled exclusively with the overt signs of glucoprivation. Rather, the delayed response suggests that the glucoprivic control comprises a constellation of stimuli assuring that food will be ingested when it becomes available, even though blood glucose has been restored by other homeostatic mechanisms.

Small changes in blood glucose concentrations are difficult or impossible to detect unless remote sampling techniques are used. The technical obstacles imposed by the remote sampling requirement have recently been overcome in two laboratories, allowing the reliable measurement of spontaneous intermeal blood glucose fluxes. Using continuous remote online blood sampling in undisturbed freely feeding rats, Louis-Sylvestre and Le Magnen (1980) found that both light-phase and dark-phase spontaneous meals were preceded by a small decline in blood glucose of about 6-8% below baseline, occurring within the 5 minutes prior to meal initiation. Thus, they showed that a fall in blood glucose was a highly reliable correlate of meal onset. They hypothesized that meal onset and the decline in blood glucose are causally related. Campfield and his coworkers, also using continuous remote sampling techniques for blood glucose, more recently obtained a very similar result (Campfield et al., 1985). They observed a consistent premeal decline in blood glucose of about 12%, occurring only during the 12 minutes prior to meal onset and at no other time. The nadir of this decline occurred at about 5.4 minutes prior to meal onset, after which blood glucose slowly returned to baseline during the ensuing 11 minutes. Thus meals were actually initiated during the rising phase of this blood glucose curve.

Campfield and his co-workers (1985) have extended this work by demonstrating that if the drop in blood glucose was attenuated by i.v. glucose infusion, meal onset was significantly delayed, even though the glucose provided only about 9% of the calories of a small meal. As these investigators point out, this finding strengthens the hypothesis that the premeal decline in blood glucose is causally related to meal initiation.

Unlike pharmacologically-induced glucoprivation, premeal blood glucose declines observed by Campfield et al. do not appear to arise from a metabolic emergency. If the food cup was covered during the premeal decline so that the meal could not occur, blood glucose returned to normal and the meal was skipped. Another meal was initiated after a normal intermeal interval and another premeal decline in blood glucose. Thus, if these premeal blood glucose declines reflect the activation of the glucoprivic control, then they appear to represent a more subtle manifestation of the control than we see during the metabolic emergency induced experimentally by 2DG or insulin and may demonstrate that in the context of the preprandial physiological environment, the glucoprivic control is extremely sensitive to glucose fluxes. Perhaps the glucoreceptors are actually primed by some other factor present during the preprandial period.

It should be pointed out that Campfield and his co-workers have not claimed that the premeal glucose declines which they observed represent the operation of the glucoprivic control of food intake. Indeed, this claim cannot be made until the receptors mediating this control are identified and their metabolic and electrophysiological responses have been studied.

XI. CONCLUSIONS

Glucoprivation is a potent stimulus for feeding, mediated by receptor cells that appear to be localized in the caudal brainstem: evidence that the control can be independently activated by forebrain or peripheral glucoreceptors is not convincing. More precise localization of glucoreceptors controlling the feeding response and a detailed characterization of their physiological and anatomical features are required to answer many of the remaining questions about this control. In its ability to be modulated by periprandial factors, the glucoprivic control is similar to other controls of feeding, but its interaction with other controls requires additional investigation. The hypothesis that glucoprivation participates in the day-to-day control of feeding has been advanced by the demonstrations that the feeding response aroused by glucoprivation is not limited to periods of overt ongoing glucoprivation, and that spontaneous meals are preceded by a decline in blood glucose concentration. However, the relative importance of glucoprivic the control in the overall regulation of food intake remains an intriguing but unanswered question.

REFERENCES

Adachi, A., and M. Kobashi. 1985. Chemosensitive neurons within the area postrema of the rat. *Neurosci. Lett.* 55: 137-140.

Alheid, G.F., L.M. McDermott, J. Kelly, A. Halaris, and S.P. Grossman. 1977. Deficits in food and water intake after knife cuts that deplete striatal DA or hypothalamic NE. *Pharmac. Biochem. Behav.* 6: 273-287.

Anderson, J., D.F. Sharman, and D.B. Stephens. 1979. An effect of haloperidol on the increased food and water intake induced in rabbits by 2-deoxy-D-glucose. *Br. J. Pharmacol.* 66: 5-6.

Baile, C.A., W. Zinn, and J. Mayer. 1971. Feeding behavior of monkeys: Glucose utilization rate and site of glucose entry. *Physiol. Behav.* 6: 537-541.

Balagura, S. and M. Kanner. 1971. Hypothalamic sensitivity to 2-deoxy-D-glucose and glucose: effects on feeding behavior. *Physiol. Behav.* 7: 251-255.

Balch, O.K., and R.C. Ritter. 1976. Brain noradrenergic participation in glucoprivic feeding evaluated using catecholamine biosynthesis inhibitors. Masters Thesis, University of Idaho, Department of Veterinary Science.

Bellin, S.I., and S. Ritter. 1981a. Disparate effects of infused nutrients on delayed glucoprivic feeding and hypothalamic norepinephrine turnover. *J. Neurosci.* 1: 1347-1353.

Bellin, S.I., and S. Ritter. 1981b. Insulin-induced elevation of hypothalamic norepinephrine turnover persists after glucorestoration unless feeding occurs. *Brain Res.* 217: 327-337.

Bellinger, H., R. Birkhahn, G. Treitley, and L. Bernardis. 1977. Failure of hepatic infusion of amino acids and/or glucose to inhibit onset of feeding in the deprived dog. *J. Neurosci. Res.* 3: 163-173.

Bellinger, L.L., and F.E. Williams. 1983. Liver denervation does not modify feeding responses to metabolic challenges or hypertonic NaCl induced water consumption. *Physiol. Behav.* 30: 463-470

Bellinger, L.L., L.L. Bernardis, and S. Brooks. 1978. Feeding responses of rats with dorsomedial hypothalamic lesions given IP 2DG or glucose. *Am. J. Physiol.* 235: R168-R174.

Betz, A.L., R. Drewes, and D.D. Gilboe. 1975. Inhibition of glucose transport into brain by phlorizin, phloretin and glucose analogues. *Biochem. Biophys. Acta* 406: 505-515.

Bird, E., C.C. Cardone, and R.J. Contreras. 1983. Area postrema lesions disrupt food intake induced by cerebroventricular infusions of 5-thioglucose in the rat. *Brain Res.* 270: 193-196.

Booth, D.A. 1972. Modulation of the feeding response to peripheral insulin,2-deoxyglucose or 3-O-methyl glucose injection. *Physiol. Behav.* 8: 1069-1076.

Brandes, J.S., and A.K. Johnson. 1978. Recovery of feeding in rats following frontal neocortical ablations. *Physiol. Behav.* 20: 763-769.

Brown, J. 1962. Effects of 2-deoxy-D-glucose on carbohydrate metabolism: review of the literature and studies in the rat. *Metabolism* 11: 1098-1112.

Campfield, L.A., P. Brandon, and F.J. Smith. 1985. On-line continuous measurement of blood glucose and meal pattern in free-feeding rats: the role of glucose in meal initiation. *Brain Res. Bull.* 14: 605-616.

Cannon, W.B., M.A. MacIver, and S.W. Bliss. 1924. Studies on the conditions of activity in endocrine glands. XIII. A sympathetic and adrenal mechanism for mobilising sugar in hypoglycaemia. *Am. J. Physiol.* 69: 46-66.

Chen, M., and R.L. Whistler. 1975. Action of 5-thio-D-glucose and its 1-phosphate with hexokinase and phosphoglucomutase. *Arch. Biochem.* 169: 392-396.

Colin-Jones, D.G., and R.I. Himsworth. 1969. The secretion of gastric acid in response to lack of metabolizable glucose. *J. Physiol. (Lond.)* 202: 97-100.

Contreras, R.J., E. Fox, and M.L. Drugovich. 1982. Area postrema lesions produce feeding deficits in the rat: effects of preoperative dieting and 2-deoxy-D-glucose. *Physiol. Behav.* 29: 875-884

Crofford, O.B., and A.E. Renold. 1965. Glucose uptake by incubated rat epididymal adipose tissue. *J. Biol. Chem.* 240: 3237-3244.

DiBattista, D. 1982. Effects of 5-thioglucose on feeding and glycemia in the hamster. *Physiol. Behav.* 29: 803-806.

DiRocco, R.J., and H.J. Grill. 1979. The forebrain is not essential for sympathoadrenal hyperglycemic response to glucoprivation. *Science* 204: 1112-1114.
Dunn, J.S., H.L. Sheehan, and N.G. McLetchie. 1943. Necrosis of islets of Langerhans produced experimentally. *Lancet* 244: 484-487.
Edwards, G.E., and R.C. Ritter. 1981. Ablation of the area postrema causes exaggerated consumption of preferred food in the rat. *Brain Res*. 216: 265-276.
Engeset, R.M., and R.C. Ritter. 1980. Intracerebroventricular 2-DG causes feeding in the absence of other signs of glucoprivation. *Brain Res*. 202: 229-233.
Epstein, A.N. 1960. Reciprocal changes in feeding behavior produced by intrahypothalamic chemical injections. *Am. J. Physiol*. 199: 969-974.
Epstein, A.N., and P. Teitelbaum. 1967. Specific loss of the hypoglycemic control of feeding in recovered lateral rats. *Am. J. Physiol*. 213: 1159-1167.
Epstein, A.N., S. Nicolaidis, and R. Miselis. 1975. The glucoprivic control of food intake and the glucostatic theory of feeding behavior. In *Neural Integration of Physiological Mechanisms and Behavior*, ed. G.J. Mogenson and F.R. Calarescu, 148-168. University Press, Toronto.
Fischer, L.J., and D.E. Richert. 1975. Pancreatic islet cell toxicity. *CRC Crit. Rev. Toxicol*. 3: 231-335.
Flynn, F.W., and H.J. Grill. 1982. Insulin elicits ingestion in decerebrate rats. *Science* 221: 188-190.
Flynn, F.W., and H. Grill. 1985. Fourth ventricular phlorizin dissociates feeding from hyperglycemia in rats. *Brain Res*. 341: 331-336.
Friedman, M.I., and J. Granneman. 1983. Food intake and peripheral factors after recovery from insulin-induced hypoglycemia. *Am. J. Physiol*. 244: R374-R382.
Friedman, M.I., I. Ramirez, G. Wade, L.I. Siegel, and J. Granneman. 1982. Metabolic and physiologic effects of a hunger-inducing injection of insulin. *Physiol. Behav*. 29: 515-518.
Frohman, L.A., E.E. Muller, and D. Cocchi. 1973. Central nervous system mediated inhibition of insulin secretion due to 2-deoxyglucose. *Horm. Metab. Res*. 5: 21-26.
Glick, Z., and J. Mayer. 1968. Hyperphagia caused by cerebral ventricular infusion of phlorizin. *Nature* 219: 1374.
Granneman, J., and M.I. Friedman. 1983. Feeding after recovery from 2-deoxyglucose injection: cerebral and peripheral factors. *Am. J. Physiol*. 244: R-383-R388.
Grossman, S.P., and L. Grossman. 1977. Food and water intake in rats after transections of fibers en passage in various portions of the tegmentum. *Physiol. Behav*. 18: 647-658.

Hawkins, R.A., and J.F. Biebuyck. 1979. Ketone bodies are selectively used by individual brain regions. *Science* 205: 325-327

Himsworth, R.L. 1968. Compensatory reactions to a lack of metabolizable glucose. *J. Physiol*. 198: 451-465.

Himsworth, R.L. 1970. Hypothalamic control of adrenalin secretion in response to insufficient glucose. *J. Physiol*. 206: 411-417.

Hoebel, B.G. 1975. Brain reward and aversion systems in control of feeding and sexual behavior. *Nebraska Symposium on Motivation: Current Theory and Research in Motivation* 22, ed. J.K. Cole and T.B. Sonderegger, 49-112.

Hokfelt, B., and S. Bydgeman. 1961. Increased adrenaline production following administration of 2-deoxy-D-glucose in the rat. *Proc. Soc. Exp. Biol. Med*. 106: 537-539.

Horton, R.S.B., B.S. Meldrum, and H.S. Bachelard. 1973. Enzymic and cerebral metabolic effects of 2-deoxy-D-glucose. *J. Neurochem*. 21: 507-520.

Houpt, T.R. 1974. Stimulation of food intake in ruminants by 2-deoxy-D- glucose and insulin. *Am. J. Physiol*. 227: 161-167.

Houpt, T.R., and H.E. Hance. 1977. Threshold levels of 2-deoxy-D-glucose for the hyperglycemic response: Dog and goat compared. *Life Sci*. 21: 513-518.

Houpt, K.A., T.R. Houpt, and W.G. Pond. 1977. Food intake control in the suckling pig: Glucoprivation and gastrointestinal factors. *Am. J. Physiol*. 232: E510-E514.

Hyde, T.A., and R.R. Miselis. 1983. Effects of area postrema/caudal medial nucleus of solitary tract lesions on food intake and body weight. *Am. J. Physiol*. 244: R577-R587.

Ipp, E., C. Garberoglio, H. Richter, A.R. Moossa, and A.H. Rubenstein. 1984. Naloxone decreases centrally induced hyperglycemia in dogs. *Diabetes* 33: 619-621.

Kadekaro, M., C. Timo-Iaria, and M.D.L. Vincentini. 1977. Control of gastric acid by the central nervous system. In *Nerves and the Gut*, ed. F.P. Brooks and P.W. Evers, 377-427. Charles B. Slack, New York.

Kanarek, R.B., R. Marks-Kaufman, and B.J. Lipeles. 1980. Increased carbohydrate intake as a function of insulin administration in rats. *Physiol. Behav*. 25: 779-782.

Kanarek, R.B., M. Salomon, and A. Khadivi. 1981. Rats with lateral hypothalamic lesions do eat following acute cellular glucoprivation. *Am. J. Physiol*. 241: R362-R369.

Kanarek, R.B., R. Marks-Kaufman, R. Ruthazer, and L. Gualtieri. 1983. Increased carbohydrate consumption by rats as a function of 2-deoxy-D- glucose administration. *Pharmacol. Biochem. Behav*. 18: 47-50.

King, B., B.A. Stamoutsos, and S.P. Grossman. 1978. Delayed response to 2-deoxy-D-glucose in hypothalamic obese rats. *Pharmacol. Biochem. Behav.* 8: 259-262.

Kow, L.-M., and D.W. Pfaff. 1985. Actions of feeding-relevant agents on hypothalamic glucose-responsive neurons, *in vitro*. *Brain Res. Bull.* 15: 509-513.

Krebs, H.A., D.H. Williamson, M.W. Bates, M.A. Page, and R.A. Hawkins.1971. The role of ketone bodies in caloric homeostasis. *Adv. Enzyme Reg.* 9: 387-409.

Larue-Achagiotis, C., and J. LeMagnen. 1980. Differential effects of 2-deoxy-D-glucose on plasma glucose, free fatty acids and feeding during the light and the dark cycle in rats. *Neurosci. Biobehav. Rev. 4, Suppl.* 1: 33-37.

Larue-Achagiotis, C., and J. LeMagnen. 1981. 2-deoxy-D-glucose inhibits food intake in rats after ventromedial hypothalamic lesions. *Physiol. Behav.* 26: 613-616.

Laszlo, J., W.R. Harlan, R.F. Klein, N. Kirschner, E.H. Estes, Jr., and M.D. Brogdonoff. 1961. The effect of 2-deoxy-D-glucose infusions on lipid and carbohydrate metabolism in man. *J. Clin. Invest.* 40: 171.

Louis-Sylvestre, J., and J. LeMagnen. 1980. A fall in blood glucose level precedes meal onset in free-feeding rats. *Neurosci. Biobehav. Rev.* 4: 13-15.

Makay, E.M., J.W. Calloway, and R.H. Barnes. 1940. Hyperalimentation in normal animals produced by protamine insulin. *J. Nutr.* 20: 59-66.

Malik, K.U., and J.C. McGiff. 1974. Relationship of glucose metabolism to adrenergic transmission in rat mesenteric arteries. *Circ. Res.* 35: 553-573.

Marshall, J.F., and J.S. Richardson. 1974. Nigrostriatal bundle damage and the lateral hypothalamic syndrome. *J. Comp. Physiol. Psychol.* 87: 808-830.

Mayer, J. 1955. Regulation of the energy intake and the body weight: The glucostatic theory and the lipostatic hypothesis. *Ann. N.Y. Acad. Sci.* 63: 15-43.

Mayer, J. 1972. In *Advances in Psychosomatic Medicine* Vol. 7, *Hunger and Satiety in Health and Disease*, ed. F. Reichsman, 322-336. S. Karger, Basel.

McCaleb, M.L., R.D. Myers, G. Singer, and G. Willis. 1979. Hypothalamic norepinephrine in the rat during feeding and push-pull perfusion with glucose, 2-DG, or insulin. *Am. J. Physiol.* 236: R312-R321.

McDermott, L.J., and S.P. Grossman. 1980. The effects of amphetamine or caffeine on the response to glucoprivation in rats with rostral zona incerta lesions. *Pharmacol. Biochem. Behav.* 12: 946-957.

McDermott, L.J., and S.P. Grossman. 1980. Responsiveness to 2-deoxy-D- glucose and insulin in rats with rostral zona incerta lesions. *Physiol. Behav.* 24: 585-592.

McDermott, L.J., G.F. Alheid, J. Kelly, A. Halaris, and S.P. Grossman. 1977. Regulatory deficits after surgical transections of three components of the MFB: Correlation with regional amine depletion. *Physiol. Behav.* 23: 1135-1140.

Mei, N. 1978. Vagal glucoreceptors in the small intestine of the cat. *J. Physiol. (Lond.)* 282: 485-506.

Miselis, R.R., and A.N. Epstein. 1971. Preoptic-hypothalamic mediation of feeding induced by cerebral glucoprivation. *Am. Zool.* 11: Abst. #31.

Miselis, R.R., and A.N. Epstein. 1975. Feeding induced by intracerebro- ventricular 2-deoxy-D-glucose in the rat. *Am. J. Physiol.* 229: 1438-1447.

Mizuno, Y., Y. Oomura, K. Hattori, T. Minami, and N. Shimizu. 1983. Glucose responding neurons in the nucleus tractus solitarius of rat. *Neurosci. Lett., Suppl.* 13: 561.

Monsereenusorn, Y., S. Kongsamut, and P.D. Pezalla. 1982. Capsaicin: a literature survey. *CRC Crit. Rev. Toxicol.* 11: 321-339.

Muller, E.E., D. Cocchi, and P. Mantegazza. 1972. Brain adrenergic system in the feeding response induced by 2-deoxy-D-glucose. *Am. J. Physiol.* 223: 945-950.

Murnane, J.M., and S. Ritter. 1985a. Intraventricular alloxan impairs feeding to both central and systemic glucoprivation. *Physiol. Behav.* 34: 609-613.

Murnane, J.M., and S. Ritter. 1985b. Alloxan-induced glucoprivic feeding deficits are blocked by D-glucose and amygdalin. *Pharmacol. Biochem. Behav.* 22: 407-413.

Niijima, A. 1980. Glucose-sensitive afferent nerve fibers in the liver and regulation of blood glucose. *Brain Res. Bull. 5, Suppl. 4*: 175-179.

Nonavinakere, V.K., and R.C. Ritter. 1983. Feeding elicited by 2-deoxyglucose occurs in the absence of reduced glucose oxidation. *Appetite* 4: 177-185.

Nonavinakere, V.K., and R.C. Ritter. 1984. Destruction of the area postrema does not abolish glucoprivic control of feeding of blood glucose. *Neurosci. Soc. Abstr.* 10: 654.

Novin, D., D.A. VanderWeele, and M. Rezek. 1973. Infusion of 2-deoxy-D- glucose into the hepatic-portal system causes eating: evidence for peripheral glucoreceptors. *Science* 181: 858-860.

O'Fallon, J., and S. Ritter. 1982. Glucoprivation-induced release of brain neurotransmitters. *Soc. Neurosci. Abstr.* 8: 712.

Olney, J.W. 1985. Excitotoxins: an overview. In *Excitotoxins*, ed. K. Fuxe, P. Roberts, and R. Schwarcz, 82-96. Macmillan, London.

Oomura, Y. 1976. Significance of glucose, insulin, and free fatty acids on the hypothalamic feeding and satiety neurons. In *Hunger: Basic Mechanisms and Clinical Implications*, ed. D. Novin, W. Wyrwicka, and G. Bray, 145-157. Raven Press, New York.

Ostrowski, N.L., N. Rowland, T.L. Foley, J.L Nelson, and L.D. Reid. 1981. Morphin antagonists and consummatory behaviors. *Pharmacol. Biochem. Behav.* 14: 549-559

Owen, O.E., A.P. Morgan, H.G. Kemp, J.M. Sullivan, M.G. Herrera, and G.F. Cahill, Jr. 1967. Brain metabolism during fasting. *J. Clin. Invest.* 46: 1589-1595.

Panksepp, J. 1974. Hypothalamic regulation of energy balance and feeding behavior. *Fed. Proc.* 33: 1150-1165.

Park, C.R., L.H. Johnson, J.H. Wright, Jr., and H. Batsel. 1957. Effect of insulin on transport of several hexoses and pentoses into cells of muscle and brain. *Am. J. Physiol.* 191: 13-18.

Patterson, J.W., A. Lazarow, and S. Levey. 1949. Alloxan and dialuric acid: Their stabilities and ultraviolet absorption spectra. *J. Biol. Chem.* 177: 187-196.

Pazur, J.H., and K. Kleppe. 1964. The oxidation of glucose and related compounds by glucose oxidase from "Aspergillus niger." *Biochemistry* 3: 578-583.

Rerup, C.C. 1970. Drugs producing diabetes through damage to the insulin secreting cells. *Pharmacol. Rev.* 22: 485-518.

Ritter, R.C., and O.K. Balch. 1978. Feeding in response t insulin butnot to 2-deoxy-D-glucose in the hamster. *Am. J. Physiol.* 234: E20-E24.

Ritter, R.C., and A.N. Epstein. 1975. Control of meal size by central noradrenergic action. *Proc. Nat. Acad. Sci. USA* 72: 3740-3743.

Ritter, R.C., and M. Neville. 1976. Hypothalamic noradrenalin turnover is increased during glucoprivic feeding. *Fed. Proc.* 35: 642.

Ritter, R.C., and P.G. Slusser. 1980. 5-thio-D-glucose causes increased feeding and hyperglycemia in the rat. *Am. J. Physiol.* 238: E141-E144.

Ritter, R.C., M. Roelke, and M. Neville. 1978. Glucoprivic feeding behavior in absence of other signs of glucoprivation. *Am. J. Physiol.* 234: E617-E621.

Ritter, R.C., P.G. Slusser, and S. Stone. 1981. Glucoreceptors controlling feeding and blood glucose: location in the hindbrain. *Science* 213: 451-453.

Ritter, S., and N.L. Pelzer. 1980. Age-related changes in norepinephrine neuron function during stress. In

Catecholamines and Stress: Recent Advances, ed. E. Usdin, R. Kvetnansky, and I.J. Kopin, 107-112. Elsevier/North-Holland, New York.
Ritter, S., and N.L. Pelzer. 1981. Orogastrically-mediated insulin release may be required for termination of delayed glucoprivic feeding by glucose. *Soc. Neurosci. Abst.* 7: 29.
Ritter, S., and R.C. Ritter. 1977. Protection against stress-induced brain norepinephrine depletion after repeated 2-deoxy-D-glucose administration. *Brain Res.* 127: 179-184.
Ritter, S., and M. Strang. 1982. Fourth ventricular alloxan injection causes feeding but not hyperglycemia in rats. *Brain Res.* 249: 198-201.
Ritter, S., C.D. Wise, and L. Stein. 1975. Neurochemical regulation of feeding in the rat: facilitation by alpha-noradrenergic, but not dopaminergic receptor stimulants. *J. Comp. Physiol. Psychol.* 88: 778-784.
Ritter, S., N.L. Pelzer, and R.C. Ritter. 1978. Absence of glucoprivic feeding after stress suggests impairment of noradrenergic neuron function. *Brain Res.* 149: 399-411.
Ritter, S., S.I. Bellin, and N.L. Pelzer. 1981. The role of gustatory and postingestive signals in the termination of delayed glucoprivic feeding and hypothalamic norepinephrine turnover. *J. Neurosci.* 1: 1354-1360.
Ritter, S., J.M. Murnane, and E.E. Ladenheim. 1982. Glucoprivic feeding is impaired by lateral or fourth ventricular alloxan injection. *Am. J. Physiol.* 243: R312-R317.
Robinson, A.S., and D.H. Williamson. 1980. Physiological roles of ketone bodies as substrates and signals in mammalian tissues. *Physiol. Rev.* 60: 143-187.
Rolls, E.T., this volume.
Roth, G.I., and W.S. Yamamoto. 1968. The microcirculation of the area postrema in the rat. *J. Comp. Neurol.* 133: 329-340.
Rowland, N. 1978. Effects of insulin and 2-deoxy-D-glucose on feeding in hamsters and gerbils. *Physiol. Behav.* 21: 291-294.
Rowland, N., and T.J. Bartness. 1982. Naloxone suppresses insulin-induced food intake in novel and familiar environments, but does not affect hypoglycemia. *Pharmacol. Biochem. Behav.* 16: 1001-1003.
Russell, P.J.D., and G.J. Mogenson. 1975. Drinking and feeding induced by jugular and portal infusions of 2-deoxy-D-glucose. *Am. J. Physiol.* 229: 1014-1018.
Sakata, T., K. Tsutsui, M. Fukushima, K. Arase, H. Kita, Y. Oomura, K. Ohki, and S. Nicolaidis. 1981. Feeding and

hyperglycemia induced by 1,5- anhydroglucitol in the rat. *Physiol. Behav.* 27: 401-405.

Sclafani, A., and D. Eisenstadt. 1980. 2-deoxy-D-glucose fails to induce feeding in hamsters fed a preferred diet. *Physiol. Behav.* 24: 641-643.

Sewell, R.D.E., and K. Jawaharlal. 1980. Antagonism of 2-deoxy-D-glucose- induced hyperphagia by naloxone: possible involvement of endorphins. *J. Pharm. Pharmacol.* 32: 148-149

Shor-Posner, G., A.P. Azar, S. Insinga, and S.F. Leibowitz. 1985. Deficits in the control of food intake after hypothalamic paraventircular nucleus lesions. *Physiol. Behav.* 35: 883-890.

Silverman, H.J. 1978. Failure of 2-deoxy-D-glucose to increase feeding in the golden hamster. *Physiol. Behav.* 21: 859-864.

Smith, G.P., and A.N. Epstein. 1969. Increased feeding in response to decreased glucose utilization in the rat and monkey. *Am. J. Physiol.* 217: 1083-1087.

Smith, G.P., and A. Root. 1969. Effect of feeding on hormonal responses to 2-deoxy-D-glucose in conscious monkeys. *Endocrinology* 85: 963-966.

Smith, G.P., S. Gibbs, A.J. Stohmayer, and P. Stokes. 1972. Threshold doses of 2-deoxy-D-glucose for hyperglycemia and feeding in rats and monkeys. *Am. J. Physiol.* 222: 77-81.

Smith, G.P., J. Gibbs, A.J. Strohmayer, A.W. Root, and P.E. Stokes. 1973. Effect of 2-deoxy-D-glucose on insulin response to glucose in intact and adrenalectomized monkeys. *Endocrinology* 92: 750-754.

Smyth, C.J., A.G. Lasichak, and L. levey. 1947. The effect of orally and intravenously administered amino acid mixtures on voluntary food consumption in normal men. *J. Clin. Invest.* 26: 439-445.

Smythe, G.A., H.S. Grunstein, J.E. Bradshaw, M.V. Nicholson, and P.J. Compton. 1984. Relationships between brain noradrenergic activity and blood glucose. *Nature* 308: 65-67.

Sokoloff, L. 1972. Circulation and energy metabolism of the brain. In *Basic Neurochemistry*, ed. G.J. Siegel, R.W. Albers, R. Katzman, and B.W. Agranoff, 388-413. Little, Brown and Co., Boston.

Sokoloff, L., M. Reivich, D. Kennedy, M.H. DesRosiers, C.S. Patlak, K.D. Pettigrew, O. Sakurada, and M. Shinohra. 1977. The [^{14}C] deoxyglucose method for measurement of local cerebral glucose utilization: theory, procedure, and normal values in the conscious and anesthetized albino rat. *J. Neurochem.* 28: 897-916.

Sols, A., and R.K. Crane. 1954. Substrate specificity of brain hexokinase. *J. Biol. Chem.* 210: 581-596.

Steffens, A.B. 1969. The influence of insulin injections and infusions on eating and blood glucose level in the rat. *Physiol. Behav.* 4: 823-828.

Stephens, C.B., and B.A. Baldwin. 1974. The lack of effect of intrajugular or intraportal injections of glucose or amino-acids on food intake in pigs. *Physiol. Behav.* 12: 923-929.

Stone, E.A. 1973. Accumulation and metabolism of norepinephrine in rat hypothalamus after exhaustive stress. *J. Neurochem.* 21: 589-601.

Stricker, E.M., and M.J. McCann. 1985. Visceral factors in the control of food intake. *Brain Res. Bull.* 14(6): 687-692.

Stricker, E.M., and N. Rowland. 1978. Hepatic versus cerebral origin of stimulus for feeding induced by 2-deoxy-D-glucose in rats. *J. Comp. Physiol. Psychol.* 92: 126-132

Stricker, E.M., M.I. Friedman, and M.J. Zigmond. 1975. Glucoregulatory feeding by rats after intraventricular 6-hydroxydopamine or lateral hypothalamic lesions. *Science* 189: 895-897.

Stricker, E.M., N. Rowland, C.F. Saller, and M.I. Friedman. 1977. Homeostasis during hypoglycemia: central control of adrenal secretion and peripheral control of feeding. *Science* 196: 79-81.

Strubbe, J.H., A.B. Steffens, and L. DeRuiter. 1977. Plasma insulin and the time pattern of feeding in the rat. *Physiol. Behav.* 18: 81-86.

Thompson, D.A., and R.G. Campbell. 1977. Hunger in humans induced by 2-deoxy-D-glucose: Glucoprivic control of taste preference and food intake. *Science* 198: 1065-1068.

Tordoff, M.G., and D. Novin. 1982. Celiac vagotomy attenuates the ingestive responses to epinephrine and hypertonic saline but not insulin, 2-deoxy-D-glucose, or polyethylene glycol. *Physiol. Behav.* 29: 605-613

Tordoff, M.G., J. Hopfenbeck, and D. Novin. 1982. Hepatic vagotomy (partial hepatic denervation) does not alter ingestive responses to metabolic challenges. *Physiol. Behav.* 28: 417-424.

Tower, D.B. 1958. The effect of 2-deoxy-D-glucose on metabolism of slices of cerebral cortex incubated in vitro. *J. Neurochem.* 3: 185-205.

Unger, R.H., and L. Orci. 1976. Physiology and pathophysiology ofglucagon. *Physiol. Rev.* 56: 778-826.

VanderWeele, D.A., D. Novin, M. Rezek, and J.D. Sanderson. 1974. Duodenal or hepatic-portal glucose perfusion: evidence for duodenally-based satiety. *Physiol. Behav.* 12: 467-473.

Wayner, M.J., A. Cott, J. Millner, and R. Tartaglione. 1971. Loss of 2-deoxy-D-glucose induced eating in recovered lateral rats. *Physiol. Behav.* 7: 881-884.

Wick, A.N., D.R. Dury, H.I. Nakada, and J.B. Wolfe. 1957. Localization of the primary metabolic block produced by 2-deoxyglucose. *J. Biol. Chem.* 224: 963-969.

Williamson, D.H., S.R. McKeown, and V. Ilic. 1975. Metabolic interactions of glucose, acetoacetate and insulin in mammary gland slices of lactating rats. *Biochem. J.* 150: 145-152.

Woods, S.C., and L.D. McKay. 1978. Intraventricular alloxan eliminates feeding elicited by 2-deoxyglucose. *Science* 202: 1209-1211.

Woods, S.C., J.R. Vasselli, E. Kaestner, G.A. Szakmary, P. Milburn, and M.V. Viliello. 1977. Conditioned insulin secretion and meal feeding in rats. *J. Comp. Physiol. Psychol.* 91: 128-133.

Woodward, G.E., and M.T. Hudson. 1954. The effect of 2-deoxy-D-glucose on glycolysis and respiration of tumor and normal tissues. *Cancer Res.* 14: 599-605.

Tamamoto, H., N. Katsuya, and H. Nakagawa. 1983. Role of catecholamine in time-dependent hyperglycemia due to 2-deoxyglucose, mannitol, and glucose. *Biomed. Res.* 4: 505-514

Yim, G.K.W., M.T. Lowry, J.M. Davis, D.R. Lamb, and P.V. Malven. 1982. Opiate involvement in glucoprivic feeding. In *The Neural Basis of Feeding and Reward*, ed. B.G. Hoebel and D. Novin, 485-498. Haer Institute for Electrophysiological Research, Brunswick, Maine.

Yin, T.H., C.L. Hamilton, and J.R. Brobeck. 1970. Effect of body fluid disturbance on feeding in the rat. *Am. J. Physiol.* 218: 1054-1059.

Zawalich, W.S. 1973. Depression of gustatory sweet response by alloxan. *Comp. Biochem. Physiol. A* 44: 903-909.

ADDENDUM

At the time this chapter went to press, Scharrer and Langhans reported that when dietary fat is relatively high, blockade of fatty acid oxidation stimulates feeding without altering blood glucose levels (*Am. J. Physiol.* 250: R1003-R1006, 1986). If borne out by further experimentation, this result may demonstrate the operation of a second control of food intake that, like the glucoprivic control, is activated by blockade of a specific metabolic pathway.

Chapter 10

THE RELATIONSHIPS AMONG BODY FAT, FEEDING, AND INSULIN

Stephen C. Woods, Daniel Porte, Jr.,
Jan H. Strubbe, and Anton B. Steffens

I. INTRODUCTION

This chapter will review the literature suggesting that body fat, or adiposity, plays a role in the regulation of food intake, and that the peptide hormone insulin is one mediator of this control. One theme of the chapter is that the absolute amount of fat in the body is a regulated variable and that deviations from ideal result in changes of average food intake until adequate fatness is restored. A second theme is that insulin provides information to the central nervous system (CNS) about the level of fatness, and that the brain in turn responds by altering meal size. The final theme is that this is but part of a complex system regulating metabolism.

II. ADIPOSITY

A. Adiposity Is Regulated

A compelling argument, documented in detail elsewhere (Bray, 1976; Woods and Porte, 1978), can be made for the concept that adiposity is regulated via a strict homeostatic negative feedback system. There appear to be powerful innate controls acting to maintain a particular level of fat, given any particular environment. Different environmental conditions, when imposed chronically, are probably associated with different regulated levels of adiposity. For example, if the average palatability of one's food were decreased, or if there were more stressors over the course of an average week, the defended level of fat might be less. Conversely, conditions could be altered such that they favored increased adiposity. The point is that the amount of fat defended varies with the ongoing environmental conditions such that there does not appear to be an absolute "set point." However, independent of the amount of fat defended or maintained, the process of regulation exists.

Experimental or therapeutic attempts to alter fat levels provide a simple demonstration of the regulation of adiposity. It is not difficult to lose weight when one eats less on a forced or voluntary diet. However, when the dieting regimen is ended, people and animals tend to overeat to a degree sufficient to regain precisely the weight that was lost (Bray, 1976; Kennedy, 1953). This return to predieting weight is, of course, a major problem for people who want to lose weight for medical or personal reasons.

The system is symmetrical. If one is forced to overeat to the point that excess weight is achieved, the process reverses when the treatment is ended such that original weight is restored through undereating (Sims et al., 1968). If adipose tissue is surgically removed in the process called lipectomy, animals overeat and regain their former fat mass. If the diet is adequate, they regain precisely the amount of fat lost at surgery (Faust et al., 1977).

It is important to note that in all of these instances the major means of returning to one's initial weight (fatness) is via changes of food intake, although this is not the only means. Changes in exercise and/or metabolic efficiency can and do also occur, but they are less important than changes of feeding unless the latter are prevented.

The existence of a rigorous negative feedback system for adiposity is consistent with the original lipostatic theory of Kennedy (1953). It implies that some signal proportional to adiposity must reach the brain and influence food intake. Although there may be many such signals, the remainder of this chapter will focus upon the possibility that the hormone insulin is a critical one.

III. INSULIN

A. Insulin and the Control of Blood Glucose

Insulin is a peptide hormone secreted from the ß-cells of the islets of Langerhans of the pancreas. Its best known and perhaps most important function in the circulation is the control of fuel flux and usage by the body. One of its actions is to cause cells to remove fuels (glucose, free fatty acids, or amino acids) from the surrounding tissue fluid and to use these fuels for instant energy or store them for future use. Glucose regulation is perhaps the most important activity, and the secretion of insulin is tightly linked with glucose levels in the blood. When plasma glucose is high (or increasing), insulin secretion is stimulated and continues until glucose levels are returned to normal. Likewise, low levels of plasma glucose cause suppression of insulin secretion below basal values.

Insulin secretion is elevated during and after meals, and this insulin has a major function in eliminating ingested nutrients from the blood. This activity is sufficiently important that animals and people anticipate meals and their consequent ingested fuels by secreting insulin before the fuel levels actually rise in the blood (Berthoud et al., 1981; Steffens, 1976; Strubbe and Steffens, 1977). This cephalic secretion of insulin may help prepare the body to curb the tide of ingested fuels (Woods, 1983), and/or it may help to stimulate appetite (Powley, 1977).

B. Insulin as an Adiposity Signal

The nature of the adiposity-related signal that reaches the brain to influence food intake is not known. What is known is that the critical information passes between animals parabiotically connected. Such animals, joined at the peritoneal cavities and flank musculature, survive relatively well and maintain food intake and body weight at near normal levels (Hervey, 1952). However, when an obese animal is parabiosed to a lean animal, the lean animal reduces its food intake and loses weight (Coleman and Hummell, 1969; Hervey, 1952; Parameswaran et al., 1977), suggesting that a signal that it is too fat reaches its body from its obese partner. The implication is that the signal must be contained in the fluids that pass between the two animals, and for the reasons documented below, we feel that insulin is a likely candidate.

There is a direct relationship between insulin levels in the plasma and adiposity (Bernstein et al., 1975; Bray, 1976; Woods et al., 1974). Obese people and animals have relatively high basal plasma insulin and lean individuals have low basal insulin. If one's weight increases, the weight gain is associated with an increase of basal plasma insulin and vice versa. The robustness of this relationship contributed to the formulation of the hypothesis that insulin might well serve as a signal to the CNS in the control of appetite (Woods and Porte, 1976; 1983).

There are several obstacles to serious consideration of insulin as a potential adiposity signal to the brain. For one thing, unlike most other tissues, brain tissue is insulin-independent. This means that the brain retains the ability to remove glucose from tissue fluid in the absence of insulin, whereas most other tissues are unable to do so. The implication is that the brain is insensitive to insulin. Second, insulin is a peptide of sufficient size (51 amino acids) that it should not easily be able to penetrate the blood-brain barrier. Blood-borne insulin would therefore seem an unlikely messenger to the brain. Finally, whereas basal plasma insulin is highly correlated with adiposity, basal insulin rarely exists unless prolonged fasting and inactivity are imposed. Insulin secretion co-varies with eating and exercise and is suppressed during stress. As a result, plasma insulin is constantly changing during normal behavior so that any information provided to the brain would necessarily be changing rapidly as well. A signal more stable than plasma insulin would presumably be involved in the regulation of appetite and adiposity.

C. Insulin in the Cerebrospinal Fluid (CSF)

Even though insulin was not thought to have a major influence upon the functioning of the brain, it was clear that the brain has a major influence over the secretion of insulin (e.g., Steffens, 1981; Steffens et al., 1972; Strubbe, 1980; Woods and Porte, 1974). While investigating this influence several years ago, we found immunoreactive insulin in the CSF of dogs (Woods and Porte, 1975). In subsequent experiments, we manipulated plasma insulin levels of anesthetized dogs and found that CSF insulin levels, obtained from the cisternum magnum, reflected plasma levels with a lag of 30 to 45 minutes, and that fluctuations of plasma insulin appeared to be damped in the CSF (Woods and Porte, 1977). We concluded that CSF insulin appeared to be an integral over time of plasma levels and thus might provide the necessary stable signal which indicates adiposity to the CNS (Porte and Woods, 1981; Woods and Porte, 1983).

More recently, we have obtained simultaneous CSF and plasma samples from awake rats. When the rats ate or were infused intravenously with glucose or insulin, CSF insulin levels rose along with plasma levels with a short time lag (15 minutes or less) (Steffens et al., in preparation). We have also found that CSF insulin levels parallel plasma levels over a normal 24-hour day (Strubbe et al., in preparation). Analogous results have been observed in the baboon (Woods et al., submitted). The point is that insulin enters the CSF from the blood and may therefore provide a relatively accurate estimate of adiposity to critical brain areas.

We have recently found a relatively high density of specific insulin binding sites in the choroid plexus of the rat (Baskin et al., 1986). Since the choroid plexus is an important interface between the contents of the blood and the CSF, the implication is that insulin may enter the CSF at this point. We have also found that when labeled insulin is administered into the CSF of rats, specific insulin binding sites can be found in the hypothalamus near regions important in the control of food intake (Baskin et al., 1983). Insulin can also interact with the brain directly from the blood without passing through the CSF, as shown by Van Houten and his colleagues, who have found that when labeled insulin is added into the carotid arteries of rats, specific insulin binding sites can be identified in circumventricular organs within 5 minutes (van Houten and Posner, 1981).

The point is that insulin has access to the brain and CSF, and that CSF insulin levels may well reflect adiposity. Consistent with this, we have found that genetically obese Zucker rats have elevated CSF insulin levels relative to their lean controls (Stein et al., 1983), and obese humans reportedly have elevated CSF insulin levels also (Owen et al., 1974).

D. Insulin in the Brain

Initial measurements of insulin concentration in brain tissue suggested that the levels were higher than those observed in the plasma (Havrankova et al., 1978b), and led to the speculation that the brain might actually synthesize insulin (Havrankova et al., 1981). There is compelling evidence against this hypothesis (Baskin et al., in press), and recent reports suggest that brain insulin concentrations are actually considerably lower than plasma levels (Baskin et al., 1983; Yalow and Eng, 1983). However, the insulin in the brain does tend to be concentrated in specific regions, including many nuclei in the anterior and ventral hypothalamus thought to be important in the control of food intake (Baskin et al., 1983; LeRoith et al., 1983).

Specific binding sites for insulin are also found within the brain, and their distribution parallels in large part the distribution of insulin content. Areas with the highest insulin concentrations (e.g., the choroid plexus and the olfactory bulb) also have the greatest concentration of insulin binding sites (Havrankova et al., 1978a; Figlewicz et al., 1985), allowing the speculation that much of the insulin which can be measured in the brain is actually bound to receptors. The important point is that insulin and insulin binding sites are located, among other places, in CNS areas important in the control of food intake and ought, therefore, to be considered as likely modulators of activity there. Consistent with this, the iontophoretic administration of insulin onto hypothalamic neurons alters their firing rate and their response to changes of glucose (Oomura and Kita, 1981). Likewise, the acute administration of insulin into either the CSF (Chen et al., 1975; Chowers et al., 1966; Woods and Porte, 1975) or the carotid arteries (Szabo and Szabo, 1983) elicits autonomic responses controlling pancreatic hormone secretion and blood glucose levels.

E. Insulin and the Reduction of Food Intake

If insulin functions as an indicator of adiposity to the brain, its local application to critical brain areas ought to elicit the same reflexes as occur when an individual is rendered overweight; i.e., the individual should eat less and lose weight. In an initial test of this hypothesis, we infused small amounts of insulin into the CSF of baboons over a 2- or 3-week interval. There was a dose-dependent decrease of caloric intake and body weight with no indication that the infused insulin had leaked into the blood and no sign of malaise in the animals (Woods et al., 1979). This phenomenon has been replicated in rats (Brief and Davis, 1984), and Nicolaidis (1981) has found that the chronic infusion of insulin directly into the hypothalamus of rats causes reduced food intake and body weight.

If CSF (and/or brain) insulin could be lowered experimentally (technically a much more difficult feat than elevating insulin levels), animals ought to interpret the decrease to mean that they are underweight and should eat more. We accomplished this by administering insulin antibodies into the hypothalamus of rats and observed an acute increase of food intake (Strubbe and Mein, 1977). Therefore, changes of insulin locally within the brain result in predictable changes of food intake and, if maintained chronically, body weight.

If endogenous insulin is important in the control of food intake, and if most insulin in the brain originates in the pancreas and travels to the brain via the blood, then chronic changes of plasma insulin ought to have the same effect on feeding as local changes of brain insulin. Experiments addressing this point are confounded by the predominant action of insulin in the blood, the lowering of blood glucose. When blood glucose is decreased to very low values as a result of peripheral insulin administration, eating is elicited (MacKay et al., 1940), presumably due to the necessity of blood glucose for brain functioning (Booth and Pitt, 1968). Likewise, if glucose utilization by tissues is compromised by the addition of the non-utilizable glucose analogue, 2-deoxyglucose, eating is also elicited (Smith and Epstein, 1969). Further, if peripheral insulin is administered chronically, the maintained increase of food intake can lead to increased body weight (Hoebel and Teitelbaum, 1966; MacKay et al., 1940), but the phenomenon is unnatural in that failure to overeat in this situation results in death by hypoglycemia (Lotter and Woods, 1977). Therefore, most experiments in which insulin has been administered peripherally have found that, if anything, food intake is increased.

Several investigations have been made in which plasma insulin is chronically elevated and hypoglycemia is prevented. Nicolaidis and Rowland (1976) reported that the addition of insulin to intravenously infused glucose suppressed food intake in rats more than glucose alone did, and we have similarly found that intravenous infusions of glucose, insulin, or both into baboons resulted in hyperinsulinemia without hypoglycemia and also reduced food intake (Woods et al., 1984). VanderWeele and his colleagues (1980) have infused small amounts of insulin subcutaneously into rats for prolonged intervals and observed decreased food intake and body weight when hypoglycemia did not occur. Therefore, increases of plasma insulin are associated with reduced feeding if hypoglycemia is prevented.

F. Brain Insulin and Obesity

Obesity presents an apparent exception to the hypothesis that elevated insulin in the blood and/or the CNS causes reduced food intake and loss of weight. Indeed, as documented above, obese individuals have hyperinsulinemia (Woods et al., 1974) and elevated levels of CSF insulin (Owen et al., 1974; Stein et al., 1983) without hypoglycemia, yet they eat more than lean controls and obviously carry more weight. The implication is that sufficient insulin to reduce appetite may not reach critical brain sites, and/or that the brain is relatively unresponsive to insulin in obese individuals.

We have found that whereas CSF insulin levels are increased in genetically obese Zucker rats, they are actually disproportionately low, given the very high levels of plasma insulin (Stein et al., 1984). It is as if the transport mechanism for passing insulin into the CSF from the blood is deficient in these animals. We have also found that brain insulin content is very low in obese Zucker rats (Baskin et al., 1985), and that they have less brain insulin binding (Figlewicz et al., 1985). It may be that the gene causing obesity in these animals is associated with a fundamental defect in the brain insulin system, since when the gene is introduced into another strain of rats, the Wistar-Kyoto, their brain insulin binding sites are also reduced (Figlewicz et al., in press). Consistent with this, we have found that whereas the infusion of insulin into the CSF of lean Zucker rats causes a reduction of food intake and body weight, infusion of insulin into the CSF of obese Zucker rats is without effect (Ikeda et al., in press).

IV. CONCLUSIONS

In this chapter, we have reviewed the evidence that the hormone insulin is a key factor in the control of food intake and adiposity. Insulin is secreted from the pancreas in direct proportion to adiposity, and this insulin gains access to receptors in the brain both directly from the blood and via the CSF. There are specific insulin binding sites in brain areas important in the control of food intake, and the local administration of insulin into the brain results in reduced feeding and loss of body weight. Finally, we have preliminary evidence for a malfunction of this brain-insulin regulatory system in the genetically obese Zucker rat.

ACKNOWLEDGEMENTS

Research mentioned in this chapter was supported by NIH grants AM 17844 and AM 12829, by the Veterans Administration, by the Regional Primate Research Center of the University of Washington (RR 00166), and by the Diabetes Research Center (AM 17047). We thank Ms. Elizabeth Rutherford for help with the manuscript.

REFERENCES

Baskin, D.G., D. Porte, Jr., K. Guest, and D.M. Dorsa. 1983. Regional concentrations of insulin in the rat brain. *Endocrinology* 112: 898-903.

Baskin, D.G., L.J. Stein, H. Ikeda, S.C. Woods, D.P. Figlewicz, D. Porte, Jr., M.R.C. Greenwood, and D.M. Dorsa. 1985. Genetically obese Zucker rats have abnormally low brain insulin content. *Life Sci.* 36: 627-633.

Baskin, D.G., B. Brewitt, D.A. Davidson, E. Corp, T. Paquette, D.P. Figlewicz, T.K. Lewellen, M.K. Graham, S.C. Woods, and D.M. Dorsa. 1986. Quantitative autoradiographic evidence for insulin receptors in the choroid plexus of the rat brain. *Diabetes* 35: 246-249.

Baskin, D.G., D.P. Figlewicz, S.C. Woods, D. Porte, Jr., and D.M. Dorsa. Insulin in the brain. *Annu. Rev. Physiol.* In press.

Bernstein, I.L., E.C. Lotter, P.J. Kulkosky, D. Porte, Jr., and S.C. Woods. 1975. Effect of force-feeding upon basal insulin levels of rats. *Proc. Soc. Exp. Biol. Med.* 150: 546-548.

Berthoud, H.R., D.A. Bereiter, E.R. Trimble, E.G. Siegel, and B. Jeanrenaud. 1981. Cephalic phase, reflex insulin secretion, neuroanatomical and physiologic characterization. *Diabetologia* 20: 393-401.

Booth, D.A., and M.E. Pitt. 1968. The role of glucose in insulin-induced feeding and drinking. *Physiol. Behav.* 3: 447-453.

Bray, G. 1976. *The Obese Patient*. Saunders, Philadelphia.

Brief, D.J., and J.D. Davis. 1984. Reduction of food intake and body weight by chronic intraventricular insulin infusion. *Brain Res. Bull.* 12: 571-575.

Chen, M., S.C. Woods, and D. Porte, Jr. 1975. Effect of cerebral intraventricular insulin on pancreatic insulin secretion in the dog. *Diabetes* 24: 910-914.

Chowers, I., S. Lavy, and L. Halpern. 1966. Effect of insulin administered intracisternally on the glucose level of the blood and the cerebrospinal fluid in vagotomized dogs. *Exp. Neurol.* 14: 383-389.

Coleman, D.L., and K.P. Hummell. 1969. Effects of parabiosis of normal with genetically diabetic mice. *Am. J. Physiol.* 217: 1298-1304.

Faust, I.M., P.R. Johnson, and J. Hirsch. 1977. Adipose tissue regeneration following lipectomy. *Science* 197: 391-393.

Figlewicz, D.P., D.M. Dorsa, L.J. Stein, D.G. Baskin, T. Paquette, M.R.C. Greenwood, S.C. Woods, and D. Porte, Jr. 1985. Brain and insulin liver binding is decreased in Zucker rats carrying the 'fa' gene. *Endocrinology* 117: 1537-1543.

Figlewicz, D.P., H. Ikeda, L.J. Stein, D.M. Dorsa, S.C. Woods, and D. Porte, Jr. Brain insulin binding is decreased in Wistar Kyoto rats carrying the 'fa' gene. *Peptides*. In press.

Havrankova, J., J. Roth, and M. Brownstein. 1978a. Insulin receptors are widely distributed in the central nervous system of the rat. *Nature* 272: 827-829.

Havrankova, J., D. Schmechel, J. Roth, and M. Brownstein. 1978b. Identification of insulin in rat brain. *Proc. Natl. Acad. Sci. USA* 75: 5737-5741.

Havrankova, J., M. Brownstein, and J. Roth. 1981. Insulin and insulin receptors in rodent brain. *Diabetologia* 20: 268-273.

Hervey, G.R. 1952. The effects of lesions in the hypothalamus in parabiotic rats. *J. Physiol. (Lond.)* 145: 336-352.

Hoebel, B.G., and P. Teitelbaum. 1966. Weight regulation in normal and hypothalamic hyperphagic rats. *J. Comp. Physiol. Psychol.* 61: 189-193.

Ikeda, H., D.B. West, J.J. Pustek, D.P. Figlewicz, M.R.C. Greenwood, D. Porte, Jr., and S.C. Woods. Intraventricular insulin reduces food intake and body weight of lean but not obese Zucker rats. *Appetite*. In press.

Kennedy, G.C. 1953. The role of depot fat in the hypothalamic control of food intake in the rat. *Proc. R. Soc. Lond. (Biol.)* 140: 579-592.

LeRoith, D., S.A. Hendricks, M.A. Lesniak, S. Rishi, K.L. Becker, J. Havrankova, J.L. Rosenzweig, M.J. Brownstein, and J. Roth. 1983. Insulin in brain and other extra-pancreatic tissues of vertebrates and nonvertebrates. *Adv. Metab. Dis.* 10: 304-340.

Lotter, E.C., and S.C. Woods. 1977. Injections of insulin and changes of body weight. *Physiol. Behav.* 18: 293-297.

MacKay, E.M., J.W. Calloway, and R.H. Barnes. 1940. Hyperalimentation in normal animals produced by protamine insulin. *J. Nutr.* 20: 59-66.

Nicolaidis, S. 1981. Lateral hypothalamic control of metabolic factors related to feeding. *Diabetologia* 20: 426-434.

Nicolaidis, S., and N. Rowland. 1976. Metering of intravenous versus oral nutrients and regulation of energy balance. *Am. J. Physiol.* 231: 661-668.

Oomura, Y., and H. Kita. 1981. Insulin acting as a modulator of feeding through the hypothalamus. *Diabetologia* 20: 290-298.

Owen, O.E., G.A. Reichard, Jr., G. Boden, and C.R. Shuman. 1974. Comparative measurements of glucose, beta-hydroxybutyrate, acetoacetate, and insulin in blood and cerebrospinal fluid during starvation. *Metabolism* 23: 7-14.

Parameswaran, S.V., A.B. Steffens, G.R. Hervey, and L. deRuiter. 1977. Involvement of a humoral factor in regulation of body weight in parabiotic rats. *Am. J. Physiol.* 232: R150-R157.

Porte, D., Jr., and S.C. Woods. 1981. Regulation of food intake and body weight by insulin. *Diabetologia* 20: 274-280.

Powley, T. 1977. The ventromedial hypothalamic syndrome, satiety, and a cephalic phase hypothesis. *Psychol. Rev.* 84: 89-126.

Sims, E.A.H., R.F. Goldman, C.M. Gluck, E.S. Horton, D.C. Keleher, and D.W. Rowe. 1968. Experimental obesity in man. *Trans. Assoc. Am. Physicians* 81: 153-170.

Smith, G.P., and A.N. Epstein. 1969. Increased feeding in response to decreased glucose utilization in the rat and monkey. *Am. J. Physiol.* 217: 1083-1087.

Stein, L.J., D.M. Dorsa, D.G. Baskin, D.P. Figlewicz, H. Ikeda, S.P. Frankmann, M.R.C. Greenwood, D. Porte, Jr., and S.C. Woods. 1983. Immunoreactive insulin levels are elevated in the cerebrospinal fluid of genetically obese Zucker rats. *Endocrinology* 113: 2299-2301.

Stein, L.J., D.P. Figlewicz, D.M. Dorsa, D.G. Baskin, D.G. Reed, D. Braget, M. Midkiff, D. Porte, Jr., and S.C. Woods. 1984. Effect of insulin infusion on cerebrospinal fluid insulin concentrations in heterozygous lean and obese Zucker rats. *Proc. North. Am. Soc. Study Obesity.*

Steffens, A.B. 1976. The influence of the oral cavity on the release of insulin in the rat. *Am. J. Physiol.* 230: 1411-1415.

Steffens, A.B. 1981. The modulatory effect of the hypothalamus on glucagon and insulin secretion in the rat. *Diabetologia* 20: 411-416.

Steffens, A.B., G.J. Mogensen, and J.A.F. Stevenson. 1972. Blood glucose, insulin, and free fatty acids after stimulation and lesions of the hypothalamus. *Am. J. Physiol.* 222: 1446-1452.
Steffens, A.B., D. Porte, Jr., and S.C. Woods. Relationship between plasma and cerebrospinal fluid insulin levels in awake rats. In preparation.
Strubbe, J.H. 1980. Hypothalamic influence on insulin secretion in rats with transplanted neonatal pancreases. *Diabetologia* 19: 318.
Strubbe, J.H., and C.G. Mein. 1977. Increased feeding in response to injection of insulin antibodies in the VMH. *Physiol. Behav.* 19: 309-313.
Strubbe, J.H., and A.B. Steffens. 1977. Blood glucose levels in portal and peripheral circulation and their relation to food intake in the rat. *Physiol. Behav.* 19: 303-308.
Strubbe, J.H., D. Porte, Jr., and S.C. Woods. Diurnal rhythms of cerebrospinal fluid insulin levels in rats. In preparation.
Szabo, A.J., and O. Szabo. 1983. Insulin injected into CNS structures or into the carotid artery: Effect on carbohydrate homeostasis of the intact animal. *Adv. Metab. Dis.* 10: 385-400.
VanderWeele, D.A., F.X. Pi-Sunyer, D. Novin, and M.J. Bush. 1980. Chronic insulin infusion suppresses food ingestion and body weight gain in rats. *Brain Res. Bull.*, Suppl. 1: 7-11.
Van Houten, M., and B.I. Posner. 1981. Cellular basis of direct insulin action in the central nervous system. *Diabetologia* 20: 255-267.
Woods, S.C. 1983. Conditioned hypoglycemia and conditioned insulin secretion. *Adv. Metab. Dis.* 10: 457-468.
Woods, S.C., and D. Porte, Jr. 1975. Effect of intracisternal insulin on plasma glucose and insulin in the dog. *Diabetes* 24: 910-914.
Woods, S.C., and D. Porte, Jr. 1976. Insulin and the set point regulation of body weight. In *Hunger: Basic Mechanisms and Clinical Implications*, ed. D. Novin, G.A. Bray, and W. Wyrwicka, 273-280. Raven Press, New York.
Woods, S.C., and D. Porte, Jr. 1977. Relationship between plasma and cerebrospinal fluid insulin levels of dogs. *Am. J. Physiol.* 233: E331-E334.
Woods, S.C., and D. Porte, Jr. 1978. The central nervous system, pancreatic hormones, feeding and obesity. *Adv. Metab. Dis.* 9: 283-312.
Woods, S.C., and D. Porte, Jr. 1983. The role of insulin as a satiety factor in the central nervous system. *Adv. Metab. Dis.* 10: 457-468.

Woods, S.C., E. Decke, and J.R. Vasselli. 1974. Metabolic hormones and regulation of body weight. *Psychol. Rev.* 81: 26-43.

Woods, S.C., E.C. Lotter, L.D. McKay, and D. Porte, Jr. 1979. Chronic intracerebroventricular infusion of insulin reduced food intake and body weight of baboons. *Nature* 282: 503-505.

Woods, S.C., L.J. Stein, L.D. McKay, and D. Porte, Jr. 1984. Suppression of food intake by intravenous nutrients and insulin in the baboon. *Am. J. Physiol.* 247: R393-R401.

Woods, S.C., D.B. West, L.J. Stein, D.P. Figlewicz, and D. Porte, Jr. Relationship between plasma and cerebrospinal fluid insulin levels in the baboon. Submitted to *Endocrinology*.

Yalow, R.S., and J. Eng. 1983. Insulin in the central nervous system. *Adv. Metab. Dis.* 10: 341-354.

Chapter 11

GUT PEPTIDES AND FEEDING BEHAVIOR: THE MODEL OF CHOLECYSTOKININN

J. Gibbs and G.P. Smith

The surface of the gastrointestinal tract, the first point of contact between food and the body, must play some role in generating the physiological events which cause eating to stop. Over the past decade, research has implicated the stomach and small intestine as gastrointestinal sites important in the production of satiety signals. In addition, several gastrointestinal peptides have been suggested to play roles as satiety signals. This chapter will examine in some detail research delineating the role of one site, the small intestine, as a generator of important satiety signals, and the possible role of one small intestinal peptide, cholecystokinin, as a physiologically relevant satiety signal. We hope that this narrow focus will not only impart information in critical relief, but will also give the reader an appreciation of the pace of progress and the difficulty of the problems. It is likely that similar progress and problems will be encountered in evaluating any other site or any other candidate in the rapidly growing list of putative satiety signals.

I. THE SATIETY ACTION OF NUTRIENTS IN THE SMALL INTESTINE

In an effort to determine the importance of the gastrointestinal tract in satiety, and to characterize its role, we used surgical techniques to limit the distribution of ingested food to particular surfaces of the gut--to the mouth, the stomach, or small intestine.

Initially, we employed chronic stainless steel cannulas, implanted in the stomach of rats, to create a sham feeding preparation. When cannulas are temporarily opened for a test meal in such a preparation, the major portion of an ingested liquid food drains immediately to the outside, preventing the accumulation of food in the stomach or the entrance of food into the small intestine. When gastric cannulas were opened at a test meal following an overnight food deprivation, the resulting sham feeding was practically uninterrupted for several hours (Young et al., 1974; Fig. 1). Although overeating might

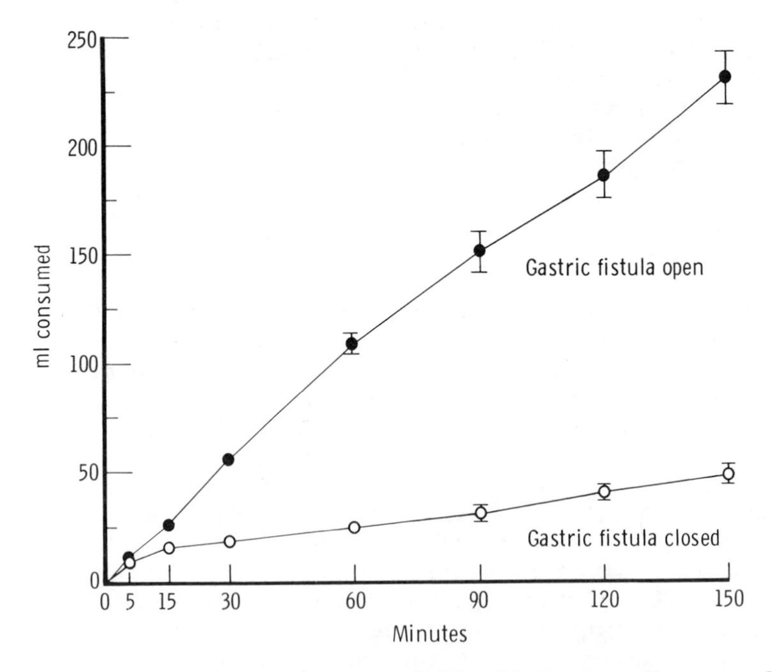

Fig. 1. Cumulative intake of liquid food by 3 rats after 17 hours of food deprivation when the gastric fistula was open (sham feeding) and when the gastric fistula was closed (real feeding). Note the very large intake that results from almost continuous sham feeding when the gastric fistula was open.

be anticipated in such circumstances, we were surprised that eating was nearly continuous for such a prolonged period. The size of the effect meant that although food was presumably smelled, tasted, and swallowed normally, satiety was absent. The same experiment in rhesus monkeys produced the same result (Gibbs et al., 1981). Thus, in both species, the massive overeating demonstrated that the contact of nutrients with the mouth and esophagus was not sufficient to produce satiety. For satiety to occur, it seemed that nutrients must accumulate in the stomach and/or enter the small intestine.

Which gut surface, the gastric or intestinal, contained the nutrient receptors important in generating normal satiety? The sham feeding preparation proved to be of further value in answering this next question. Since sham feeding after overnight food deprivation provides a bioassay for satiety--a situation in which satiety does not occur spontaneously--it can be used to test putative satiety signals. First, we asked whether nutrients delivered directly to the small intestine would induce satiety in this bioassay. The answer was clear. Nutrients introduced directly into the small intestine through chronic duodenal cannulas stopped sham feeding (Liebling et al., 1975). The effect was dose-related, and has been replicated in rats (Reidelberger et al., 1983) and in rhesus monkeys (Gibbs et al., 1981; see Fig. 2).

The potency of nutrients in the small intestine was surprising: Reidelberger et al. (1983) showed that satiety can be induced in the sham-feeding rat by the intra-duodenal infusion of a volume of liquid food that does not exceed the volume which empties into the intestine from the stomach during a normally consumed meal. Note that in the sham feeding situation, the inhibitory effect of food in the small intestine cannot depend upon gross distention of the stomach, because distention cannot occur when gastric cannulas are open.

In addition to reductions in the amount of liquid food consumed, important behavioral changes emerged during these tests. When food was delivered to the small intestine, rats and rhesus monkeys displayed the same sequence of behaviors that they regularly display at the end of a normal meal--the cessation of eating, followed by alternating periods of grooming and exploratory behaviors, and then apparent sleep (Antin et al., 1975; Fig. 2). Thus, nutrients in the intestine, even in the absence of gastric distention, are behaviorally potent: they not only stop further nutrient intake, but they elicit the full behavioral sequence characteristic of normal satiety. We have referred to this complex action of nutrients at this site as "intestinal satiety" (Smith et al., 1974).

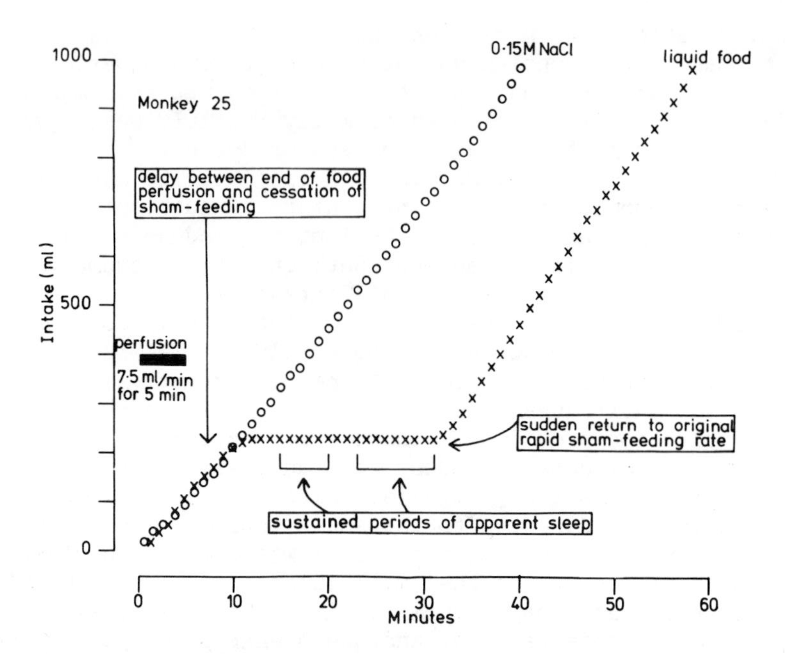

Fig. 2. Cumulative intake in a monkey after intraduodenal infusion of 0.15M NaCl or liquid food during sham feeding.

II. IN SEARCH OF THE MECHANISMS OF INTESTINAL SATIETY

How do nutrients produce intestinal satiety? There are several possibilities: they could release hormones from the small intestine into the circulation that would in turn act as satiety signals; they could activate extrinsic or intrinsic intestinal nerves; they could exert their effects directly, following absorption into the portal or systemic circulations; or they could involve all of these possibilities.

For one nutrient, fat, there is now powerful evidence from two laboratories that the small intestine itself generates a potent satiety signal, and that absorbed fats play no role in the control of meal size. In rats, Greenberg et al. (1986) have shown that a potent dose-related suppression of sham feeding is produced by delivery of a lipid emulsion to the lumen of the duodenum. The same emulsion delivered into the hepatic portal vein was without effect. Similarly, in humans Welch et al. (1985) have shown a significant suppression in food intake at a test meal after ileal infusions of a lipid; equivalent intravenous infusions failed to reduce meal size.

These studies limit the possibilities, at least for fats, to the stimulation of intestinal nerves and/or the release of intestinal hormones.

The behavioral results of the experiment shown in Figure 2 directed our attention to hormones. Examination of the results of every test on each rhesus monkey involved in this study revealed a significant delay, usually five to eight minutes, between the end of the nutrient infusion into the intestine and the cessation of sham feeding. This latency suggested that intestinal satiety was not being mediated solely by neural elements, and that some humoral mechanism was involved.

Earlier work had noted that intestinal extracts could decrease food intake (Maclagan, 1937; Ugolev, 1960; Schally et al., 1967; Sjödin, 1972). In 1972, building on these studies, we tested the three gastrointestinal hormones then available in partially purified form, in order to determine if one of them would inhibit food intake. These hormones, all peptides, were gastrin (from gastric mucosa), secretin, and cholecystokinin (from duodenal mucosa).

III. EXOGENOUS CHOLECYSTOKININ AND SATIETY

Gastrin and secretin had no apparent satiety effect in rats, but cholecystokinin (CCK) showed remarkable potency. Intraperitoneal (i.p.) injections of a 10% pure preparation of CCK produced a dose-related reduction in both solid and liquid food intakes (Gibbs et al., 1973a). Additional observations suggested that this gut peptide was exerting a specific satiety action when it reduced food ingestion: First, rats were not incapacitated by the injections, and they showed only those behaviors usually seen at the close of a meal. Second, CCK did not prevent or delay feeding, but it did shorten the duration of feeding; thus, CCK advanced the time of meal termination, the characteristic action of a satiety signal. Finally, doses of CCK that produced a marked reduction of liquid food intake after a period of food deprivation failed to reduce drinking when rats were offered water after a period of water deprivation. This differential effect suggested that CCK was not acting simply by producing some subclinical discomfort, since the peptide was able to dissociate two different motivated behaviors (eating and drinking) that employed the identical motor act--lapping a liquid from a spout (Gibbs et al., 1973a).

Further work demonstrated a degree of chemical specificity in showing that the biologically active synthetic C-terminal octapeptide of CCK (CCK-8) matched the properties of the impure preparation, while desulfated CCK-8, a biologically weak analogue, failed to inhibit food intake. It was important that CCK and CCK-8 were potent inhibitors of sham feeding, as well as real feeding, both in rats (Gibbs et al., 1973b) and in rhesus monkeys (Falasco et al., 1979). Figure 3 illustrates dose-related inhibitions of sham feeding in rats produced by 10% pure CCK and by CCK-8.

Thus, the use of the sham feeding paradigm revealed strong parallels between the satiating effects of nutrients delivered to the small intestine and the intraperitoneal administration of CCK-8. Both manipulations inhibited sham feeding in a dose-response manner, and both elicited the characteristic behavioral display of satiety. These parallels strongly suggested, but did not prove, that endogenous CCK released by the action of ingested nutrients on the mucosa of the upper small intes-

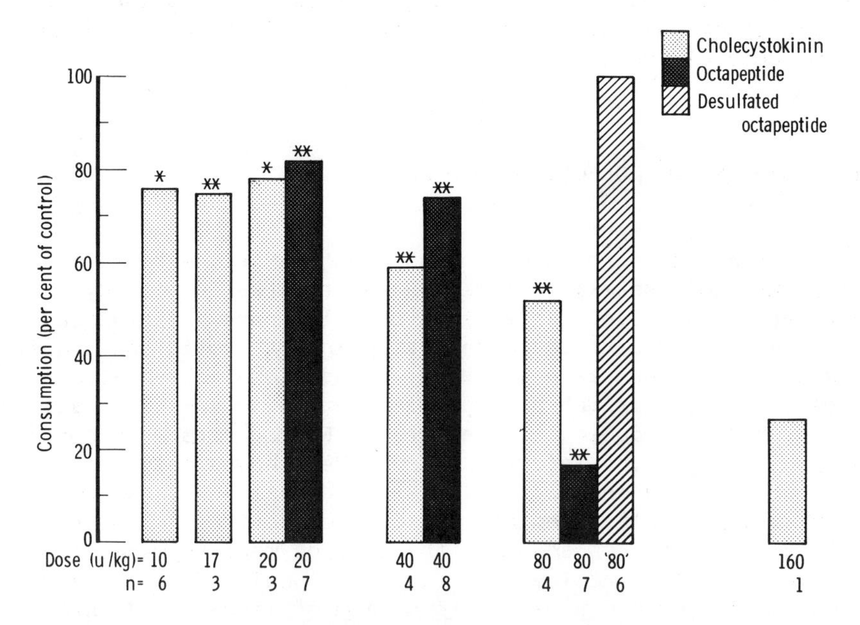

Fig. 3. Consumption of solid food (expressed as percentage of control consumption) during first 30 min of the test period following intraperitoneal injection of cholecystokinin (CCK) [in light bars]; octapeptide (CCK-8) [dark bars]; and caerulein [hatched bars] in various doses. Caerulein doses were calculated by assuming that caerulein was 15 times as potent by weight as CCK. 1 microgram of CCK-8 is equivalent to approximately 20 units. Mean control intakes ranged 2.8-4.3 gm. Statistical differences from saline control at $*p < 0.01$ and $**p < 0.001$. (From Gibbs et al., 1973b, with permission).

tine plays a role in the regulation of meal size. We will return to the evidence for and against this suggestion.

CCK reduced food intake in mice (McLaughlin and Baile, 1981; Parrott and Batt, 1980; Strohmayer and Smith, 1981), chickens (Savory and Gentle, 1980), rabbits (Houpt et al., 1978), sheep (Grovum, 1981), and pigs (Anika et al., 1981), as well as in rats and monkeys.

Two groups of investigators have observed a satiating effect of CCK in humans. In these studies, carried out under double-blind conditions, slow intravenous (i.v.) infusions of CCK-8 or the CCK-like peptide ceruletide produced inhibitions of test meal size ranging from 15 to 50% in lean or obese humans (Kissileff et al., 1981; Pi-Sunyer et al., 1982; Stacher et al., 1982; for a thorough review see Stacher, 1986). Two features of these human studies were notable. The first was that the doses of exogenous CCK or exogenous ceruletide required to inhibit food intake in humans were small, in the low nanogram per kilogram dose range. The second was that the effective doses of exogenous CCK failed to produce significant side effects (Smith, 1984). These features are consistent with a physiological role for endogenous CCK in satiety.

IV. ON THE PERIPHERAL MECHANISM OF ACTION OF CHOLECYSTOKININ

Because a systemic injection of CCK produces a large change in the behavior of feeding, it must, directly or indirectly, produce large changes in the neural function of the brain. How are these changes effected? Through what pathway(s) does the satiety information provided by CCK reach the brain? The major pathway appears to be afferent fibers of the abdominal vagal system. This assertion is based on four pieces of evidence. First, bilateral subdiaphragmatic vagotomy markedly reduces or abolishes the satiating effect of peripherally administered CCK (Smith et al., 1981; Lorenz and Goldman, 1982; Morley et al., 1982). Second, the critical lesion involves the afferent fibers (Smith et al., 1985). Third, Niijima (1983) reported electrophysiological evidence that intraportal or intraperitoneal injections of CCK stimulated gastric vagal afferents and inhibited hepatic vagal afferents. This differential pattern of the electrophysiological effects of CCK on abdominal vagal afferents may be relevant to the differential effect of selective gastric or hepatic vagotomy on the satiating effect of CCK--gastric vagotomy reduces or abolishes it, but hepatic vagotomy does not (Smith et al., 1981). Fourth, small-diameter sensory neurons in the afferent vagus are the important fibers, because capsaicin, a neurotoxin which is relatively selective for such neurons, produces a major attenuation of CCK-induced satiety when it is applied directly to the vagal trunks (South and Ritter, 1985). Capsaicin has been shown to deplete peptides such as substance P, somatostatin, vasoactive intestinal polypeptide, and CCK-8 from sensory nerves, and all of these peptides are housed in the vagus. Nevertheless, it remains unknown which, if any, of these peptides are important constituents of the particular afferent fibers on which the satiety action of exogenous CCK depends.

This apparently compelling evidence for mediation of the satiating effect of CCK by afferent fibers of the abdominal vagus has been challenged recently by Kraly (1984), who reported that lesioning the gastric and major celiac vagal branches (the hepatic and accessory celiac branches were left intact) did not change the satiating effect of impure CCK or of CCK-8 in doses up to 4.5 μg/kg in _sham-feeding_ rats. This contrasted with the results discussed above that were observed during _real feeding_.

To investigate this matter further, we recently administered CCK-8 (2-8 μg/kg) to rats after total abdominal vagotomy during real feeding and sham feeding. The preliminary results are that abdominal vagotomy abolished the satiating effect of 2 and 4 μg/kg and markedly reduced, but did not abolish, the satiating effect of 8 μg/kg. These results were observed in both real and sham feeding tests (Joyner et al., 1986).

Thus, as a working hypothesis, we conclude that afferent vagal fibers mediate the satiating effect of doses of CCK-8 equal to or less than 4 μg/kg, but that there is some unidentified, nonvagal pathway that mediates a small part of the response to larger doses (8 μg/kg). Two possible candidates for this nonvagal pathway are fibers relaying through the area postrema and the visceral afferent fibers of the spinal cord. There is some evidence that lesions of the area postrema (AP) decrease the satiating effect of CCK (Vander Kooy, 1984; R.C. Ritter and colleagues, see this volume), but the AP lesions decrease the response to low doses of CCK rather than the high dose, and the AP lesions usually encroach on the primary projection site of the abdominal vagal afferents. To our knowledge, no one has evaluated the role of the AP or the visceral afferents of the spinal cord for the mediation of the satiating effect of relatively high doses of CCK-8 in the _vagotomized_ rat.

If the afferent vagal fibers mediate a significant part of the satiating effect of CCK, how does peripherally administered CCK activate them? There are two possibilities. The first is by direct stimulation of specific receptors that move distally along the vagus and are present in all the abdominal branches

(Zarbin et al., 1981; Moran et al., 1986a). The other possibility is by indirect stimulation of the vagus due to the contractions of pyloric smooth muscle produced by activation of the receptors that cluster there (Smith et al., 1982). In this case, the vagus is activated by sensing the change in muscle contractility or tension produced by CCK through its mechanoreceptors (Davison, 1986).

In an attempt to distinguish between these two possibilities, Falasco et al. (1984) removed the pyloric sphincter surgically and anastamosed the remaining stomach to the duodenum. Such rats appeared to respond normally to CCK. This result suggests that CCK-8 is acting directly on the vagal receptors in this situation, but a study of the stomach in the rats that had the sphincter removed revealed some CCK receptors on gastric smooth muscle near the gastroduodenal anastamosis. Thus, although the experiment shows that the CCK receptors in the pyloric sphincter are not necessary for the satiating effect of CCK, the results do not distinguish clearly between the vagal receptors and the gastric receptors as the site of CCK's action in these animals.

The presence of CCK receptors in the pyloric sphincter and CCK's well-known effect of decreasing gastric emptying led McHugh (1979) to suggest that CCK could activate the vagus by distending the stomach as a result of the decreased emptying produced by CCK. As a demonstration of this possibility, Moran and McHugh (1982) showed that a dose of CCK-8 that had no satiating effect when administered to a monkey with an empty stomach did have a significant satiating effect when the stomach was preloaded with 0.15M saline. This is clear evidence for synergistic interaction, but the experiment did not evaluate the role of the vagus in this effect, and it did prove that distention was the critical event.

Miceli (1985) has also obtained some evidence for a synergism between stomach distention and CCK, observing that in the hamster, a lower dose of CCK was required to reduce real feeding than was required to reduce sham feeding. This evidence is weak because (1) sham feeding prevents the emptying of food into the duodenum as well as preventing gastric distention, and (2) Joyner et al. (1986) observed no difference in the dose required to inhibit real and sham feeding in the rat.

Perhaps the best evidence for vagal fibers mediating the synergism between gastric distention and CCK is the electrophysiological results of Dockray et al. (1985). They recorded extracellularly from neurons in the nucleus of the solitary tract (NST) and other areas of the dorsomedial medulla. They identified subsets of cells that responded to gastric distention by an increase, a decrease, or by no change in their spontaneous firing rate, and found that administration of intravenous CCK-8 produced the same effect on firing rate in each subset of cells--increase, decrease, or no change--that gastric distention itself produced. These investigators suggested "CCK-8 might act directly on vagal afferent fibers that relay to the CNS information from gastric mechanoreceptors about the volume of the gastric contents." (Dockray et al., 1985).

Although the matter is not settled, we believe that current evidence suggests that CCK can activate the vagus by a change in tension without a change in intragastric pressure (Davison, 1986) and without distention (sham feeding), but that CCK and some gastric event related to volume (possibly distention) interact synergistically. The specification of the gastric and the function of vagal afferent fibers in this synergism remains to be demonstrated in more detail.

V. THE BRAIN AND SYSTEMIC CHOLECYSTOKININ

The first synapse for the central processing of information carried by abdominal vagal afferents stimulated by CCK occurs at the NST. An extensive lesion of the commissural and medial subnuclei of NST decreased the satiating effect of CCK of 4 μg/kg or less, but not of a larger dose (Edwards et al., 1985). Since this NST lesion had less effect than the total abdominal vagotomy in the study of Joyner et al. (1986), we conclude that some of the vagal afferent terminals transmitting the peripheral signal of CCK's action have not been damaged by the NST lesion.

Since CCK is in vagal afferent fibers, it is possible that CCK is released from the central end of these fibers when peripherally administered CCK activates their peripheral ends. Crawley (1985) tested this possibility by injecting CCK directly into the NST, but there was no change in food intake.

Crawley and her colleagues, however, have found two central lesions that abolish the satiating effect of peripherally administered CCK. First, bilateral midbrain transections that interrupted ascending pathways from NST to hypothalamus abolished the inhibition of food intake produced by systemically-administered CCK (Crawley et al., 1984). Second, Crawley and Kiss (1985) reported that bilateral knife cut lesions of the paraventricular nucleus of the hypothalamus had the same effect. Others have reported that electrolytic lesions of the dorsomedial nucleus of the hypothalamus attenuate the effect of peripherally administered CCK (Bellinger and Bernardis, 1984).

At this point, we should note that the logical and tempting interpretation--tracing a neurally mediated, necessary and sufficient satiety pathway from peripheral circulating CCK via vagus to hypothalamus--must be resisted. The reason is that rats with chronic decerebration (achieved by a complete knife cut directed from the supracollicular level dorsally to the brainstem ventrally, thus eliminating neural communication between NST and hypothalamus) not only satiate in response to food (Grill and Norgren, 1978), but show an inhibition of food intake after systemic CCK-8 (Grill et al., 1983). Further work is required to understand these paradoxical results.

VI. CENTRAL CHOLECYSTOKININ AND SATIETY

CCK intrinsic to neurons in the brain may also play a role in the satiating effect of ingested food. In 1975, Vanderhaeghen and his colleagues discovered a peptide in the vertebrate central nervous system that reacted with antibodies to gastrin, a gut peptide structurally quite similar to CCK (Vanderhaegen et al., 1975). In rapid succession, Dockray (1976), Muller et al. (1977), and the Vanderhaeghen group (Robberecht et al., 1978) showed that the source of this gastrin-like immunoreactivity in brain was due in large part to CCK, not gastrin; Rehfeld (1978) and Beinfeld et al. (1981) provided comprehensive and compelling evidence that brain CCK was heterogeneous in molecular form and in distribution. Finally, Williams and his colleagues have demonstrated specific binding sites for CCK in brain (Saito et al., 1980).

Behavioral applications followed quickly. Baile, Della-Fera, and their colleagues, in a series of publications since 1979, produced strong support for their conclusion that in the sheep, CCK of brain origin, not peripheral origin, had a satiating effect. These studies utilized intracerebroventricular (i.c.v.) infusions of very small doses of CCK (e.g., Della-Fera and Baile, 1979) or of CCK antagonists (e.g., Della-Fera et al., 1981; for an extensive recent review of the field by this group, see Baile et al., 1986).

In species other than sheep, repeated early attempts to inhibit feeding by the introduction of CCK into the cerebral ventricles were negative. The notable exception was the work of Maddison (1977), who compared the i.p. and i.c.v. routes for their potencies and temporal characteristics in reducing the rate of an operant response for food in the rat, and found the effective central dose of an impure preparation of CCK to be well below the effective peripheral dose. Later, Parrott and Baldwin (1981) achieved an inhibition of food intake by CCK-8 in pigs by use of the cerebroventricular route. More recently,

Ritter and Ladenheim (1984) demonstrated a dose-related suppression of food intake in the rat after doses of CCK-8 in the low nanogram range directly into the fourth cerebral ventricle. Also in the rat, Zhang Bula, and Stellar (1986) demonstrated a chemically specific inhibition of food intake (i.e., induced by CCK-8 but not by desulfated CCK-8), and a dose-related inhibition of running speed, as an index of motivation for food reward, after injections of low doses into the third ventricle.

The chemically specific inhibition demonstrated by Zhang et al. was surprising. It has been thought that all brain CCK receptors were approximately as avid in binding desulfated CCK-8 as sulfated CCK-8, and were thus quite different from peripheral CCK receptors, which bind sulfated CCK-8 with much greater affinity. The demonstration of a satiety effect of sulfated CCK-8, but not of desulfated CCK-8, constituted unequivocal behavioral evidence that the brain must harbor colonies of peripheral CCK receptors. Moran and his colleagues have recently discovered some of these colonies--in the area postrema, nucleus of the solitary tract, interpenduncular nuclei, and posterior hypothalamic nuclei (Moran et al., 1986b). Their roles in satiety remain to be determined.

In the studies by Maddison and by Ritter and Ladenheim, the effective doses were well below those required for a satiety effect when the peptide was administered systemically; this is an important point, since i.c.v. injected peptides, including CCK-8, can readily and rapidly pass from cerebrospinal fluid to blood (Passaro et al., 1982). Another important point considered by these investigators was whether CCK-8 was behaviorally specific when it reduced food intake after i.c.v. administration: Ritter and Ladenheim found that administration of CCK-8 into the fourth cerebral ventricle produced an inhibition of food intake after food deprivation, but not an inhibition of water intake after water deprivation. Zhang et al. found that their ventricular CCK-8 produced an inhibition of running speed for food reward in hungry rats, but not for water reward in the same rats when they were thirsty. Attention to behavioral specificity, always important, is perhaps never more crucial than when treatments are administered to the hypothalamus, brainstem, or their neighboring ventricular spaces, since manipulations of these areas can readily produce changes in cardiovascular, respiratory, or other autonomic functions that may nonspecifically inhibit ongoing behavior.

Investigators have recently begun to employ microinjections of CCK-8 directly into brain tissue in attempts to discover the sites at which the i.c.v. administered peptide acts. Some of these reports have been flawed by lack of attention to the points made above--for example, the use of doses of CCK-8 so large that they would be effective even if given peripherally.

There are notable exceptions to this criticism. Using paraventricular nucleus lesions in a study that complemented the findings of Crawley and Kiss, Faris and Olney (1985) reported that direct bilateral injections of CCK-8 into the PVN, in doses as low as 4.4 pmol, bilaterally inhibited food intake in rats. Identical injections into the ventromedial nucleus had no effect. Dorfman et al. (1985) did the complementary experiment by injecting benzotript, an antagonist of CCK, into the PVN and observed a significant increase of food intake. There is also evidence that medial hypothalamic injections of CCK inhibit food intake elicited by intrahypothalamic injections of norepinephrine (McCaleb and Myers, 1980).

It is clear that the analysis of the function of central CCK in the process of satiation is just beginning and that it is likely to be most productive.

VII. FORMULATION

Summarizing and interpreting the studies reviewed thus far, we can make the following statements:

First, there is strong evidence from animal and human studies that a potent preabsorptive satiety signal is released when ingested nutrients contact the surface of the small intestine.

Second, there is strong evidence from animal and human studies that the systemic administration CCK (an endogenous peptide known to be released into circulation when ingested nutrients enter the small intestine) produces the behaviors and the experience of satiety.

Third, the full expression of systemically-administered exogenous CCK in producing satiety requires a population of afferent, capsaicin-sensitive, abdominal vagal neurons and the nucleus of the solitary tract. In the chronic decerebrate rat, this appears to be enough neural organization, but in the intact rat an ascending pathway through the midbrain and the paraventricular nucleus of the hypothalamus are also involved.

Fourth, there is increasing evidence that very low doses of exogenous CCK--amounts too small to produce satiety effects if given peripherally--will produce satiety when they are delivered directly into cerebral ventricles or brain tissue; this evidence strongly suggests, but does not prove, that endogenous brain CCK plays an important role in satiety.

What is still lacking, of course, are links between these four statements that would provide a compelling and coherent plan of the mechanisms involved, from gut through brain, when exogenous CCK and/or ingested food produces satiety.

VIII. PROBLEMS FACING THE CHOLECYSTOKININ SATIETY HYPOTHESIS

Substantial progress has been made in establishing the satiety action of exogenous CCK in the periphery and in the brain, and in identifying some pieces of the peripheral and central machinery important to this action. Nevertheless, major problems remain.

The first problem is whether the release of endogenous CCK can exert a satiety action. Based on early bioassay evidence of CCK release by the L-isomer, but not by the D-isomer, of phenylalanine in the dog (Meyer and Grossman, 1972), we used these two isomers in studies comparing their potencies to inhibit food intake. As predicted, we demonstrated a potent satiating effect of L-phenylalanine, but not of D-phenylalanine, after intragastric administration in rhesus monkeys (Gibbs et al., 1976) or after intraduodenal administration in sham-feeding rats (Lew et al., 1983). Unfortunately, circulating levels of endogenous CCK in response to the two isomers were not measured in either experiment. This was a costly omission, for it now appears that there are important species differences concerning which nutrients release CCK into circulating blood: fats are potent stimulants of circulating CCK bioactivity in humans (Liddle et al., 1985) but not in rats; furthermore, recent evidence in the rat demonstrates that proteins are potent stimulants of circulating CCK bioactivity, but amino acids, including L-phenylalanine, are not (Liddle et al., in press).

Other investigators have used trypsin inhibitors (which bind luminal trypsin and thereby prevent its negative feedback action on CCK release--Green and Lyman, 1972) in an effort to increase circulating levels of endogenous CCK. In an encouraging report, the intragastric administration of the trypsin inhibitor aprotinin sharply decreased test meal size in obese and lean Zucker rats (McLaughlin et al., 1983). This result is what would be predicted if endogenous CCK levels were raised by trypsin inhibitor, but unfortunately circulating CCK levels were not measured in this study. In contrast, we have now demonstrated sharply increased circulating levels of endogenous CCK after intragastric administration of soybean trypsin inhibitor, but these increased levels of CCK were not associated with a decrease in food intake (Greenberg et al., 1985).

It is clear that future studies employing this strategy, if they are to be compelling, must track changes in CCK and changes in feeding behavior simultaneously.

A second problem remains, that of determining whether CCK has a physiologically relevant satiety action when it is administered exogenously or when it is released endogenously as nutrients are ingested.

Some investigators have compared the doses of exogenous CCK required to reduce food intake with the doses required to produce effects thought to be physiological, such as an increase in pancreatic enzyme secretion. One recent study using the rat found that the doses of exogenous CCK-8 required to reduce sham feeding were five times greater than those required to drive pancreatic amylase output maximally (Reidelberger and Solomon, 1986). Another study using the dog showed that the doses of exogenous CCK-8 required to reduce a short (7.5 minute) period of sham feeding by 50% were three to four times those required to increase pancreatic protein output by 50% (Pappas et al., 1985). These two studies suggest that the circulating levels of CCK-8 likely to prevail after a meal are not sufficient to produce satiety. However, it cannot be the case that CCK is the only satiety signal acting at a meal. It is more likely that CCK interacts with other satiety signals released by the action of nutrients along the gut surface. Recall that Moran and McHugh (1982) have demonstrated a potentiation of the satiating effect of a very low dose of exogenous CCK-8 (250 pmol/kg/h) by the addition of a small gastric load of physiological saline in the rhesus monkey. It is hazardous to categorize putative satiety signals as pharmacological or physiological--particularly on the basis of tests employing exogenous administration of the candidates--until all of the important signals are identified and their interactions quantified.

A strategy promising a better penetration into this problem would be the use of highly selective and potent antagonists to block the action of endogenous CCK at test meals. If such an agent significantly increased food intake and if its specificity under the test conditions could be demonstrated, these results would constitute compelling evidence for a physiological satiety action of CCK released by food ingestion. Although the ideal agents are not yet available, the recent availability of reasonably selective CCK antagonists has allowed tests of this proposal to begin. Proglumide, a competitive antagonist of the in vitro actions of CCK on preparations of gut viscera (e.g., Hahne et al., 1981) has been employed in the effort to determine whether blockade of the endogenous CCK released by food in the gastrointestinal tract would increase food intake at test meals. Four studies have attempted to increase meal size by prior administration of proglumide; none of them have been successful (Collins et al., 1983; Shillabeer and Davison, 1984; Crawley et al., 1986; and Jerome et al., 1986).

On the other hand, there have been two reports of proglumide antagonizing the satiating effect of food stimuli. Collins et al. (1985) showed that proglumide partially reversed the inhibition of sham feeding produced by large, but not small, intraduodenal loads of liquid food in rats. Shillabeer and Davison have demonstrated (1984) and replicated (1985) a proglumide-induced blockade of the satiety effects of a food preload on a subsequent test meal in rats. Unfortunately, Schneider et al. (1986) could not replicate the effect of proglumide on an oral preload.

A different tactic was employed by McLaughlin et al. (1985). These investigators administered serum that contained antibodies raised against CCK-8 (but which in fact bound both radioiodinated gastrin and radioiodinated CCK-33 in tests of binding capacity). They administered this serum, or, as the control, serum that did not contain antibodies, to lean and obese Zucker rats prior to a test meal of solid food. Both lean and obese rats showed significant increases in meal size on days when they received serum containing antibodies to gastrin and CCK. The effect was present after a short fast preceding the test meal, but not after a prolonged fast. Unfortunately, the authors were unable to demonstrate an effect after administration of a larger volume of antiserum, and they did not employ the ideal control--serum preabsorbed with CCK.

Our conclusion is that the role of endogenous CCK in satiety remains unproven. Nevertheless, despite these generally negative results, the problem needs further testing with the newer and more potent antagonists of CCK now available (Niederau et al., 1986).

IX. CONCLUSION

Substantial progress has been made in analyzing the hypothesis that CCK plays a role in satiety. The case for a satiety action of peripherally administered exogenous CCK is strong, and advances in understanding the mechanism of its action are clear, if far from complete. Two major problems remain: whether endogenous CCK can induce satiety, and whether endogenous CCK can induce satiety under the physiological conditions that obtain at the close of a meal. Finally, several interesting reports suggest that centrally administered CCK may have a satiety action, and these observations offer clear and difficult experimental challenges.

In addition to CCK, several gut peptides--bombesin, pancreatic glucagon, somatostatin, and others--have been implicated in satiety, but the experimental case for each is far less developed than that for CCK. Critical treatments of the evidence supporting each of these peptides is beyond the scope of this chapter; they can be found in several available reviews (e.g., Gibbs and Smith, 1984; Morley et al., 1985). It will be interesting to see how these other candidates survive the intense experimental scrutiny which CCK has undergone, and to what extent the experimental trail for CCK serves as a model for their understanding.

ACKNOWLEDGEMENTS

This chapter was written while the authors were supported by NIMH RSDA MH70874 (JG) and NIMH RSA MH 00149 (GPS). We thank Jane Magnetti for her expert assistance in preparing the manuscript.

REFERENCES

Anika, S.M., T.R. Houpt, and K.A. Houpt. 1981. Cholecystokinin and satiety in pigs. *Am. J. Physiol.* 240: R310-R318.

Antin, J., J. Gibbs, J. Holt, R.C. Young, and G.P. Smith. 1975. Cholecystokinin elicits the complete behavioral sequence of satiety in rats. *J. Comp. Physiol. Psychol.* 89: 784-790.

Baile, C.A., C.L. McLaughlin, and M.A. Della-Fera. 1986. Role of cholecystokinin and opoid peptides in control of food intake. *Physiol. Rev.* 66: 172-234.

Beinfeld, M.C., D.K. Meyer, R.L. Eskay, R.T. Jensen, and M.J. Brownstein. 1981. The distribution of cholecystokinin immunoreactivity in the central nervous system of the rat as determined by radioimmunoassay. *Brain Res.* 212: 51-57.

Bellinger, L.L., and L.L. Bernardis. 1984. Suppression of feeding by cholecystokinin but not bombesin is attenuated in dorsomedial hypothalamic nuclei lesioned rats. *Peptides* 5: 547-552.

Collins, S., D. Walker, P. Forsyth, and L. Belbeck. 1983. The effects of proglumide on cholecystokinin-, bombesin-, and

glucagon-induced satiety in the rat. *Life Sci.* 32: 2223-2229.
Collins, S.M., K.L. Conover, P.A. Forsyth, and H.P. Weingarten. 1985. Endogenous cholecystokinin and intestinal satiety. *Am. J. Physiol.* 249: R667-R680.
Crawley, J.N. 1985. Neurochemical investigation of the afferent pathway from the vagus nerve to the nucleus tractus solitarius in mediating the "satiety syndrome" induced by systemic cholecystokinin. *Peptides* 6, 133-137.
Crawley, J.N., and J.Z. Kiss. 1985. Paraventricular nucleus lesions abolish the inhibition of feeding induced by systemic cholecystokinin. *Peptides* 6: 927-935.
Crawley, J.N., J.Z. Kiss, and E. Mezey. 1984. Bilateral midbrain transections block the behavioral effects of cholecystokinin on feeding and exploration in rats. *Brain Res.* 232: 316-321.
Crawley, J.N., J.A. Stiver, D.W. Hommer, L.R. Skirboll, and S.M. Paul. 1986. Antagonists of central and peripheral behavioral actions of cholecystokinin octapeptide. *J. Pharmacol. Exp.* 236: 320-330.
Davison, J.S. 1986. Activation of vagal gastric mechano-receptors by cholecystokinin. *Gastroenterology* 90: 1388.
Della-Fera, M.A., and C.A. Baile. 1979. Cholecystokinin octapeptide: continuous picomole injections into the cerebral ventricles of sheep suppress feeding. *Science* 206: 471-473.
Della-Fera, M.A., C.A. Baile, B.S. Schneider, and J.A. Grinker. 1981. Cholecystokinin antibody injected in cerebral-ventricles stimulates feeding in sheep. *Science* 212: 687-689.
Dockray, G.J. 1976. Immunochemical evidence of cholecysto-kininlike peptide in brain. *Nature* 264: 568-570.
Dockray, G.J., H. Desmond, R.J. Gayton, A.C. Jonsson, H. Raybould, K.A. Sharkey, A. Varro, and R.G. Williams. 1985. Cholecystokinin and gastrin forms in the nervous system. In *Neuronal Cholecystokinin*, ed. J.J. Vanderhaeghen and J.N. Crawley, 32-43. The New York Academy of Sciences, New York.
Dorfman, D.B., D. Schwartz, and B.G. Hoebel. 1985. Feeding induced by benzotript in the PVN: Further evidence that CCK receptors in the PVN inhibit feeding. Program of Satellite Symposium on Mechanisms of Appetite and Obesity.
Edwards, G.L., E.E. Ladenheim, and R.C. Ritter. 1985. Dorsal hindbrain participation in cholecystokinin-induced satiety. *Soc. Neurosci. Abstr.* 11: 37.
Falasco, J.D., G.P. Smith, and J. Gibbs. 1979. Cholecystokinin suppresses sham feeding in the rhesus monkey. *Physiol. Behav.* 23: 887-890.

Falasco, J.D., K.M.S. Joyner, J. Gibbs, and G.P. Smith. 1984. Pyloric binding sites for cholecystokinin are not necessary for its satiety effect. *Soc. Neurosci. Abstr.* 10: 532.

Faris, P.L., and J.A. Olney. 1983. Suppression of food intake in rats by microinjection of cholecystokinin (CCK) to the paraventricular nucleus (PVN). *Soc. Neurosci. Abstr.* 11: 39.

Gibbs, J., and G.P. Smith. 1984. The neuroendocrinology of postprandial satiety. In *Frontiers in Neuroendocrinology*, ed. L. Martini and W.F. Ganong, 223-245. Raven Press, New York.

Gibbs, J., R.C. Young, and G.P. Smith. 1973a. Cholecystokinin decreases food intake in rats. *J. Comp. Physiol. Psychol.* 84: 488-495.

Gibbs, J., R.C. Young, and G.P. Smith. 1973b. Cholecystokinin elicits satiety in rats with open gastric fistulas. *Nature* 245: 323-325.

Gibbs, J., J.D. Falasco, and P.R. McHugh. 1976. Cholecystokinin-decreased food intake in rhesus monkeys. *Am. J. Physiol.* 230: 15-18.

Gibbs, J., S.P. Maddison, and E.T. Rolls. 1981. Satiety role of the small intestine examined in sham-feeding rhesus monkeys. *J. Comp. Physiol. Psychol.* 95: 1003-1015.

Green, G.M., and R.L. Lyman. 1972. Feedback regulation of pancreatic enzyme secretion as a mechanism for trypsin inhibitor-induced hypersecretion in rats. *Proc. Soc. Exp. Biol. Med.* 140: 6-12.

Greenberg, D., G.P. Smith, J. Gibbs, J.D. Falasco, R.A. Liddle, and J.A. Williams. 1985. Soybean trypsin inhibitor increases plasma cholecystokinin (CCK) but does not decrease food intake. *Soc. Neurosci. Abstr.* 11: 557.

Greenberg, D., D.C. Becker, J. Gibbs, and G.P. Smith. 1986. Intraportal administration of fats fails to elicit satiety. *Soc. Neurosci. Abstr.* In press.

Grill, H.J., and R. Norgren. 1978. Chronic decerebrate rats demonstrate satiation, but not bait shyness. *Science* 201: 267-269

Grill, H.J., D. Ganster, and G.P. Smith. 1983. CCK-8 decreases sucrose intake in chronic decerebrate rats. *Soc. Neurosci. Abstr.* 9: 903.

Grovum, W.L. 1981. Factors affecting the voluntary intake of food by sheep. 3. The effect of intravenous infusions of gastrin, cholecystokinin and secretin on motility of the reticulo-rumen and intake. *Br. J. Nutr.* 45: 183-201.

Hahne, W.H., R.T. Jensen, G.F. Lemp, and J.D. Gardner. 1981. Proglumide and benzotript: members of a new class of cholecystokinin antagonists. *Proc. Natl. Acad. Sci. USA* 78: 6304-6308.

Houpt, T.R., S.M. Anika, and N.C. Wolff. 1978. Satiety effects of cholecystokinin and caerulein in rabbits. *Am. J. Physiol.* 235: R23-R28.
Jerome, C., R. Campbell, M. Lew, J. Gibbs, and G.P. Smith. 1986. Evidence against the CCK hypothesis: proglumide does not increase food intake in rats. In *Program, IXth International Conference on the Physiology of Food and Fluid Intake*, p. 38.
Joyner, K., J. Gibbs, and G.P. Smith. 1986. Effect of CCK-8 on sham feeding in vagotomized rats. In *Program, IXth International Conference on the Physiology of Food and Fluid Intake*, p. 39.
Kissileff, H.R., F.X. Pi-Sunyer, J. Thornton, and G.P. Smith. 1981. Cholecystokinin-octapeptide (CCK-8) decreases food intake in man. *Am. J. Clin. Nutr.* 34: 154-160.
Kraly, F.S. 1984. Vagotomy does not alter cholecystoknin's inhibition of sham feeding. *Am. J. Physiol.* 246: R829-R831.
Lew, M.F., J. Gibbs, and G.P. Smith. 1983. Intestinal satiety is elicited by L- but not by D-phenylalanine. *Soc. Neurosci. Abstr.* 9: 900.
Liddle, R.A., I.D. Goldfine, M.S. Rosen, R.A. Taplitz, and J.A. Williams, 1985. Cholecystokinin bioactivity in human plasma: molecular forms, responses to feeding, and relationship to gallbladder contraction. *J. Clin. Invest.* 75: 1144-1152.
Liddle, R.A., G.M. Green, C.K. Conrad, and J.A. Williams. Proteins but not amino acids, carbohydrates or fats stimulate cholecystokinin secretion in the rat. *Am. J. Physiol.* In press.
Liebling, D.S., J.D. Eisner, J. Gibbs, and G.P. Smith. 1975. Intestinal satiety in rats. *J. Comp. Physiol. Psychol.* 89: 955-965.
Lorenz, D.N., and S.A. Goldman. 1982. Vagal mediation of the cholecystokinin satiety effect in rats. *Physiol. Behav.* 29: 599-604.
Maclagan, N.F. 1937. The role of appetite in the control of body weight. *J. Physiol.* 90: 385-394.
Maddison, S. 1977. Intraperitoneal and intracranial cholecystokinin depress operant responding for food. *Physiol. Behav.* 19: 819-824.
McCaleb, M.L., and R.D. Myers. 1980. Cholecystokinin acts on the hypothalamic "noradrenergic system" involved in feeding. *Peptides* 1: 47-49.
McHugh, P.R. 1979. Aspects of the control of feeding: application of quantitation in psychobiology. *Johns Hopkins Med. J.* 144: 147-155.
McLaughlin, C.L., and C.A. Baile. 1981. Obese mice and the

satiety effects of cholecystokinin, bombesin and pancreatic polypeptide. *Physiol. Behav.* 26: 433-437.

McLaughlin, C.L., S.R. Peikin, and C.A. Baile. 1983. Food intake response to modulation of secretion of cholecystokinin in Zucker rats. *Am. J. Physiol.* 244: R676-R685.

McLaughlin, C.L., C.A. Baile, and F.C. Buonomo. 1985. Effect of CCK antibodies on food intake and weight gain in Zucker rats. *Physiol. Behav.* 277: 277-282.

Meyer, J.H., and M.I. Grossman. 1972. Comparison of D- and L-phenylalanine as pancreatic stimulants. *Am. J. Physiol.* 222: 1058-1063.

Miceli, M.O. 1985. Role of gastric distention in cholecystokinin's satiety effect in golden hamsters. *Physiol. Behav.* 35: 945-953.

Moran, T.H., and P.R. McHugh. 1982. Cholecystokinin suppresses food intake by inhibiting gastric emptying. *Am. J. Physiol.* 242: R491-R497.

Moran, T.H., G.P. Smith, A.M. Hostetler, and P.R. McHugh. 1986a. Transport of CCK receptors in abdominal vagal branches. *Program, IXth International Conference on the Physiology of Food and Fluid Intake*, p. 55.

Moran, T.H., P.H. Robinson, M.S. Goldrich, and P.R. McHugh. 1986b. Two brain cholecystokinin receptors: implications for behavioral actions. *Brain Res.* 362: 175-179.

Morley, J.E., A.S. Levine, J. Kniep, and M. Grace. 1982. The effect of vagotomy on the satiety effects of neuropeptides and naloxone. *Life Sci.* 30: 1943-1947.

Morley, J.E., A.S. Levine, B.A. Gosnell, J.E. Mitchell, D.D. Krahn, and S.E. Nizielski. 1985. Peptides and feeding. *Peptides* 6: 181-192.

Muller, J.E., E. Straus, and R.S, Yalow. 1977. Cholecystokinin and its COOH-terminal octapeptide in the pig brain. *Proc. Natl. Acad. Sci. USA* 74: 3035-3037.

Niederau, C., M. Niederau, J.A. Williams, and J.H. Grendell. 1986. New proglumide-analogue CCK receptor antagonists: very potent and selective for peripheral tissues. *Am. J. Physiol.* 251: G856-G860

Niijima, A. 1983. Glucose-sensitive afferent nerve fibers in the liver and their role in food intake and blood glucose regulation. In *Vagal Nerve Function: Behavioral and Methodological Considerations*, ed. J.G. Kral, T.L. Powley and C. McC. Brooks, 207-220. Elsevier Science Publishers, New York.

Pappas, T.N., R.L. Melendez, K.M. Strah, and H.T. Debas. 1985. Cholecystokinin is not a peripheral satiety signal in the dog. *Am. J. Physiol.* 249: G733-G738.

Parrott, R.F., and B.A. Baldwin. 1981. Operant feeding and drinking in pigs following intracerebroventricular injection of synthetic cholecystokinin octapeptide. *Physiol. Behav.* 26: 419-422.

Parrott, R.F., and R.A.L. Batt. 1980. The feeding response of obese mice (genotype, ob ob) and their wild-type littermates to cholecystokinin (pancreozymin). *Physiol. Behav.* 24: 751-753.

Passaro, E. Jr., H. Debas, W. Oldendorf, and T. Yamada. 1982. Rapid appearance of in intraventricularly administered neuropeptides in the peripheral circulation. *Brain Res.* 241: 333-340.

Pi-Sunyer, X., H.R. Kissileff, J. Thornton, and G.P. Smith. 1982. C-terminal octapeptide of cholecystokinin decreases food intake in obese men. *Physiol. Behav.* 29: 627-630.

Rehfeld, J.F. 1978. Immunochemical studies on cholecystokinin. II. Distribution and molecular heterogeneity in the central nervous system and small intestine of man and hog. *J. Biol. Chem.* 253: 4022-4030.

Reidelberger, R.D., and T.E. Solomon. 1986. Comparative effects of CCK-8 on feeding, sham feeding, and exocrine pancreatic secretion in rats. *Am. J. Physiol.* 251: R97-R105.

Reidelberger, R.D., T.J. Kalogberis, P.M.B. Leung, and V. Mendel. 1983. Postgastric satiety in the sham-feeding rat. *Am. J. Physiol.* 244: R872-R881.

Ritter, R.C., and E.E. Ladenheim. 1984. Fourth ventricle infusion of cholecystokinin suppresses feeding in rats. *Soc. Neurosci. Abstr.* 10: 652.

Robberecht, P., M. Deschodt-Lanckman, and J.J. Vanderhaeghen. 1978. Demonstration of biological activity of brain gastrin-like peptidic material in the human: Its relationship with the COOH-terminal octapeptide of cholecystokinin. *Proc. Natl. Acad. Sci. USA* 75: 524-528.

Saito, A., H. Sankaran, I.D. Goldfine, and J.A. Williams. 1980. Cholecystokinin receptors in the brain: characterization and distribution. *Science* 208: 1155-1156.

Savory, C.J., and M.J. Gentle. 1980. Intravenous injections of cholecystokinin and caerulein suppress food intake in domestic fowls. *Experientia* 36: 1191-1192.

Schally, A.V., T.W. Redding, H.W. Lucien, and J. Meyer. 1967. Enterogastrone inhibits eating by fasted mice. *Science* 157: 210-211.

Schneider, L.H., J. Gibbs, and G.P. Smith. 1986. Proglumide fails to increase food intake after an ingested preload. *Peptides* 7: 135-140

Shillabeer, G., and J.S. Davison. 1984. The cholecystokinin antagonist, proglumide, increases food intake in the rat. *Regul. Pept.* 8: 171-176.

Shillabeer, G., and J.S. Davison. 1985. The effect of vagotomy on the increase in food intake induced by the cholecystokinin antagonist, proglumide. *Regul. Pept.* 12: 91-99.
Sjödin, L. 1972. Influence of secretin and cholecystokinin on canine gastrin secretion elicited by food and by exogenous gastrin. *Acta Physiol. Scand.* 85: 110-117.
Smith, G.P. 1984. The therapeutic potential of cholecystokinin. *Int. J. Obes.* 8, Suppl. 1, 35-38.
Smith, G.P., J. Gibbs, and R.C. Young. 1974. Cholecystokinin and intestinal satiety in the rat. *Fed. Proc.* 33: 1146-1149.
Smith, G.P., C. Jerome, B.J. Cushin, R. Eterno, and K.J. Simansky. 1981. Abdominal vagotomy blocks the satiety effect of cholecystokinin in the rat. *Science* 213: 1036-1037.
Smith, G.T., T.H. Moran, J.T. Coyle, M.J. Kuhar, T.L. O'Donohue, and P.R. McHugh. 1982. Anatomic localization of cholecystokinin receptors to the pyloric sphincter. *Am. J. Physiol.* 246: R127-R130.
Smith, G.P., C. Jerome, and R. Norgren. 1985. Afferent axons in abdominal vagus mediate satiety effect of cholecystokinin in rats. *Am. J. Physiol.* 249: R638-R641.
South, E.H., and R.C. Ritter. 1985. Attenuation of CCK-induced satiety by systemic, intraventricular or perivagal capsaicin administration. *Soc. Neurosci. Abstr.* 11: 37.
Stacher, G. 1986. The effects of cholecystokinin and caerulein on human eating behavior and pain sensation. *Psychoneuroendocrinology* 11: 39-48.
Stacher, G., H. Steinringer, G. Schmierer, C. Schneider, and S. Winklehner. 1982. Cholecystokinin octapeptide decreases intake of solid food in man. *Peptides* 3: 133-136.
Strohmayer, A.M., and G.P. Smith. 1981. Cholecystokinin inhibits food-intake in genetically obese (C57Bl-6J-ob) mice. *Peptides* 2: 39-43.
Ugolev, A.M. 1960. The influence of duodenal extracts on general appetite. *Doklady Akademii Nauk SSSR* 133: 632-634.
Vanderhaeghen, J.J., J.C. Signeau, and W. Gepts. 1975. New peptide in the vertebrate CNS reacting with antigastrin antibodies. *Nature* 257: 604-605.
Van der Kooy, D. 1984. Area postrema: Site where cholecystokinin acts to decrease food intake. *Brain Res.* 295: 345-347.
Welch, I., K. Saunders, and N.W. Read. 1985. Effect of ileal and intravenous infusions of fat emulsions on feeding and satiety in human volunteers. *Gastroenterology* 89: 1293-1297.

Young, R.C., J. Gibbs, J. Antin, J. Hold, and G.P. Smith. 1974. Absence of satiety during sham feeding in the rat. *J. Comp. Physiol. Psychol.* 87: 795-800.

Zarbin, M.A., J.K. Wamsley, R.B. Innis, and M.J. Kuhar. 1981. Cholecystokinin receptors: presence and axonal flow in the rat vagus nerve. *Life Sci.* 29: 697-705.

Zhang, D.M., W. Bula, and E. Stellar. 1986. Brain cholecystokinin as a satiety peptide. *Physiol. Behav.* 36: 1183-1186.

Index

E

F

G

H

W